Climate Justice in Tourism

ASPECTS OF TOURISM
Series Editors: **C. Michael Hall** (*Massey University, New Zealand*) and **Dallen J. Timothy** (*Arizona State University, USA*)

Aspects of Tourism is an innovative, multifaceted series, which comprises authoritative reference handbooks on global tourism regions, research volumes, texts and monographs. It is designed to provide readers with the latest thinking on tourism worldwide and in so doing will push back the frontiers of tourism knowledge. The series also introduces a new generation of international tourism authors writing on leading edge topics.

The volumes are authoritative, readable and user-friendly, providing accessible sources for further research. Books in the series are commissioned to probe the relationship between tourism and cognate subject areas such as strategy, development, retailing, sport and environmental studies. The publisher and series editors welcome proposals from writers with projects on the above topics.

All books in this series are externally peer-reviewed.

Full details of all the books in this series and of all our other publications can be found on http://www.channelviewpublications.com, or by writing to Channel View Publications, BLOCK, The Fairfax, Pithay Court, Bristol, BS1 3BN, UK.

ASPECTS OF TOURISM: 104

Climate Justice in Tourism

Edited by

Freya Higgins-Desbiolles, Raymond Rastegar and Roshis Krishna Shrestha

CHANNEL VIEW PUBLICATIONS
Bristol • Jackson

DOI https://doi.org/10.21832/HIGGIN0091
Library of Congress Cataloging in Publication Data
A catalog record for this book is available from the Library of Congress.
Names: Higgins-Desbiolles, Freya editor | Rastegar, Raymond editor |
   Shrestha, Roshis Krishna editor
Title: Climate Justice in Tourism/Edited by Freya Higgins-Desbiolles,
   Raymond Rastegar and Roshis Krishna Shrestha.
Description: Bristol; Jackson: Channel View Publications, [2026] |
   Series: Aspects of Tourism: 104 | Includes bibliographical references.
   | Summary: 'This book provides a comprehensive exploration of climate
   justice in tourism, critically examining how tourism contributes to and
   is impacted by the global climate crisis. It interrogates dominant
   growth-centric, colonial and anthropocentric tourism models, and
   calls for radical transitions toward just, decolonised and regenerative
   futures' – Provided by publisher.
Identifiers: LCCN 2025047934 (print) | LCCN 2025047935 (ebook) |
   ISBN 9781836460084 hardback | ISBN 9781836460091 paperback |
   ISBN 9781836460114 pdf | ISBN 9781836460107 epub
Subjects: LCSH: Tourism – Environmental aspects – Case studies |
   Climate justice – Case studies
Classification: LCC G156.5.E58 C55 2026 (print) | LCC G156.5.E58 (ebook)
LC record available at https://lccn.loc.gov/2025047934
LC ebook record available at https://lccn.loc.gov/2025047935

British Library Cataloguing in Publication Data
A catalogue entry for this book is available from the British Library.

ISBN-13: 978-1-83646-008-4 (hbk)
ISBN-13: 978-1-83646-009-1 (pbk)

Channel View Publications
UK: BLOCK, The Fairfax, Pithay Court, Bristol, BS1 3BN, UK.
USA: Ingram, Jackson, TN, USA.

Authorised Representative: Easy Access System Europe - Mustamäe tee 50, 10621 Tallinn, Estonia, gpsr.requests@easproject.com.

Website: https://www.channelviewpublications.com
Bluesky: @multi-ling-mat.bsky.social
X: Channel_View
Facebook: https://www.facebook.com/channelviewpublications
Blog: https://www.channelviewpublications.wordpress.com

Copyright © 2026 Freya Higgins-Desbiolles, Raymond Rastegar, Roshis Krishna Shrestha and the authors of individual chapters.

All rights reserved. No part of this work may be reproduced in any form or by any means without permission in writing from the publisher.

The policy of Multilingual Matters/Channel View Publications is to use papers that are natural, renewable and recyclable products, made from wood grown in sustainable forests. In the manufacturing process of our books, and to further support our policy, preference is given to printers that have FSC and PEFC Chain of Custody certification. The FSC and/or PEFC logos will appear on those books where full certification has been granted to the printer concerned.

Typeset by Riverside Publishing Solutions.

Contents

Figures and Tables — vii
Contributors — xi
Acronyms — xix
Foreword — xxiii
Jeremy Smith

Introduction: Climate Change, Polycrisis and the Climate Justice Imperative in Tourism — 1
Freya Higgins-Desbiolles, Raymond Rastegar and Roshis Krishna Shrestha

Part 1: The Challenge of Climate Change for the Tourism Sector, Destinations and Communities

1 Tourism in a Warming World: Who Emits and Who Pays the Price? — 24
   Ya-Yen Sun, Futu Faturay and Wanru Zhou

2 Climate (In)Justice in Tourism: Meanings and Perspectives from Industry and Academic Leaders — 40
   Shenshen He, Bobbie Chew Bigby, Freya Higgins-Desbiolles and Bryan Grimwood

3 Towards a Just Transition in Aviation Climate Policy: Implications for Tourism — 60
   Gerben Broekema, Vishal Babajee, Ramón Fisac García and Daniel Scott

Part 2: Governance (In)Justices in Tourism

4 Stay Grounded: Re-Imagining Tourism for a Grounded Climate-Just Future — 75
   Angel Sulub and Daniela Subtil Fialho

5 Climate Justice in Nepal's Tourism Sector: Confronting Water Scarcity and Systemic Inequities — 101
   Nirmal Mani Dahal and Sudhan Subedi

6 Tourism Mobilities, Climate Mobilities and
   Mobility Injustice in the US Virgin Islands 120
   *Mimi Sheller, Leah Trotman, Greg Guannel
   and Kim Waddell*

Part 3: Case Studies in Climate (In)Justice in Tourism

7 Carbon Offsetting and Rights of Tourism Hosting
   Communities in Kenya's Conservancies 142
   Judy Kepher Gona and Lucy Atieno

8 The Nexus of Environmental Justice and Potential
   for Sustainable Tourism under Colonial Occupation
   in Palestine 167
   Mazin B. Qumsiyeh and Andrea Bibee

9 Affective Solidarity in Melting Destinations:
   Stepping Forward, Standing with and Staying
   Connected to Climate Justice 186
   Monica Nadegger and Carina Ren

Part 4: Imagining More Just Tourism Futures

10 The Regenerative Vanua Stewardship Framework:
    Ol Vanuas blong yumi oli no ol showgrounds blong
    ol turis! (Our Vanuas are not Destinations) 205
    *Kehana Andrews, Jerry Spooner,
    Laurana Rakau-Tokataake, Eva Addinsall
    and Cherise Addinsall*

11 'No Climate Justice without Gender Justice' and Racial
    Justice: A Critical Feminist Analysis 227
    Freya Higgins-Desbiolles

12 Deep Adaptation, Climate Justice and Tourism Futures 246
    Freya Higgins-Desbiolles

Conclusion: Advancing Climate Justice –
Pathways Forward for More Just Tourism Futures 271
*Roshis Krishna Shrestha, Raymond Rastegar and
Freya Higgins-Desbiolles*

Index 286

Figures and Tables

Figures

I.1	The 2023 update to the Planetary boundaries	3
I.2	Understanding climate justice in tourism: A multidimensional framework	10
1.1	Tourism expenditure share (left) and per capita tourism expenditure (right) for four income groups in 2019	27
1.2	Carbon footprint share (top left), carbon intensity (bottom left), and per capita tourism carbon footprint (right) for four income groups in 2019	30
1.3	Global map of tourism demand changes by 2050	33
1.4	Climate change impacts on tourism demand by country and income group in 2050	34
3.1	GDP and government debt as % to GDP development in 2020 and 2021 vs. 2019	66
3.2	Economic growth vs tourism balance pre-COVID to COVID	66
3.3	Potential increase in a long-haul ticket price with full implementation of aviation climate policy measures	69
4.1	Amount of CO_2 emissions and type of flights that CORSIA may cover	79
4.2	The graph on the left shows how, between 1970 and 2020, efficiency improvements have slowed down over time (downward curve), whereas the rate of air traffic growth has increased over time. The graph on the right shows sustained growth of aviation CO_2 emissions, closely linked to air-traffic growth and despite efficiency gains (SG, 2021a)	80
4.3	Illustration showing how (indirect) land-use changes associated with biofuel production can lead to higher GHG emissions than using fossil fuels	81
4.4	Graphic shows how current growth rate of air traffic and corresponding emissions will, before 2035, overrun the CO_2 budget for aviation to stay below	

	the 1.5°C threshold temperature increase. If the planned 'solutions' are not delayed, only in 2035 could aviation emissions start to fall below 2019 levels (SG, 2022)	82
4.5	Map of Airport-related Injustice and Resistance documents socioenvironmental conflicts resulting from airport expansion projects worldwide	84
4.6	Residents in Tulum protest the new international airport and military airbase in Tulum	91
6.1	An overhead shot of the West Indian Company Dock (cruise ship port) in downtown Charlotte Amalie, St. Thomas, USVI	128
6.2	An overhead shot of The Ann E. Abramson Marine Facility (cruise ship port) in Frederiksted, St. Croix	129
6.3	A 1954 orthophotograph of Crown Bay in St. Thomas, USVI	130
6.4	A 2000 orthophotograph of Crown Bay in St. Thomas, USVI	130
7.1	Conservancy land resource use system, centring interactions and outcomes of key action situations, based on components of Ostrom's SES framework	150
7.2	Mobile boma in one of the conservancies visited for field research. File photo taken by Judy Kepher Gona	151
8.1	Students at a marginalised school holding our climate change brochures that articulate challenges and solutions	174
8.2	Children visiting one of the exhibit areas of the natural history museum and learning about the geography of Palestine with issue of desertification	174
8.3	Two children competing in a game we created that teaches them about conservation and sustainability	175
9.1	Demonstration against uranium mining	195
9.2	Warm winter weather and melting snow days in the Tyrolean Alps	197
10.1	Regenerative Vanua Stewardship Framework (RVSF)	216
12.1	Community participation in planning for Yitpi Yartapuultiku Aboriginal Cultural Place, Port Adelaide, South Australia, 17 November 2022	260
12.2	Berkana Institute's 'Two Loops Model' of Transition between Systems	263

Tables

3.1	Intended and unintended effects of climate change and other environmental impact measures reductions	70
7.1	Survey statements with noted disagreement from respondents	155

Contributors

Editors

Freya Higgins-Desbiolles is an Adjunct in Tourism Management, Business Unit, Adelaide University (Australia); Adjunct Professor with the Centre for Research and Innovation in Tourism, Taylor's University (Malaysia); and Adjunct Associate Professor with Department of Recreation and Leisure Studies, University of Waterloo (Canada). She has worked with industry, community and non-profits on projects that address community in tourism, human rights and justice. She has introduced ideas such as 'tourism as a social force', 'socialising tourism' and defining tourism by the local community to tourism's study and practice.

Raymond Rastegar is a researcher with the Department of Tourism and Marketing, Griffith University, where his scholarly work focuses on justice, sustainability transitions and environmental conservation. His research explores how sustainability transitions can foster inclusive and equitable futures at local, national and global levels. He is particularly interested in developing theoretical and empirical foundations that integrate justice into sustainability transitions, offering insights to guide tourism communities and inform policy responses.

Roshis Krishna Shrestha obtained his PhD in Business and Economics at the Research School of Management at the Australian National University in December 2023. He joined the School of Hotel and Tourism Management, Hong Kong Polytechnic University, as a Research Assistant Professor in January 2024. Before embarking on his academic career, Roshis gained over five years of managerial experience in various sectors in Nepal, including automotive and consumer durables industries. His research interests include Indigenous tourism, sustainable tourism, culture and heritage tourism and rural tourism as applied to his native Nepal. Roshis has special interests in ethnographic methods and grounded theory.

Authors

Eva Addinsall is a postgraduate student at the University of Melbourne in the School of Culture and Communication. She holds an undergraduate degree in Politics and International Relations and is the Intercultural Communications and Engagement Advisor for Regenerative Vanua.

Cherise Addinsall is a Senior Research Fellow in Regenerative Agritourism at the University of the Sunshine Coast and the Project Leader for the Australia Centre for International Agriculture Research (ACIAR) that funded Our Vanua Project and Regenerative Agritourism Vanuatu project.

Kehana Andrews is a Ni-Vanuatu woman from the island of Malo in the Sanma Province. She is an Adjunct Fellow within the School of Science, Technology and Engineering at the University of Sunshine Coast, Australia, and holds a degree in Tourism Management at the University of the South Pacific. She is the Regenerative Vanua Project Manager of the Australian Centre for International Agricultural Research (ACIAR) SSS/2021/120 *Our Vanua Project*.

Lucy Atieno is currently pursuing PhD studies at Bremen University in Germany. She applies participatory research techniques to explore gender, tourism and climate change. Her academic journey is complemented by professional experience in sustainable tourism consulting in East Africa, where she focused on enhancing local linkages to tourism value chains. In 2024, she contributed to the report *Sustainable Blue Tourism in the Western Indian Ocean: Trends, Challenges, and Policy Pathways*, helping to shape policy guidance for the region's blue tourism development.

Vishal Babajee is a seasoned executive with 20+ years of experience across diverse corporate roles in Aviation, complemented by a background in economic and financial research within Banking and as Economist in the Ministry of Economic Development & Financial Services in Mauritius. He holds a Master's degree in Strategy & Organisation Consulting and a BSc (Hons) in Economics. Vishal is a guest speaker at ESCP Business School and has contributed to Master's programmes in International Business & Diplomacy, Strategy & Digital Business, Hospitality & Tourism Management, Big Data and Business Analytics and Strategy and Organisation Consulting. His work bridges the practitioner–academic gap in Strategy and Transformation, Sustainable Growth and Economic Development.

Andrea Bibee graduated from the University of Oregon in 2013 with her Juris Doctor and Master's degrees. She has spent her career focused on civil and human rights law and policy. She met Professor Qumsiyeh during a visit to the West Bank in 2024 and is honoured to collaborate. Andrea is an organising member of Corvallis Palestine Solidarity group and frequently hosts public events on historical, legal and ethical issues related to the ongoing illegal occupation of Palestine. She lives in Oregon with her three children.

Bobbie Chew Bigby is a Postdoctoral Research Fellow at the University of Waterloo (Canada), an Adjunct Research Fellow at Nulungu Research Institute (Australia) and a Lecturer at NYU Tulsa (United States). Bobbie's research explores the ways that tourism can be a tool for cultural resurgence among Indigenous and post-conflict communities. Bobbie looks beyond tourism's economic potential alone and seeks to understand how tourism can help people (re)connect with culture, community and Country (living lands and waters).

Gerben Broekema is an aviation strategist with 25+ years of experience in the aviation sector both in consulting roles at McKinsey & Company and Broekema Aviation Advisory Services as well as in corporate roles at Royal Schiphol Group where he held the position of Head of Group Strategy & International Development until 2017. He holds a Master's degree in Human and Economic Geography with a specialisation in transport geography and economics. Gerben has a passion for researching and shaping the role of aviation in society and developing a perspective on how to achieve a cost effective and 'just' transition. He is a leading expert in (future) electric Regional Air Mobility.

Nirmal Mani Dahal is a highly accomplished researcher from Nepal. Holding a PhD from the Chinese Academy of Sciences, his research expertise lies in the intersection of climate change, drought and agricultural impacts. He has contributed several publications to peer-reviewed journals and given presentations at international conferences in the field. Beyond his research, he actively participates in Nepali professional organisations and serves as visiting faculty at Nepal Open University. Currently, he is a Research Assistant at the School of Hotel and Tourism Management, The Hong Kong Polytechnic University, where he continues to contribute to innovative projects.

Futu Faturay has been a Postdoctoral Research Fellow at the UQ Business School since 2023. Previously, he worked as a policy analyst at the Fiscal Policy Agency, Ministry of Finance Indonesia. He holds a PhD in Sustainability Science from the University of Sydney (2019) and a Master's in economics from Georgia State University, USA (2012). Futu has authored multiple academic publications on input–output analysis, climate change impact assessment and tourism-related carbon emissions inventories.

Daniela Subtil Fialho is a network coordinator at Stay Grounded (SG) and has been mainly working with frontline groups resisting airport expansion and associated tourism development in the Global South, as well as coordinating SG's working group on tourism and aviation.

She studied international development as well as sustainable resource management with a focus on climate change and land use. In the last 15 years, Daniela has been an activist with Amnesty International and an organiser within the climate justice movement: with *Climáximo* in Portugal, *Ende Gelände* in Germany and in the wider European network. She is an anti-imperialist and anti-capitalist militant for climate justice and the liberation of oppressed peoples.

Ramón Fisac García holds a MSc in Industrial Engineering (2007) and a PhD in Management Engineering (2014). He is Executive Director of the Master's in Hospitality and Tourism Management. His fields of research are innovative business models for sustainable tourism, behavioural transformation in the tourism industry and the socioeconomic impacts of tourism in destinations. For three years he worked as a sustainability analyst at ACCIONA, where he designed and implemented the Corporate Training Program on Sustainability. During the last decade, he has been lecturing for postgraduate programs in different national Universities and international business schools (Mexico, Indonesia).

Bryan Grimwood is a Professor in the Department of Recreation and Leisure Studies at the University of Waterloo, Canada. His research examines ethical and political dimensions of tourism, leisure and cultural livelihoods.

Greg Guannel is the Director of the Caribbean Green Technology Centre at the University of the Virgin Islands. The Centre has focused on civil and coastal infrastructure resilience, renewable energy technology and policy and water resource management. Major publications include the US Virgin Islands' Hazard Mitigation and Resilience Plan, the territory's Climate Vulnerability Assessment and the islands' first ever estimate of Social Vulnerability. Dr Guannel has a Master's in Public Works from the Ecole Supérieure des Travaux Public, a Master's in Coastal and Ocean Engineering from Texas A&M University and a PhD in Civil Engineering from Oregon State University.

Shenshen He is a Doctoral student in Recreation and Leisure Studies at the University of Waterloo, specialising in Sustainable Tourism Management and Indigenous Cultural Heritage Preservation. Her research focuses on the sociocultural and environmental impacts of tourism, with a background in Tourism Administration from The George Washington University. She has contributed to various research projects, including the Caribbean Tourism Organisations Project, and has hands-on experience in destination development and Green Globe Certification auditing. Her work aims to integrate sustainability and community-based solutions into tourism practices and policy.

Judy Kepher Gona is a leading ecotourism and sustainable tourism expert with over 20 years' experience in Africa. As Founder and Principal Consultant of Sustainable Travel & Tourism Agenda (STTA), she has advanced responsible tourism, community empowerment and environmental conservation. Judy pioneered Kenya's first sustainable tourism certification scheme, championed youth and women's voices, supported Micro, Small and Medium Enterprises (MSMEs), and led community conservation models. Recognised among Africa's Top 100 Women in Tourism, she has judged global sustainability awards and contributed to policy discussions on ethical tourism.

Monica Nadegger is a Postdoctoral Researcher at the LMU University of Munich (Innovation and Entrepreneurship Centre) and the MCI – The Entrepreneurial School (Department for Tourism and Leisure Business). Her research is guided by CCO (communicative constitution of organisation) and feminist new materialist perspectives, and she focuses on materiality, resistance and alternative organising. Her most recent work explores the constitution of resistance and more-than-human relations in the winter tourism industry through non-representational, auto/ethnographic and post-human approaches. Her research includes articles published in *Organisation Studies*, *Organisation*, *Culture and Organisation*, *Annals of Tourism Research*, *Journal of Sustainable Tourism* and *Leisure Studies*, as well as several book chapters.

Mazin B. Qumsiyeh is founder and volunteer director of the Palestine Institute for Biodiversity and Sustainability, Bethlehem University, Palestine (see palestinenature.org for PIBS' 10 year booklet and annual reports). He previously served at US universities including Tennessee, Duke and Yale. Qumsiyeh has published over 250 scientific papers, over 30 book chapters, hundreds of articles and several books on topics ranging from the environmental impact of colonisation to environmental and climate justice to cultural heritage to human rights to biodiversity conservation to cancer. He is a laureate of the Paul K. Feyerabend Foundation award, the Takreem award, Peace Seeker of the Year award and was nominated for the Nobel Peace Prize 2025.

Laurana Rakau-Tokataake is a Ni-Vanuatu woman from the Island of Aneityum, in the province of Tafea. She is the Operations Manager of Regenerative Vanua, a Pasifika not-for-profit which strengthens, connects and regenerates Oceanic Vanuas. She holds a degree in Tourism Management from the University of the South Pacific.

Carina Ren is an Associate Professor at Aalborg University, Denmark, and head of her university's Arctic research platform, AAU Arctic. Inspired by relational and feminist material thinking, Carina explores how tourism interferes with other fields of the social, exploring the practices, events and technologies through which tourism is developed, organised and valued. She has published widely in tourism and Arctic studies and is the co-editor of books such as *Tourism Encounters and Controversies: Ontological Politics of Tourism Development* (Ashgate, 2015), *Co-Creating Tourism Research: Collaborative Ways of Knowing* (Routledge, 2017) and *Collaborative Research Methods in the Arctic* (Routledge, 2020).

Daniel Scott is a Distinguished Professor and Research Chair in the Department of Geography and Environmental Management at the University of Waterloo. He has worked extensively on sustainable tourism for 25 years, with a focus on the transition to a low carbon tourism economy and adaptation to the complex impacts of a changing climate. He has advised and led projects for a wide range of government agencies and tourism organisations around the world, including the United Nations World Tourism Organisation, United Nations Environment Programme, World Bank, European Tourism Commission, World Travel and Tourism Council, International Olympic Committee and the OECD. He has also been a contributor to multiple UN Intergovernmental Panel on Climate Change assessment and special reports.

Mimi Sheller is Inaugural Dean of The Global School at Worcester Polytechnic Institute, in Massachusetts. Sheller is an interdisciplinary social scientist with work in Caribbean Studies, Mobilities Research and Social Theory. Sheller was founding co-editor of the journal *Mobilities* and past President of the International Association for the History of Transport, Traffic and Mobility. She has published more than 150 articles and chapters, and recent books include *Advanced Introduction to Mobilities* (Edward Elgar, 2021); *Island Futures: Caribbean Survival in the Anthropocene* (Duke University Press, 2020); and *Mobility Justice: The Politics of Movement in an Age of Extremes* (Verso, 2018).

Jeremy Smith works at the intersection of tourism, climate and systems change, helping the tourism and travel sector towards a just transition. His focus is on writing, advising, strategy and convening. He co-founded Tourism Declares a Climate Emergency and co-authored the Glasgow Declaration on Climate Action in Tourism. Since then, he has worked closely with destinations, regional authorities and international organisations to support climate action planning rooted in justice, accountability and care.

Jerry Spooner is a Ni-Vanuatu man from the Islands of Ambae, in the Penama Province. He is the Executive Director of Regenerative Vanua, a Pasifika not-for-profit which strengthens, connects and regenerates Oceanic Vanuas. He holds a degree in Business Studies at the University of the South Pacific.

Sudhan Subedi is a Senior Officer at Nepal Tourism Board (NTB). He has more than 20 years of experience in the tourism sector. He has keen interest in the social and cultural aspects of tourism. With his extensive experiences, he has contributed significantly to promoting and managing Nepal's tourism endeavours. He holds a Master's degree in Sociology from Tribhuvan University, Nepal. He is also a Lecturer in Tourism Studies at the Tribhuvan and Mid-West University. He is keen on the growth and sustainability of the industry and pens articles on tourism-related issues regularly in different media.

Angel Sulub is the coordinator for Stay Grounded's regional network in Latin America & the Caribbean, Permanecer en la Tierra (PET). He is a Mayan activist involved in the defence of the rights of Indigenous peoples. He is also a Mayan delegate in the National Indigenous Congress of Mexico. Prior to PET, he was coordinator of the Centro Comunitario Maya U Kúuchil K Ch'i'ibalo'on, a space for the exchange of community knowledge. Angel has been active against the Tren Maya, a project for territorial reorganisation in the southeast of Mexico that involves the increase of tourism and the construction of airport infrastructure. He has also been actively involved in the shaping of El Sur Resiste, a space of struggle and resistance against mega-projects in eight states in the south of Mexico.

Ya-Yen Sun is an Associate Professor at the University of Queensland, Australia. Her interest is in tourism sustainability, focusing on economic impacts and environmental footprinting. In addition, she also works on the environmental perspectives of travel behaviour, quantifying the tourism carbon footprint and tourism water footprint. She successfully constructed and analysed tourism impacts for individual countries (Taiwan, China, Japan, United States and New Zealand) and provided the first detailed estimate of global travel's impact on greenhouse gas emissions.

Leah Trotman received a MSc in Health and International Development from the London School of Economics and Political Science and an MA in Caribbean and Latin American Studies from University College London. Over the past five years, her research and practical work have centred on the impacts of (un)natural disasters (and other climate change-related events) on health and development in the Caribbean and Latin America. Leah is the US Virgin Islands' first Marshall Scholar and a 2020 Truman Scholar. She currently works at the Virgin Islands Housing Finance Authority (VIHFA) as a National Environmental Policy Act Specialist.

Kim Waddell is Principal Investigator and Project Director for the Virgin Islands Established Program for Stimulating Competitive Research – a National Science Foundation-supported research capacity building program based at the University of the Virgin Islands focused on marine ecosystem function under climate change, and STEM education opportunities for underrepresented students. Kim was the project lead for the new US Virgin Islands Hazard Mitigation and Resilience Plan. He received his PhD in Biological Sciences from the University of South Carolina and his BA in Environmental Studies from the University of California, Santa Cruz.

Wanru Zhou is a PhD candidate at the Business School, University of Queensland, Australia. Her research focuses on climate change issues in tourism, with a particular interest in assessing the socioeconomic impacts of climate change on tourism and evaluating the environmental and economic consequences of tourism activities. She employs advanced econometric modelling techniques to provide evidence-based insights for sustainable tourism planning and climate policy.

Acronyms

ACA	Annapurna Conservation Area
ACIAR	Australian Centre for International Agricultural Research
AD	Anno Domini, Latin for in the year of the Lord; based on system for counting years based on the estimated birth of Jesus Christ
AMOC	Atlantic Meridional Ocean Circulation
ASAL	Arid and Semi-Arid Lands
ATG	Alternative Tourism Group of Palestine
B&Bs	Bed and Breakfast
BIPOC	Black, Indigenous and People of Colour
CBDR	Common But Differentiated Responsibilities
CBT	community-based tourism
CCAs	Community Conserved Areas
CCAN	Caribbean Climate Adaptation Network
CEKA	Cyril E. King Airport
CO_2	carbon dioxide
COP	Conference of the Parties
CORSIA	Carbon Offsetting and Reduction Scheme for International Aviation
COVID-19	corona virus disease 2019
CPAR	Critical Participatory Action Research
DBA	destination-based accounting
EIT	Equality in Tourism
ESBs	Earth Systems Boundaries
ETS	European Trading Scheme
EU	European Union
FIFA	Federation Internationale de Football Association
GCF	Green Climate Fund
GDP	Gross Domestic Product
GHG	Greenhouse Gas Emissions
GLOFs	Glacial Lake Outburst Floods
GNI	Gross National Income

GNP	Gross National Product
Gt	Gigatonne
HLITOA	Holy Land Incoming Tour Operators Association
HVO	hydrotreated Vegetable Oils
IA	Inuit Ataqatigiit
IATA	International Aviation Transport Association
ICAO	International Civil Aviation Organization
ICJ	International Court of Justice
IDB	Inter-American Development Bank
ILO	International Labor Organization
IMF	International Monetary Fund
INE	Instituto Nacional de Estadistica
IPCC	Intergovernmental Panel on Climate Change
IWRM	Integrated Water Resources Management
KPIs	key performance indicators
LAC	Latin America & the Caribbean
LAPA	Local Adaptation Plan of Action
LGBTQI+	Lesbian, Gay, Bisexual, Transgender, Intersex and Queer/Questioning, with the '+' indicating inclusion of other identities not explicitly listed
MEU	Mobile Educational Unit
MMWCA	Maasai Mara Wildlife Conservancies Association
MNCC	Malvatumauri National Council of Chiefs
MST	Measuring the Sustainability of Tourism
NAACP	National Association for the Advancement of Coloured People
NAP	National Adaptation Plan
NASA	National Aeronautics and Space Administration
NBSAP	National Biodiversity Strategy and Action Plan
NDA	National Designated Authority
NDCs	Nationally Determined Contributions
NGO	Non-governmental Organisation
NOAA	National Oceanic and Atmospheric Administration
NPO	Non-Profit Organisation
NZE	Net Zero Emissions
PIBS	Palestine Institute for Biodiversity and Sustainability
RBA	residence-based accounting
RVSF	Regenerative Vanua Stewardship Framework
SAFs	Sustainable Aviation Fuels
SARS	Severe Acute Respiratory Syndrome
SDGs	UN Sustainable Development Goals
SESC	Serviço Social de Comércio

SG	Stay Grounded
SIDS	Small Island Developing States
SPADES	South Pacific Action for Development Strategy
STT	St Thomas, US Virgin Islands IATA code
SUVs	Suburban Utility Vehicles
TAAF	Tourism Alert and Action Forum
TID	Tourism-inbound destinations
TOC	Tourism-outbound countries
TPCC	Tourism Panel on Climate Change
UNCTAD	United Nations Conference on Trade and Development
UNDP	United Nations Development Program
UNEP	United Nations Environment Programme
UNFCCC	United Nations Framework Convention on Climate Change
UNGA	United Nations General Assembly
UN Tourism	United Nations Tourism Organization
UNWTO	United Nations Tourism Organization – now UN Tourism
USD	United States dollar
USVI	United States Virgin Islands
VAT	Value Added Taxes
WEF	World Economic Forum
WMO	World Meteorological Organisation
WTTC	World Travel and Tourism Council
ZET	Zero Emissions Tourism

Foreword

Many years ago, I stayed in a forest lodge in South Africa whose website included a page titled *Don't Come If....* It wasn't a list of booking conditions or a novel way to limit numbers. It was a gentle provocation: a way of interrogating the dynamic between the hosts, the place and the guest. A way of asking us travellers to consider the expectations and assumptions we'd brought with us. It challenged us to not treat this unfamiliar place as just somewhere else to consume. It reframed the act of travel as a relationship rather than a transaction.

To borrow that framing here: don't read this book if you're looking for reassurance. Don't expect a smooth journey through replicable case studies or a comforting checklist of best practices to implement in your own operations. Don't come in search of the business case for sustainability, distilled into KPIs and carbon metrics. This book doesn't offer any of that.

It is, instead, a necessarily discomforting read. It starts here, with the fact that I have been asked to write the foreword. After all, as a white, European, anglophone male of relative privilege, I have long benefited from many of the iniquities, imbalances and entitlements this book dissects. And while I have spent many years reading and writing about the challenges of developing a more equitable tourism industry, the only lived experience that I bring is that I've travelled a lot.

My foreword is not here to appropriate the ideas that belong to the writers in the following pages. I am, however, the first person to read all their finished essays. This means I got to read criticism of some of my own climate action work. It made me uncomfortable. I felt defensive. But it was useful and necessary, and far better to be challenged than be reassured that everything I am doing or believe in is going fine.

At the heart of all these chapters are many similarly challenging explorations of climate justice in tourism. They are diverse in origin, voice and emphasis, yet united in their refusal to endorse our industry's dominant narrative. These are stories and viewpoints many of us in the sector choose not to hear. That is reason enough to listen.

Too often, many of us imagine justice as a kind of convenient accounting exercise. We look to balance out our harms with an appropriate amount of offsets. We try to erase our footprints with financial contributions to tree planting or community projects.

Is this justice? Or does this framing contain its own injustices, and avoid deeper questioning? Who decides that a certain amount of investment in community projects vindicates X tonnes of carbon emissions? Who has the power to make these decisions? And who is heard only when convenient? Who gets to write the forewords, deliver the keynotes, run the workshops? And who is thanked for their perspective but left out of the actual decision making? Who do I mean – and who do you hear – in my repeated use of the word 'We'?

Whoever we are, this book asks us to consider whether our sector's talk of a 'just transition' is matched by the willingness to confront the complexity of what this work really entails. We talk of developing tourism that is more Responsible. Or even Regenerative. But what of Reparative? Is such a concept even possible? And once again, who would decide?

We have to go further still. We need to consider that justice is not owed only to people. It is owed to land, to rivers, to all that our colonisation, capitalism and consumption have treated as inert or ownable. Climate action that does not embrace this wider kinship and reckoning is not fully justice. It is responsibility deferred. Again.

We need to consider all of this – and more – if we are truly to engage with climate justice. And then, just as we feel overwhelmed with the scale of the challenge, I'm afraid we need to consider the other side of the coin.

What do we need to let go?

Climate justice doesn't mean doing things better so that Business as Usual can continue. In some other cases, climate justice requires us to stop some aspects of our industry and relinquish some of our power.

For those of us long accustomed to the freedoms of travel, this might feel like a deeply personal loss or threat. We've built careers cherishing tourism's much heralded benefits of mobility, leisure, autonomy and access. Yet we live in a world where these freedoms are not universally shared, never have been and never will be. How long can we sustain the fantasy of a sector that sells freedom of movement, while millions are displaced or denied the right to remain on or access their ancestral land?

What remains then, if we begin to let go of the illusions of endless growth and endlessly manufactured joy? Perhaps something quieter yet more essential. Perhaps a form of tourism that moves at a slower pace, that honours limits and prioritises shared rest and real connection over private luxury. Perhaps a way of travelling that does not treat the world as a theme park of 'authentic' experiences, but as a living system of relationships that includes us, but does not exist for us.

This book does not reach a tidy, reassuring conclusion. It does not attempt to bail out our industry with a set of strategies or the latest sustainable buzzword. Instead, it offers us a rare place to listen to many voices who live daily with the consequences of what most of us choose to ignore.

Of course, listening is not enough. What matters is how we then act.

Jeremy Smith
Co-founder of Tourism Declares
a Climate Emergency
Co-author of the Glasgow Declaration
on Climate Action in Tourism

Introduction: Climate Change, Polycrisis and the Climate Justice Imperative in Tourism

Freya Higgins-Desbiolles, Raymond Rastegar and Roshis Krishna Shrestha

Climate justice is concerned with the inequities and injustices that underpin the phenomenon of climate change. This introduction opens a book that presents the first comprehensive treatment of climate justice in tourism, examining the ways tourism both contributes to climate injustices and has potential to counteract such tendencies.

This chapter provides a context for discussions of climate justice, demonstrating strategies and actions to date are failing to rise to the challenge and seriousness of the predicament that we confront. It introduces a method of 'critical climate justice praxis' as an approach that combines critical reflection and transformative action to address the systemic root causes of climate injustices. It additionally outlines a multidimensional framework on justice to assist in identifying the ways certain peoples are privileged or marginalised as a tool for delving deeper into climate justice dynamics. The chapter also addresses certain key dynamics, including between reformist and radical approaches and degrowth and decolonisation. The chapter closes with an overview of the book's chapters which collectively provide a comprehensive insight into climate justice in tourism.

Introduction

'We do not want that dreaded death sentence, and we have come here today to say, "Try harder, try harder", because our people, the climate army, the world, the planet needs our actions now, not next year, not in the next decade'. Prime Minister of Barbados, Mia Mottley, speech to the Conference of the Parties (COP) 26

It has taken decades for climate change to enter the public consciousness. This is despite the fact that it is likely that the US oil and gas industry's largest trade association had known since at least the 1950s that 'fossil fuel products could lead to global warming with "dramatic environmental effects before the year 2050"' (Supran *et al.*, 2023: 1). Focus on concerns of climate justice is only now emerging; however, its meaning is not widely or well understood, particularly by those who are shielded by privilege and power. And yet climate justice is one of the most important issues of our lives. Climate justice is a class issue (Glen & Clarke, 2024), a gender issue, a 'race' issue, an ecological issue, a decolonising issue and even a spiritual issue. The terms are contested, with 'climate change' being argued to be too benign a phrase to activate people; other descriptors include global warming, global heating, climate crisis, global boiling and even climate catastrophe.

Climate justice focuses attention on the differential, uneven and disproportionate impacts of climate change, as well as the injustices and inequities that accompany efforts to mitigate and adapt to climate change. Sultana argued:

> The goals are to reduce marginalization, exploitation, and oppression, and enhance equity and justice. Applying a climate justice approach is an intentional process that involves carefully analyzing who is excluded or marginalized by climate change processes as well as any adaptation or mitigation interventions pursued. A climate justice approach focuses on who benefits, who loses out, in what ways, where, and why. It is an explanatory tool that helps better explain the relationships at different scales that co-create and maintain injustices. (2022a: 118–119)

In thinking climate justice in tourism, there are additional compounding injustices that enter the equation. This includes the nature of a multinational corporate dominated industry that is built on a model of wealth extraction from destinations to which they hold few obligations in terms of taxes and regulations and on which they try to externalise the environmental and social costs. Many of these destinations have been incorporated into the global tourism political economy in vulnerable conditions of tourism dependency that have been in part established through ongoing legacies of imperialism and settler colonialism. Additionally, there are peoples in communities that have been oppressed, racialised and/or marginalised who experience added layers of difficulty in the face of climate crisis and impeded access to support and crisis relief. As well, there are those who are currently not well represented at the table to assert their rights and demand justice in climate related decision making, including non-human kin and future generations.

The climate crisis is compounded by accompanying crises that result from human activities exceeding fixed planetary boundaries, that include not only climate change but also biosphere integrity, ocean acidification, land system change and others. The 2023 planetary boundaries framework update found that six of the nine boundaries (see Figure I.1) were transgressed 'suggesting that Earth is now well outside of the safe operating space for humanity' (Richardson, 2023: 1). It is the urgency that this truth engenders that can be read in Prime Minister Mottley's words opening this introduction and compels much of the work gathered in this edited volume. This reality has sparked the use of the term 'polycrisis' to describe the complex, interwoven, overlapping and compounding nature of these multiple crises we now face (which include in addition to these ecological crises and thresholds, associated social, financial and political crises; see Bianchi & Milano, 2024).

This book is an attempt to make sense of the complex interface between climate change and tourism, with tourism being both a catalyst to the crisis but also a victim of its impacts. But more specifically, not all stakeholders are equally implicated in the causation nor are all stakeholders equally impacted. This is why climate change is a climate

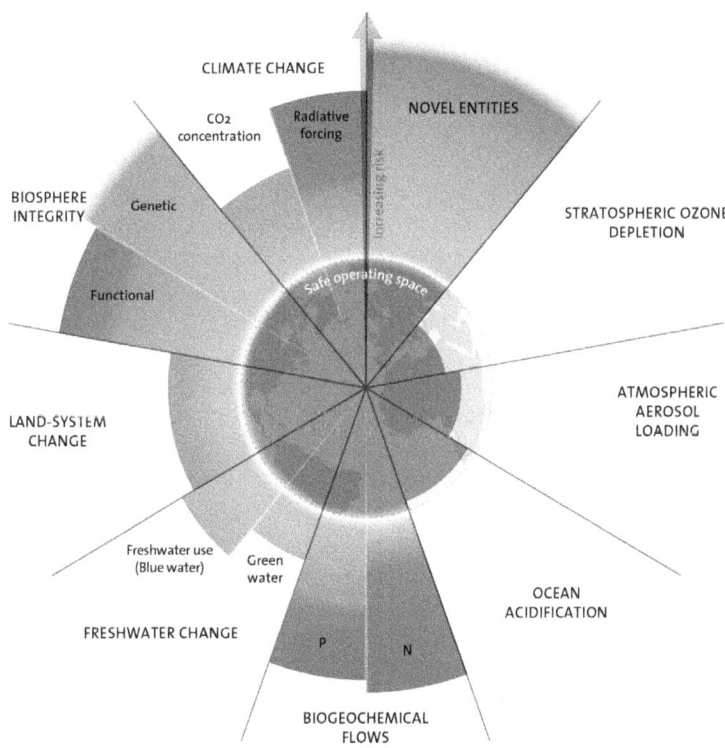

Figure I.1 The 2023 update to the Planetary boundaries. Licensed under CC BY-NC-ND 3.0. Source: Azote for Stockholm Resilience Centre, based on analysis in Richardson *et al.* (2023).

justice issue that is deserving of considered attention. This chapter introduces the book's key pillar of 'critical climate justice praxis' as an approach that combines critical reflection and transformative action to address the systemic root causes of climate injustices.

Setting a Context for Climate Justice in Tourism Discussions

In establishing a context for the discussion, it is first essential to realise that the fossil fuel industry was alerted early on to the role that fossil fuel emissions would play in climate warming. For example, investigative journalists in 2015 found '[…] internal company documents showing that Exxon scientists have been warning their executives about "potentially catastrophic" anthropogenic (human-caused) global warming since at least 1977' (Supran *et al.*, 2023: 1). In the intervening years, climate scientists and others have been developing rigorous climate assessments and models to inform states so that responsible policies and strategies can be developed to guide requisite action.

At the moment, climate obligations are set by the United Nations Framework Convention on Climate Change (UNFCCC) and more recently by the terms of the 2015 Paris Agreement. The UNFCCC came into force in 1994, with universal membership, 198 countries having ratified the Convention, agreeing to collectively address human-induced climate change. The Paris Agreement marked a milestone as a multilateral binding agreement that brings all nations together to combat climate change and to adapt to its effects. The 'overarching goal' of the Paris Agreement is to limit 'the increase in the global average temperature to well below 2°C above pre-industrial levels' and pursue efforts 'to limit the temperature increase to 1.5°C above pre-industrial levels' (UNFCCC, n.d.). A key feature of this agreement is the approach that nations develop self-determined pledges to reduce their emissions through Nationally Determined Contributions. 'Rather than being binding on States individually, this is a collective obligation, or rather, aspiration […] that is obtained through the submission of nationally determined contributions (NDCs) by individual Parties that they "intend to achieve" (Paris Agreement, art. 4(2)' (Klerk, 2024: 1). Thus, the key weakness is '[…] absence of legally binding human rights provisions or any complaints mechanisms empowered to offer remedies to affected populations, with the Paris Agreement implementation mechanism carefully framed as being "facilitative", "non-adversarial and non-punitive" […]' (Venn, 2023: 3). These mechanisms are clearly not sufficiently effective, as the World Meteorological Organisation (WMO) reported: 'Global carbon emissions from fossil fuels reached a record high in 2024 and there is still "no sign" that the world has reached a peak' (WMO, 2024).

Climate justice has emerged from grassroots action and social justice and environmental justice movements (Sultana, 2022a). For example, in 2019, a group of 27 students from the University of the South Pacific undertaking a university assignment collaborated on a campaign to persuade the leaders of the Pacific Islands Forum to petition the International Court of Justice on climate change and human rights. As a result, Vanuatu tabled on the agenda of the 2022 Pacific Island Forum to formally endorse the request for an advisory opinion from the International Court of Justice on the obligations of States with regards to climate change and human rights (Pacific Islands Students Fighting Climate Change, n.d.). This was unanimously endorsed. This led to a request from the United Nations (UN) General Assembly in March 2023 for the International Court of Justice (ICJ) to consider what are the international legal obligations of states to address climate change and what legal consequences might they incur for acts and omissions causing harm to states, particularly Small Island Developing States (SIDS), as well as 'peoples and individuals of the present and future generations' (ICJ, 2023). The ICJ handed down its unanimous opinion on 23 July 2025, finding that the production and consumption of fossil fuels 'may constitute an internationally wrongful act', for which a nation state is responsible (McVey & Savaresi, 2025). This allows for the possibility for the big emitters to be successfully sued for reparations. The opinion is not legally binding but carries a weight as authoritative legal interpretation.

More tangibly, states may hold obligations under other international legal instruments and treaties that they have signed, including the United Nations Convention on the Rights of the Child and the United Nations Convention on the Law of the Sea, which entail binding obligations. In terms of the former, the UN Committee on the Rights of the Child issued new guidance in 2023 that calls for governments to take action to protect children from the climate crisis (UN, 2023b). In 2024, the International Tribunal for the Law of the Sea handed down a finding that greenhouse gases constitute marine pollution, obligating states to act (Scott, 2024). However, there are clear failings for international legal instruments to meet the challenges that climate change presents, with 'climate refugees' being one of the most significant weaknesses. As Dioszegi noted: 'millions of climate refugees remain stuck in legal limbos and gaps, as existing legal frameworks fail to acknowledge new realities presented by climate change and tailor protections for people forcibly displaced by climate disasters' (2025: n.p.).

Climate change exacerbates already profound inequalities between people (see Hickel *et al.* (2022) on how the Global North has 'drained' $10 trillion per year from the Global South). It was only in the 2022 International Panel on Climate Change (IPCC) report that the impacts of colonialism, both historic and ongoing, were recognised as one of the catalysts to the current climate crisis (IPCC, 2022). Rickards highlighted:

> [...] climate justice illuminates two broad human groups. The first is those who have benefited from fossil fuel- and colonialism-enabled economic development and now sit in positions of privilege. Compared to others, this group is well placed to adapt to the negative side effects of the development trajectory they have helped generate and have largely benefited from – side effects that include, but are not limited to, a disrupted climate…[and] the second group: the much larger and more diverse population who have long been, and continue to be, exploited and sacrificed in the development processes that have birthed climate change. (2020: 295)

It is these injustices and inequities that have inspired demand for reparations and the Loss and Damage Fund. But the 'finance gap' for adaptation has been described as an 'adaptation emergency' in the 2023 *Global Adaptation Gap Report* from the United Nations Environment Programme (UNEP), which estimated the Global South will require US$215 billion to US$387 billion per year this decade. However, the report noted that public multilateral and bilateral adaptation finance flows to these countries declined by 15 per cent to US$21 billion in 2021, despite pledges made at COP26 in Glasgow to provide some US$40 billion per year in adaptation finance support by 2025 (UNEP, 2023).

The politicisation of climate change policy and action is also a threatening development. This politicisation has gathered pace since 2005, with debates about the annual negotiations for the United Nations Framework Convention on Climate Change (UNFCCC), especially as Global South nations began to demand climate justice (Warlenius, 2017: 131). Most recently, it was reported that the US National Institutes of Health began 'mass terminations of research grants that fund active scientific projects because they no longer meet "agency priorities", including threatening work focused on environmental justice and climate change' (Kozlov & Mallapaty, 2025: n.p.). Such efforts seek to shield industry from the governmental regulation that climate action requires, to terminate progress towards gender and racial equity and to stave off wealth redistribution in the interests of justice.

Why climate change is a justice issue: Inequities in impact, voice and response

Climate change is an escalating global crisis that magnifies existing inequalities, disproportionately harming those least responsible, especially marginalised communities and ecosystems. Climate justice, as an integrative framework, demands more than technical fixes or carbon accounting, it necessitates a fundamental reckoning with how multiple forms of injustice coalesce across scales, systems and species. In the context of tourism, this reckoning is long overdue. The current climate crisis is not only the product of excessive carbon emissions but also of systemic injustices embedded in colonial histories, neoliberal economic logics and anthropocentric worldviews that have long shaped tourism development (Bigby *et al.*, 2024; Higgins-Desbiolles, 2008, 2020; Jamal & Higham, 2020). The sector's continued prioritisation of economic growth, often at the expense of social and ecological wellbeing, has exacerbated inequalities and left vulnerable communities and ecosystems increasingly exposed to climate impacts. Despite tourism's rhetoric of sustainability, the failure to embed justice into climate policy and practice reflects a deep epistemic and political failure (Becken & Rastegar, 2025; Rastegar & Becken, 2024).

At the foundation of climate justice lies the environmental justice movement, which emerged as a grassroots response to environmental racism and the disproportionate siting of environmental hazards in poor and racially marginalised communities. Environmental justice exposed how environmental burdens, such as pollution, waste and hazardous development, were unevenly distributed across racial, class and geographic lines. This foundational movement reframed environmentalism from a conservationist agenda to one of recognition, distributive and procedural fairness (Agyeman *et al.*, 2002). Over time, this critique evolved into what is now recognised as climate justice, a framework that applies environmental justice principles to the global, intergenerational and systemic impacts of climate change. As climate change continues to amplify environmental degradation, resource scarcity and displacement, it reproduces the very injustices the environmental justice movement sought to combat. Therefore, environmental justice must be seen not only as a precursor to climate justice but as its epistemological and political root (Rastegar *et al.*, 2023). As the recent *Climate Justice in Tourism* report notes, climate injustice occurs when those least responsible for emissions are the most exposed to their consequences and the least resourced to respond (Bigby *et al.*, 2024).

Recognition justice, an essential pillar of climate justice, is routinely violated in tourism-related climate policy, which often ignores the lived experiences and knowledges of Indigenous peoples and marginalised groups. In Australia, the ongoing exclusion of Aboriginal communities from climate governance related to the Great Barrier Reef illustrates the colonial continuity of policymaking (Lopez, 2019). While these communities possess deep ecological knowledge and cultural ties to the reef, they are frequently sidelined in decisions that affect their territories and livelihoods. Similarly, the rise of 'green gentrification' in urban tourism destinations such as Łódź Stare Polesie (Poland) and Leipzig's inner east (Germany) reveals how climate adaptation strategies, such as the greening of public spaces, are co-opted by market forces that displace low-income and minority communities under the guise of environmental progress (Haase *et al.*, 2022). These examples demonstrate that climate initiatives devoid of recognition justice reproduce the very inequalities they purport to address.

Procedural justice, which demands meaningful participation in decision making, remains largely aspirational in tourism's climate agenda. The global push toward net zero, for instance, has largely been shaped by powerful actors in the Global North, with minimal engagement from those most affected by climate change. The UNWTO's Tourism Shared Pathways initiative and the Glasgow Declaration reference inclusivity and collaboration yet provide little evidence of mechanisms for genuine stakeholder involvement from the Global South (Rastegar & Ruhanen, 2023; UNWTO, 2021). The *Climate Justice in Tourism* report confirms this gap, noting that 'there is little evidence in industry plans that the organisation sought the perspectives of communities most closely impacted by climate change during the plan's creation' (Bigby *et al.*, 2024: 15). This procedural void perpetuates a form of climate colonialism.

Distributive justice, the most frequently invoked yet shallowly applied concept in climate discourse, also warrants critical interrogation. The assumption that tourism's economic benefits can be redistributed to offset its environmental harms is a myth that obscures deeper injustices. In SIDS, tourism is promoted as a development pathway, yet these same destinations bear the brunt of tourism-induced emissions from long-haul travel emissions for which they are not responsible (TPCC, 2023). The case of the Maldives is emblematic: while international tourists enjoy luxury resorts on climate-vulnerable atolls, local communities face housing insecurity, limited access to basic services and increased exposure to rising sea levels. Moreover, distributive justice remains skewed within tourist-receiving countries. In cities like Barcelona and Venice, local residents are displaced or priced out by short-term rentals and cruise tourism, while the costs of infrastructure maintenance, waste management and climate adaptation are socialised across the population (e.g. WEF, 2023). These dynamics make clear that, without structural redistribution of power and resources, distributive justice remains aspirational (Higgins-Desbiolles, 2020).

Restorative justice, perhaps the most neglected dimension, demands an acknowledgement of historical wrongs and proactive measures to repair the damage inflicted on communities and ecosystems (Rastegar, 2025). Yet tourism climate policy remains largely ahistorical, failing to account for the legacy of colonial extraction and forced displacement that underpins many iconic tourism destinations (Higgins-Desbiolles, 2018). For example, the so-called blue economy, often hailed as a sustainable solution for climate-vulnerable destinations, frequently masks and reproduces existing injustices under the guise of innovation and sustainability (Rastegar, 2022). In coastal areas of the Global South, such as the Maldives and Seychelles, tourism growth has led to the enclosure of beachfronts and marine ecosystems for exclusive tourist use, restricting access for local fishers and communities while reinforcing sociospatial segregation (Rastegar & Becken, 2024; TPCC, 2023). Similarly, in Panama, rising sea levels and heat have forced communities reliant on tourism to abandon their homes and traditional livelihoods, revealing the limits of adaptation strategies that fail to protect both cultural continuity and climate resilience (Delacroix & Zamorano, 2024). These examples highlight how tourism-led climate responses often exacerbate existing vulnerabilities rather than repair them. Scholars have cautioned that growth-driven development paradigms, particularly when paired with unsustainable tourism practices, risk reinforcing the 'genocide–ecocide nexus', a dynamic in which economic progress unfolds through the simultaneous erasure of cultural identities and the degradation of ecological systems (Dunlap, 2021). Genuine restorative justice would involve reparative actions such as land return, revenue redistribution and the legal empowerment of Indigenous governance systems.

Ecological justice, which challenges the anthropocentrism of dominant tourism models, is critical to any meaningful climate response. Despite the sector's reliance on natural environments, nonhuman species are routinely treated as scenery or commodities (Danby *et al.*, 2019). Wildlife tourism continues to thrive on the exploitation of animals for entertainment, from elephant rides in Southeast Asia to marine life enclosures in the Caribbean. More insidiously, climate adaptation strategies in tourism destinations often prioritise human infrastructure and economic assets, while neglecting the ecological systems they depend on, disrupting habitats, biodiversity and nonhuman life (Mayer-Pinto *et al.*, 2019). The mass mortality of coral reefs due to warming seas has profound implications not only for marine species but also for Indigenous cosmologies that view reefs as ancestral kin (Great Barrier Reef Marine Park Authority, n.d.). The call for a multispecies justice framework demands a radical shift in tourism ethics, from dominion over nature to co-existence with it (Fennell & Sheppard, 2021; Rastegar, 2022).

Cosmopolitan justice further complicates the picture by exposing the uneven geographies of climate risk and responsibility. The global mobility asymmetry, where a small, affluent segment of the global population accounts for the majority of tourism emissions, undermines any claim to fairness. As India expands its aviation sector to meet rising middle-class demand, it faces criticism for contributing to global emissions (Rastegar & Becken, 2024). Yet such critiques often obscure the fact that more than 80% of the world's population has never flown, and that high-frequency fliers in the Global North continue to dominate global air travel (Gössling & Humpe, 2020). Climate policies that seek to restrict mobility must grapple with this hypocrisy. It is unjust to limit the mobility aspirations of populations historically denied access to leisure travel while allowing elites to maintain carbon-intensive lifestyles. True cosmopolitan justice demands global frameworks that not only curb emissions equitably but also ensure mobility rights are not sacrificed for planetary health in a way that reproduces privilege.

At its core, climate change is a force multiplier of inequality. Addressing it through tourism requires recognising that climate justice is not a technical fix, but a political agenda to be contested, co-created and transformed. It requires more than tokenistic inclusion or the rebranding of existing practices. It necessitates confronting uncomfortable truths: that tourism, as currently constituted, is complicit in the production of social, ecological and epistemic harm. Here, we may need to embrace a 'local turn', one that holistically engages local communities, ecosystems and intergenerational perspectives, to more meaningfully reconnect people, place and planet (Higgins-Desbiolles & Bigby, 2023). It necessitates embracing justice as a performative action, evolving process, one rooted in solidarity, accountability and transformation (Figure I.2). Only through such a lens can the tourism sector transition toward futures that are not only sustainable, but just.

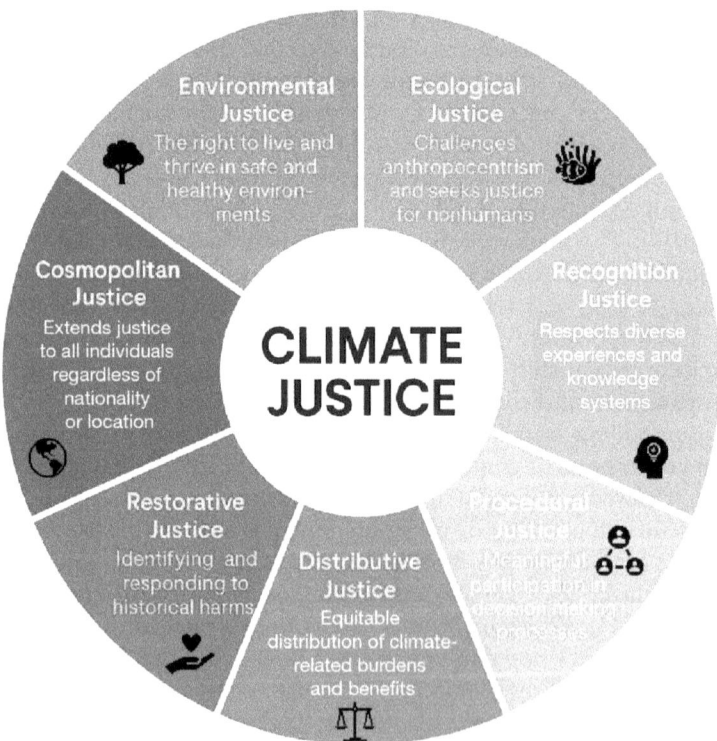

Figure I.2 Understanding climate justice in tourism: A multidimensional framework

Reformism and Radicalism – Degrowth and Decolonisation

'We're resisting climate colonialism, where rich dirty corporations profit from our destruction'. Brianna Fruean, Samoan climate activist and member of the Pacific Climate Warriors delegation at Cop26 (2021: n.p.)

The threat of climate change and wider environmental crises that have resulted from human activities and excessive demands on a finite environment indicate that action is needed. There are debates concerning whether reform of the current economic system to a form of 'green capitalism' (see Beuret & Bloom, 2025) and use of technologies (such as renewable energy technologies, 'sustainable' aviation fuels and artificial intelligence) will be sufficient to manage the situation or instead whether we need fundamental and radical transformations of our current systems (see Ritter & Thaler, 2022). Following a radicalism line, another essential debate arises concerning pathways of degrowth and decolonisation, which exposes fault lines between the Global North and Global South.

There are a variety of reformist strategies to effect decarbonisation in the global economy. Well-known approaches include transitioning away from fossil fuels to renewable energy sources, implementing energy efficiencies, electrification, sustainable land use reforms and carbon capture and storage. There are new and emerging reform proposals that are less well known. One example is Hickel and Stevenson's recommendation to use 'credit guidance' as a way to align finance with the goals of the green transition. This describes states' central banks using their abilities to guide credit to more socially and ecologically sound activities. This would be accomplished by limiting or stopping 'credit that commercial banks and other financial institutions direct to destructive sectors (say, fossil fuels and SUVs) while increasing credit flows to more beneficial sectors (such as renewables and other green technologies)' (Hickel & Stevenson, 2024: n.p.).

However, a growing number of analysts are demonstrating that sustainability and climate justice are not attainable in the current capitalist system (e.g. Bianchi & Milano, 2024; Fletcher *et al.*, 2021; Hickel, 2023), largely because of its requirement for continual consumption to drive economic growth, which is what underpins corporate profits and capitalist accumulation. Such an economic system is not compatible with a finite environment which is burdened with meeting resource demands and acting as a sink for the many pollutions and wastes it produces in the process, including carbon emissions. Hickel (2023) argues that social and ecological justice are intimately entangled and as we address ecological crises, such as climate change, we must do so with attention to social justice outcomes; that is addressing the grave inequalities that capitalism catalyses. Hickel states: 'No political program that promises to analyze and resolve the ecological crisis can hope to succeed if it does not also simultaneously [...] analyze and resolve the social crisis' (2023: n.p.). However, this is impossible under this capitalist system because: 'Capital wields the power to mobilize our collective labor and our planet's resources for whatever it wants, determining what we produce, under what conditions, and how the surplus we generate shall be used and distributed' (Hickel, 2023: n.p.). Rather than the world capitalist production system, he advocates a 'post capitalist' transition. This particularly entails recognising: '[...] socially necessary production that clearly needs to increase for social progress, and the destructive and less-necessary forms of production that urgently need to be scaled down. This is the revolutionary world-historical objective that faces our generation' (Hickel, 2023: n.p.). A critical question is whether tourism, or certain forms of tourism, can be justified at this critical juncture of polycrisis or whether it falls into the category of Hickel's 'less-necessary' activities (see Fletcher *et al.*, 2021; Higgins-Desbiolles, 2022b). Some of the chapters in this edited volume will grapple with these issues.

Degrowth is an agenda of voluntary societal shrinking of production and consumption with the goal of achieving social and ecological sustainability (Hall, 2009). With the exceeding of planetary boundaries and especially widescale climate change, degrowth is an approach that makes logical sense. The degrowth movement in tourism has proposed strategies to slow the growth trajectory in tourism and achieve a 'steady state' approach, foster slow tourism and advocate for more conscious consumerism in tourism (Higgins-Desbiolles, 2018). Like the wider degrowth movement, degrowth in tourism has evolved to take aim at capitalist forms of wealth extraction as a key catalyst to unsustainability and injustice, and so currently tends to advocate anti-capitalist or 'post-capitalist' futures (e.g. Fletcher *et al.*, 2021; Higgins-Desbiolles, 2018). Degrowth in tourism advocates have also turned to diverse knowledges and lifeways to draw on local sociocultural-economic systems to counter hegemonic capitalism, for example Buen Vivir in Latin America (e.g. Chassagne & Everingham, 2021).

However, degrowth in tourism may not be uniformly implemented globally with just outcomes, when many Global South countries and SIDS have yet to attain the level of development, wealth and wellbeing for their peoples they aspire to. Those Global South countries that are highly tourism dependent may want greater growth in tourism to meet their development goals. They also might argue that it is a climate injustice if climate action policies force degrowth on them by discouraging, for example, long-haul aviation through greater taxation, limits to expansion and promotion of flight shame (see Chapters 3 and 4). The ecological goal of reducing carbon emissions confronts a social inequity of unfulfilled development goals. This was part of the point that Hickel (2023) was addressing in his advocacy of a socioecologically aware post-capitalist transition which could chart a just and equitable pathway to universal wellbeing and fair sharing of the benefits of development and the burdens of climate change. To arrive at such a possibility, decolonisation and decolonisation in tourism will be vital work to undertake.

Decolonisation today refers to 'the process of returning sovereign control of indigenous lands to indigenous nations, and concurrently, revitalizing indigenous lifeways' (Grimwood *et al.*, 2024: 903). Decolonising tourism is an emerging approach that seeks to expose: the imperial and colonial roots of tourism and as well as their 'ongoingness' (Higgins-Desbiolles, 2022a); the ways Global South communities have been enculturated in Western-centric development models that have fostered a colonised mentality (see Ayikoru, 2024); the system of tourism dependency that advantages hegemonic tourism industrial interests and privileged tourists; and the wealth of Indigenous, local and pluriversal epistemologies, ontologies and lifeways that continue to exist despite colonial assault that are sustaining or are awaiting revival for colonised or formerly colonised peoples.

A decolonising approach to climate change exposes the roots of today's predicament in the colonial and extractive ideologies that are built on racial, gender and class hierarchies at their core. As we have discussed, the cascading and compounding impacts of climate change are suffered highly unevenly, with the most severe impacts too often striking those who have least contributed to the crisis and are least resourced to adapt. Because efforts to address climate change are enacted in a context of already existing gross inequities and injustices, decolonising work is required in our climate responses at all scales. Grimwood *et al.* noted:

> The responsibility for enacting and making tangible decolonial futures must be collectively shouldered, and no longer presumed to be of concern only to colonized communities [...] With decolonization comes the potential to transform tourism and begin healing the traumas and scars it has inflicted upon lands and life. (2024: 903)

A climate just approach would respect sovereignty and self-determination of communities. Climate impacted communities, including Indigenous communities, SIDS and Global South communities are refusing a victim designation being imposed on them – 'We are not just victims to this crisis. We are not drowning, we are fighting. Now the world needs to hear us and follow our lead' (Fruean, 2021: n.p.). This is an important act of resistance against ongoing imperialism and extractive logics and a vital strike at the root cause of the climate crisis we confront (see Chapter 10).

Terminology

The topic of climate justice and climate justice in tourism is very new and emergent. As a result, some of the concepts and terms may not be widely known. This section briefly outlines some of these terms that are essential to understand, many of which are used in the subsequent chapters.

Climate debt: Climate debt is '[…] Basically the idea that climate change is caused by rich people while mainly harming people that are poor, and therefore, the former should take the burden of mitigation and adaptation costs [… it] is at the very core of climate justice' (Warlenius, 2017: 132).

Climate colonialism: Climate colonialism refers to the ongoing legacies of colonialism that see the Global North and Global South differentially impacted by climate change. It also refers to the Global North's efforts to continue to exploit the Global South for its climate agendas, such as through carbon-offsetting projects. Sultana noted: 'Climate coloniality reproduces the hauntings of colonialism and imperialism through climate impacts in the post-colony (located primarily in the tropics and subtropics where climate-induced disasters and shifts have been prevalent for some time' (2022b: 3).

Climate justice: Climate justice entails the recognition that climate change has differential, uneven and disproportionate impacts and that efforts to mitigate and adapt to climate change can bring marginalisation, injustice and oppression to vulnerable people. This concept has a particular focus on unjust relations between the nations of the Global North and Global South.

Climate just tourism taxes: Recognising that tourist consumption in destinations can bring injustices and too little benefit, destinations can implement 'climate just tourist taxes' to ensure adequate compensation for negative tourism impacts, to fund relief efforts for climate-induced disasters and to protect social and ecological wellbeing in tourism destinations. An example is Hawai'i's new 'green fee' which was described as intended to address the growing impacts of climate change (Graham, 2025).

Climate washing: Climate washing or climate-related greenwashing refers to misleading or false reporting of climate actions by business, governments and other entities. Climate washing '[…] communicates a message that exaggerates or misrepresents climate credentials through advertising, branding, labelling or reporting' (Schuijers, 2023: n.p.).

Common but differentiated responsibilities (CBDR): The CBDR principle recognises that Global North nations, due to their industrialisation and higher emissions, bear a greater historical responsibility for climate change than the Global South. In the Kyoto Protocol, this meant 'Annex 1' countries, mostly 'developed' nations, were obligated to commit to quantified emissions reductions and additional measures on technology transfers and adaptation finance owed to 'developing' nations (see de Lucia, 2007).

Conscientisation: Paulo Freire's concept of 'conscientisation' (conscientização) refers to mutual, critical awareness work that the oppressed undertake to understand their conditions and leads to a determination that 'the concrete situation that begets oppression must be transformed' (1970: 50).

Critical climate justice praxis: Critical climate justice praxis is a framework that combines critical reflection and transformative action to address the systemic root causes of climate injustices. Sultana's (2022a) analysis referenced Paulo Freire, who described praxis as 'reflection and action upon the world in order to transform it' (Freire, 1970: 51, cited in Sultana, 2022a: 119). Climate justice has catalysed actions and collaborations by social justice and environmental movements.

Critical feminist climate justice: This approach spotlights the unequal impacts of both climate change and the responses to climate change on women, diverse genders, peoples made marginalised and vulnerable, future generations and our more than human relations. It employs an intersectional and structural justice approach to addressing the climate crisis. The solutions and responses it offers are not solely focused on empowering women, but rather advancing the collective struggle for justice.

Disaster capitalism: Disaster capitalism refers to exploitative economic practices in which corporations, governments or elites capitalise on crises triggered by climate disasters, conflicts or socioeconomic breakdown to advance profit-making agendas (Klein & Sproat, 2023).

Environmental justice: Environmental justice offers a social justice analysis of environmental decision making, revealing how environmental racism results in Black, Indigenous and marginalised communities being negatively impacted by government decision making, including in climate change policies.

Gender justice: Gender justice refers to achieving fairness and equity for all genders. A gender just approach would recognise that different genders may have different needs and circumstances and seeks to ensure equitable access to resources and opportunities based on those needs. Critical gender justice is concerned with the systemic inequalities and supports solidarity across movements.

Global South: 'The use of the phrase 'Global South' marks a shift from a focus on development or cultural difference toward an emphasis on geopolitical power relations… The term Global South functions as more than a metaphor for underdevelopment. It references an entire history of colonialism, neo-imperialism, and differential economic and social change through which large inequalities in living standards, life expectancy, and access to resources are maintained' (Dados & Connell, 2012: 12–13).

Inequality: Inequality is a multidimensional concept encompassing 'economic inequalities such as inequalities in income, wealth, wages and social protection, as well as social and legal inequalities where different groups are discriminated, excluded or otherwise denied full equality' (UN, n.d.).

Intersectionality: Intersectionality describes systems of inequality, including class hierarchies, racism and supremacy, sexism, heterosexism, cisgenderism, ableism and border policing, '[…] constitute one another, meaning that they construct one another and interact to create institutions and differential social positions […] Social institutions and positions are therefore shaped by multiple, mutually constituting, divisions operating simultaneously. Applying intersectionality, in both theory and practice, therefore means engagement with the interrelationship of these systems of inequality' (Christoffersen & Emejulu, 2023: 633).

Intragenerational and intergenerational equity: Key justice concerns are ensuring equity in allocations of resources and opportunities within the current generation (intragenerational equity) and also ensuring future generations have access to the resources they need for their wellbeing (intergenerational equity).

Just transition: A just transition refers to the concept of '[…] ensuring that no one is left or pushed behind in the transition to low-carbon and environmentally sustainable economies and societies' (UN, 2023a: 1).

Loss and Damage Fund: This describes official funding arrangements for assisting Global South countries that are particularly vulnerable to the adverse effects of climate change. The 'Santiago Network' connects vulnerable countries with providers of technical assistance, knowledge and resources to address climate risks.

Marginalisation: Marginalisation describes 'A state that prevents individuals or groups from full participation in social, economic, and political life and from asserting their rights' (WFTO, 2015: 4).

Mobility justice: 'Mobility justice is an overarching concept for thinking about how power and inequality inform the governance and control of movement, shaping the patterns of unequal mobility and immobility in the circulation of people, resources, and information' (Sheller, 2018: 30–31).

Racial capitalism: Racial capitalism refers to interconnected system of racialised and colonial exploitation that enables the process of capital accumulation. 'According to this framework, capitalism, as we know it today, would not have been possible if not for imperialism, colonialism, racial slavery, expropriation, and superexploitation. Capital accumulation would not be possible today if not for these ongoing logics' (Edwards, 2021: 2).

Reparations: Reparations pertain to compensation for the injustices of climate change paid by the wealthy Global North (most responsible for emissions) to the nations of the Global South (least responsible and most impacted by climate change). Known officially in UN channels as Loss and Damage Funding.

Tourism dependency: Tourism dependency describes a situation where a state or community becomes overly reliant on tourism, leading to vulnerability when tourism suffers a downturn. Climate change will add to the risk of being tourism dependent.

Trade-offs: Climate justice trade-offs include balancing climate action with other societal goals like economic development and equitable distribution of resources, often leading to difficult choices.

Vulnerability: 'The conditions determined by physical, social, economic, and environmental factors or processes which increase the susceptibility of an individual, a community, assets or systems to the impacts of hazards' (UNDRR, 2017).

These are some of the terms and concepts that we need to begin to engage with the complexities of climate justice in tourism. Climate justice vocabulary helps in revealing underlying issues of power, privilege, structural injustice and various forms of supremacism. Words matter and so we should also be aware of and avoid the use of deficit-based language in talking about those who are impacted by climate injustice. Calling them 'victims', describing them as 'vulnerable' and advocating 'saviour' interventions obscures the nature of the structural injustices that have made them vulnerable, marginalised and/or racialised. It is intended to obscure the culpability of the privileged in the climate crisis and their need to take responsibility and make reparations for their historical and ongoing exploitation and wealth extraction. This edited volume offers many rich insights into these dynamics so that climate (in)justice can be better understood and addressed.

Overview of Book

This book examines the complex relationship between tourism and climate justice through 12 rigorously researched and/or theorised chapters that are offered from tourism experts and actors from diverse parts of the world. The book is organised into four thematic parts that assist in unpacking, contextualising, interrogating and making sense of the meaning and value of climate justice in tourism futures.

The first part, 'The Challenge of Climate Change for the Tourism Sector, Destinations and Communities', discusses the asymmetric impacts of climate change on tourism. Chapter 1, by Ya-Yen Sun *et al.*, investigates the intersection of climate justice and global tourism, analysing disparities in emissions contributions and climate impacts. It accentuates the urgency of embedding equity into climate agreements to break the cycle where high emitters profit while vulnerable nations bear disproportionate costs. Chapter 2, by He *et al.*, amplifies this critique through interviews with global tourism professionals, demonstrating how carbon-intensive travel habits disproportionately impact many Global South countries that bear the brunt of climate change. It also critiques superficial industry commitments to sustainability, rooted in a lack of accountability and transparency, reflecting consequences of an absence of structural justice. Chapter 3, by Broekema *et al.*, shifts the focus to aviation policies, examining the socioeconomic implications of aviation climate policies through the lens of a 'just transition'. Crucially, it provides critical reflective questions for policymakers to align decarbonisation goals with socioeconomic justice. Together, these chapters call to depart from the illusion of tourism as a 'win-win' for development and sustainability, reframing it as a significant hurdle in achieving climate justice.

The second part, 'Governance (In)Justices in Tourism', dives deeper into systemic issues that deepen climate inequalities. Chapter 4, by Sulub and Subtil Fialho, frames air travel as a pillar of global inequality by engaging with the Stay Grounded network's advocacy for aviation degrowth and climate justice. Drawing on the case of the Mayan people in Mexico, this chapter advocates for a just transition, calling for drastic reduction in air traffic, reparation of climate debts and the prioritisation of Indigenous land rights. Chapter 5, by Dahal and Subedi, directs us to the mountain tourism economy of Nepal, where communities face the dual threat of glacial floods and prolonged droughts – both symptoms of a climate-induced water crisis characterised by extremes: either 'too much' or 'too little' water. It also demonstrates that while traditional community-based water management models offer a viable solution, their effectiveness is undermined by power asymmetries between large operators and small-scale lodges. Chapter 6, by Sheller *et al.* explores mobility justice in the US Virgin Islands, showing how historical colonialism, extractive industries and contemporary tourism infrastructure perpetuate unequal mobilities. It also reveals how racialised hierarchies and climate disruptions exacerbate stalled mobilities, driving brain drain and uneven recovery. Together, these chapters call for the critical reflection of contemporary approaches to tourism governance, urging transformative actions via degrowth, reparations and equitable transitions to prioritise climate justice.

The third part, 'Case Studies in Climate (In)Justice in Tourism', engages the issues of climate injustice by examining cases from Kenya, Palestine, Greenland and Tyrol. Chapter 7, by Gona and Atieno, examines the unequal power relationships among carbon developers, tourism investors and pastoral communities in Kenya's nature conservancies. It critiques the 'triple-win' narrative of carbon markets, arguing that carbon markets perpetuate colonial approaches to resource management. Finally, it advocates for governance reforms centring Indigenous land stewardship, transparent benefit sharing and intergenerational equity. Chapter 8, by Qumsiyeh and Bibee, confronts tourism under Israeli occupation in Palestine, where Israel's apartheid policies and resource monopolies exacerbate climate vulnerabilities, water scarcity and habitat loss. It, however, suggests that tourism can be a decolonial tool that fosters equitable practices. Chapter 9, by Nadegger and Ren, takes the context of two distinct melting destinations, i.e. Greenland and Tyrol, to demonstrate how affective solidarity emerges from emotional and material injustices. Crucially, it highlights how climate justice movements contest structural inequalities while navigating tensions between growth and ecological limits. Together, these cases illustrate how climate injustice is embodied and contested, emphasising the importance of Indigenous self-determination and offering lessons in collective resilience and agency.

The final section, 'Imagining More Just Tourism Futures', charts radical pathways for climate justice and socioecological transformation. Chapter 10, by Andrews *et al.*, recentres Indigenous sovereignty through Vanuatu's Regenerative Vanua Stewardship Framework, a holistic model that advances regenerative intercultural hosting through stewardship of self, others, place and nature. It then challenges tourism's extractive patterns, emphasising Indigenous leadership and epistemic justice to empower custodians for leading climate resilient futures. Chapter 11, by Higgins-Desbiolles, calls for the need to integrate critical feminist justice thinking into climate discourse, asking to go beyond reformist approaches to proactively dismantle structural systems of oppression deep rooted in patriarchy, colonialism and racism. It concludes with radical recommendations, suggesting delinking from capitalist tourism, prioritising climate reparations and reimagining tourism through a justice lens. Chapter 12, also by Higgins-Desbiolles, applies Jem Bendell's Deep Adaptation Framework to reflect on how tourism education and practice can be reformed in the context of possible societal and tourism collapse due to polycrisis. It advocates radical rethinking of tourism practices, including adoption of regenerative practices, relinquishment of growth-centric models and restoration of Indigenous-led economies.

Together, these chapters offer novel insights into climate justice in tourism at a timely moment when attention to climate change issues in tourism is growing. With a lens of justice and equity, incremental reformism and technological fixes ring hollow in our response to the global challenge we confront. The chapters gathered here collectively accomplish two key critical tasks for the climate justice in tourism agenda: firstly, they critically unpack the nature of injustices in tourism that arise from or are exacerbated by the climate crisis we face, and secondly, they offer a constructive agenda for just transitions to a more climate just future.

Looking forward, we note the hopeful insight that social tipping points may counterbalance the current power configuration that sees the hegemony of fossil fuel interests and corporate agendas pushing business as usual (Steffen & Morgan, 2021: 237). However, such social tipping points depend on us, and so this edited volume is intended to play its role in fostering a positive social tipping point in tourism. This collaborative project on climate justice in tourism is based on a shared commitment to promote a climate just transition and we hope you will value joining us as you read the chapters that follow.

References

Agyeman, J., Bullard, R.D. and Evans, B. (2002) Exploring the nexus: Bringing together sustainability, environmental justice and equity. *Space and Polity* 6 (1), 77–90. https://doi.org/10.1080/13562570220137907.

Ayikoru, M. (2024) Pragmatic arguments for decolonising tourism praxis in Africa. *Tourism Geographies* 26 (6), 917–934. https://doi.org/10.1080/14616688.2024.2335955.

Becken, S. and Rastegar, R. (2025) Advancing climate justice in tourism: A critical evaluation of the TPCC Stocktake. *Annals of Tourism Research* 113, 103962. https://doi.org/10.1016/j.annals.2025.103962.

Beuret, N. and Bloom, P. (2025) Climate capitalism won't save us. *The Ecologist*. See https://theecologist.org/2025/jun/09/climate-capitalism-wont-save-us (accessed June 2025).

Bianchi, R.V. and Milano, C. (2024) Polycrisis and the metamorphosis of tourism capitalism. *Annals of Tourism Research* 104, 103731. https://doi.org/10.1016/j.annals.2024.103731.

Bigby, B.C., Smith, J. and Higgins-Desbiolles, F. (2024) Climate justice in tourism: An introductory guide. Travel Foundation. See https://www.thetravelfoundation.org.uk/wp-content/uploads/2024/07/Climate_Justice_Tourism_v5.pdf (accessed June 2025).

Chassagne, N. and Everingham, P. (2021) Buen vivir. In F. Higgins-Desbiolles, A. Doering and B.C. Bigby (eds) *Socialising Tourism: Rethinking Tourism for Social and Ecological Justice* (pp. 214–228). Routledge.

Christoffersen, A. and Emejulu, A. (2023) 'Diversity within': The problems with 'intersectional' white feminism in practice. *Social Politics: International Studies in Gender, State & Society* 30 (2), 630–653. https://doi.org/10.1093/sp/jxac044.

Dados, N. and Connell, R. (2012) The Global South. *Contexts* 11 (1), 12–13. https://doi.org/10.1177/1536504212436479.

Danby, P., Dashper, K. and Finkel, R. (2019) Multispecies leisure: Human-animal interactions in leisure landscapes. *Leisure Studies* 38 (3), 291–302. https://doi.org/10.1080/02614367.2019.1628802.

Delacroix, M. and Zamorano, J. (2024) Panama prepares to evacuate first island in face of rising sea levels. *AP News*. See https://apnews.com/article/panama-island-guna-climate-change-f368711649ff6986ea25a79534405a84 (accessed June 2025).

De Lucia, V. (2007) Common but differentiated responsibility. In C.J. Cleveland (ed.) *Encyclopedia of Earth*. Environmental Information Coalition. See https://www.ipcc.ch/apps/njlite/srex/njlite_download.php?id=5114 (accessed 15 June 2025).

Dioszegi, R. (2025) No refuge for climate refugees: How international environmental law and human rights law can guide Lex Specialis. *Sec Jure*. See https://www.secjure.nl/2025/02/25/no-refuge-for-climate-refugees-how-international-environmental-law-and-human-rights-law-can-guide-lex-specialis/ (accessed June 2025).

Dunlap, A. (2021) The politics of ecocide, genocide and megaprojects: Interrogating natural resource extraction, identity and the normalization of erasure. *Journal of Genocide Research* 23 (2), 212–235. https://doi.org/10.1080/14623528.2020.1754051.

Edwards, Z. (2021) Racial capitalism and COVID-19. *Sustainable Human Development*, August. See https://jussemper.org/Resources/Economic%20Data/Resources/Zophia EdwardsRacialCapitalism.pdf (accessed June 2025).

Fennell, D. and Sheppard, V. (2021) Tourism, animals and the scales of justice. *Journal of Sustainable Tourism* 29 (2–3), 314–335. https://doi.org/10.1080/09669582.2020.1768263.

Fletcher, R., Blanco-Romero, A., Blázquez-Salom, M., Cañada, E., Murray Mas, I. and Sekulova, F. (2021) Pathways to post-capitalist tourism. *Tourism Geographies* 25 (2–3), 707–728. https://doi.org/10.1080/14616688.2021.1965202.
Freire, P. (1970) *Pedagogy of the Oppressed*. Continuum.
Fruean, B. (2021) Pacific islanders aren't just victims – we know how to fight the climate crisis. *The Guardian* (Online). See https://www.theguardian.com/commentisfree/2021/nov/02/pacific-islanders-fight-climate-crisis-cop26 (accessed April 2025).
Glen, S. and Clarke, J. (2024) Climate change is a class issue. Published by the Authors. See https://climateandclass.net/wp-content/uploads/2024/10/climate-change-is-a-class-issue-3.pdf (accessed January 2025).
Gössling, S. and Humpe, A. (2020) The global scale, distribution and growth of aviation. *Global Environmental Change* 65, 102194. https://doi.org/10.1016/j.gloenvcha.2020.102194.
Graham, D. (2025) Hawaii adds 'green fee' to hotel stays to combat climate change. *Skift* (Online), 29 May. See https://skift.com/2025/05/28/hawaii-adds-green-fee-to-hotel-stays-to-combat-climate-change/ (accessed May 2025).
Great Barrier Reef Marine Park Authority (n.d.) Great Barrier Reef traditional owners. *Reef Authority*. See https://www2.gbrmpa.gov.au/learn/traditional-owners/great-barrier-reef-traditional-owners#/ (accessed May 2025).
Grimwood, B.S.R., Lee, E. and Higgins-Desbiolles, F. (2024) Unsettling geographies of tourism. *Tourism Geographies* 26 (6), 899–916. https://doi.org/10.1080/14616688.2024.2402997.
Haase, A., Koprowska, K. and Borgström, S. (2022) Green regeneration for more justice? An analysis of the purpose, implementation, and impacts of greening policies from a justice perspective in Łódź Stare Polesie (Poland) and Leipzig's inner east (Germany). *Environmental Science & Policy* 136, 726–737. https://doi.org/10.1016/j.envsci.2022.08.001.
Hall, C.M. (2009) Degrowing tourism: Décroissance, sustainable consumption and steady-state tourism. *Anatolia* 20 (1), 46–61. https://doi.org/10.1080/13032917.2009.10518894.
Hickel, J. (2023) The double objective of democratic ecosocialism. *The Monthly Review* (Online). See https://monthlyreview.org/2023/09/01/the-double-objective-of-democratic-ecosocialism/ (accessed June 2025).
Hickel, J., Dorninger, C., Wieland, H. and Suwandi, I. (2022) Imperialist appropriation in the world economy: Drain from the global South through unequal exchange, 1990–2015, *Global Environmental Change* 73, 102467. https://doi.org/10.1016/j.gloenvcha.2022.102467.
Hickel, J. and Stevenson, C. (2024) How to force capitalism to stop climate change. *Foreign Policy* (Online). See https://foreignpolicy.com/2024/08/16/climate-change-central-banks-credit-guidance/ (accessed June 2025).
Higgins-Desbiolles, F. (2008) Justice tourism and alternative globalisation. *Journal of Sustainable Tourism* 16 (3), 345–364. https://doi.org/10.1080/09669580802154132.
Higgins-Desbiolles, F. (2018) Sustainable tourism: Sustaining tourism or something more? *Tourism Management Perspectives* 25, 157–160. https://doi.org/10.1016/j.tmp.2017.11.017.
Higgins-Desbiolles, F. (2020) Socialising tourism for social and ecological justice after COVID-19. *Tourism Geographies* 22 (3), 610–623. https://doi.org/10.1080/14616688.2020.1757748.
Higgins-Desbiolles, F. (2022a) The ongoingness of imperialism: The problem of tourism dependency and the promise of radical equality. *Annals of Tourism Research* 94, 103382. https://doi.org/10.1016/j.annals.2022.103382.
Higgins-Desbiolles, F. (2022b) Subsidiarity in tourism and travel circuits in the face of climate crisis. *Current Issues in Tourism* 26 (19), 3091–3101. https://doi.org/10.1080/13683500.2022.2116306.
Higgins-Desbiolles, F. and Bigby, B.C. (eds) (2023) *The Local Turn in Tourism: Empowering communities*. Channel View Publications.

ICJ (2023) Request for advisory opinion. See https://www.icj-cij.org/sites/default/files/case-related/187/187-20230412-app-01-00-en.pdf. (accessed May 2025).

IPCC (2022) Summary for policymakers. See https://www.ipcc.ch/report/ar6/wg2/downloads/report/IPCC_AR6_WGII_SummaryForPolicymakers.pdf (accessed June 2025).

Jamal, T. and Higham, J. (2020) Justice and ethics: Towards a new platform for tourism and sustainability. *Journal of Sustainable Tourism* 29 (2–3), 143–157. https://doi.org/10.1080/09669582.2020.1835933.

Klein, N. and Sproat, K. (2023) Why was there no water to fight the fire in Maui? *The Guardian* (Online), 18 August. See https://www.theguardian.com/commentisfree/2023/aug/17/hawaii-fires-maui-water-rights-disaster-capitalism (accessed August 2023).

Klerk, B. (2024) To lex specialis, or not to lex specialis? The Paris-UNCLOS nexus in the ITLOS advisory opinion on climate change. *NCLOS blog*. See https://site.uit.no/nclos/2024/06/28/nclos-blog-series/ (accessed May 2025).

Kozlov, M. and Mallapaty, S. (2025) Exclusive: NIH to terminate hundreds of active research grants. *Nature* (Online). See https://archive.is/gz3wW#selection-871.0-878.0 (accessed May 2025).

Lopez, J.R. (2019) Australia: As the Great barrier Reef shrinks, Torres Strait Islanders have everything to lose. In P. Grant (ed.) *Minority and Indigenous Trends 2019: Focus on climate justice*. Minority Rights Group (pp. 115–118). See https://minorityrights.org/publications/minority-and-indigenous-trends-2019/ (accessed June 2025).

Mayer-Pinto, M., Dafforn, K.A. and Johnston, E.L. (2019) A decision framework for coastal infrastructure to optimize biotic resistance and resilience in a changing climate. *Bioscience* 69 (10), 833–843. https://doi.org/10.1093/biosci/biz092.

McVey, M. and Savaresi, S. (2025). The ICJ advisory opinion on climate change: A business and human rights perspective. *OpinioJuris*. See https://opiniojuris.org/2025/08/04/the-icj-advisory-opinion-on-climate-change-a-business-and-human-rights-perspective/ (accessed September 2025).

Pacific Islands Students Fighting Climate Change (n.d.) Our journey. See https://www.pisfcc.org/ourjourney (accessed May 2025).

Rastegar, R. (2022) Towards a just sustainability transition in tourism: A multispecies justice perspective. *Journal of Hospitality and Tourism Management* 52, 113–122. https://doi.org/10.1016/j.jhtm.2022.06.008.

Rastegar, R. (2025) Regenerative justice and tourism: How can tourism go beyond restoration? *Annals of Tourism Research* 111, 103896. https://doi.org/10.1016/j.annals.2025.103896.

Rastegar, R. and Becken, S. (2024) Embedding justice into climate policy and practice relevant to tourism. *Journal of Sustainable Tourism* 33 (10), 2011–2028. https://doi.org/10.1080/09669582.2024.2377720.

Rastegar, R., Higgins-Desbiolles, F. and Ruhanen, L. (2023) Tourism, global crises and justice: Rethinking, redefining and reorienting tourism futures. *Journal of Sustainable Tourism* 31 (12), 2613–2627. https://doi.org/10.1080/09669582.2023.2219037.

Rastegar, R. and Ruhanen, L. (2023) Climate change and tourism transition: From cosmopolitan to local justice. *Annals of Tourism Research* 100, 103565. https://doi.org/10.1016/j.annals.2023.103565.

Richardson, K., Steffen, W., Lucht, W., *et al.* (2023) Earth beyond six of nine planetary boundaries. *Science Advances* 9, 1–16. DOI:10.1126/sciadv.adh2458.

Rickards, L. (2020) Using and interrogating privilege to progress climate justice. *Planning Theory & Practice* 21 (2), 293–321. https://doi.org/10.1080/14649357.2020.1748959.

Ritter, E. and Thaler, G.M. (2022) Technical reform or radical justice? Environmental discourse in non-governmental organizations. *Environment and Planning E: Nature and Space* 6 (3), 2071–2095. https://doi.org/10.1177/25148486221119750.

Schuijers, L. (2023) Capitalising on climate anxiety: what you need to know about 'climate-washing'. *The Conversation* (Online). See https://theconversation.com/capitalising-on-climate-anxiety-what-you-need-to-know-about-climate-washing-202507 (accessed May 2025).

Scott, K. (2024) A new ruling says countries – including NZ – must take action on climate change under the law of the sea. *The Conversation* (Online). See https://theconversation.com/a-new-ruling-says-countries-including-nz-must-take-action-on-climate-change-under-the-law-of-the-sea-230420 (accessed May 2025).

Sheller, M. (2018) Theorizing mobility justice. *Tempo Social* 30 (2), 17–34. https://doi.org/10.11606/0103-2070.ts.2018.142763.

Steffen, W. and Morgan, J. (2021) From the Paris Agreement to the Anthropocene and Planetary Boundaries Framework: An interview with Will Steffen. *Globalizations* 18 (7), 1298–1310. https://doi.org/10.1080/14747731.2021.1940070.

Sultana, F. (2022a) Critical climate justice. *The Geographical Journal* 188, 118–124. DOI: 10.1111/geoj.12417.

Sultana, F. (2022b) The unbearable heaviness of climate coloniality. *Political Geography* 99, 102638. https://doi.org/10.1016/j.polgeo.2022.102638.

Supran, G., Rahmstorf, S. and Oreskes, N. (2023) Assessing Exxon Mobil's global warming projections. *Science* 379, eabk0063. eabk0063. DOI:10.1126/science.abk0063.

TPCC (2023) Tourism and climate change stocktake 2023. S. Becken, and D. Scott (eds). See https://tpcc.info/ (accessed May 2025).

Tourism Panel on Climate Change (2023) Tourism and Climate Change Stocktake 2023. [Eds. Becken, S. & Scott, D.]. See https://tpcc.info/ (accessed October 2025).

UNWTO (2021) Launching of the Glasgow declaration: A commitment to a decade of climate action in tourism. See https://www.unwto.org/event/cop-26-launch-of-the-glasgow-declaration-a-commitment-to-a-decade-of-climate-action-in-tourism (accessed May 2025).

United Nations Office for Disaster Risk Reduction (UNDRR) (2017) The Sendai framework terminology on disaster risk reduction. 'Vulnerability'. See https://www.undrr.org/terminology/vulnerability (accessed May 2025).

UN (n.d.) Inequalities. See https://unsceb.org/topics/inequalities#:~:text=Inequality%20has%20multiple%20dimensions%20and,or%20otherwise%20denied%20full%20equality (accessed May 2025).

UN (2023a) Just transition. See https://www.un.org/development/desa/dpad/wp-content/uploads/sites/45/CDP-excerpt-2023-1.pdf (accessed May 2025).

UN (2023b) New UN guidance affirms children's right to a clean, healthy environment. See https://news.un.org/en/story/2023/08/1140122 (accessed May 2025).

UNEP (2023) As climate impacts accelerate, finance gap for adaptation efforts at least 50% bigger than thought. Press release. See https://www.unep.org/news-and-stories/press-release/climate-impacts-accelerate-finance-gap-adaptation-efforts-least-50 (accessed April 2025).

UNFCCC (n.d.) The Paris Agreement. See https://unfccc.int/process-and-meetings/the-paris-agreement (accessed May 2025).

Venn, A. (2023) Rendering international human rights law fit for purpose on climate change. *Human Rights Law Review* 23 (1), ngac034. https://doi.org/10.1093/hrlr/ngac034.

Warlenius, R. (2017) Decolonizing the atmosphere: The climate justice movement on climate debt. *The Journal of Environment and Development* 27 (2), 131–155. https://doi.org/10.1177/1070496517744593.

WEF (2023) What is overtourism and how can we overcome it? See https://www.weforum.org/agenda/2023/10/what-is-overtourism-and-how-canwe-overcome-it/ (accessed May 2025).

WFTO (2015) Defining the marginalised. See https://wfto-asia.com/wp-content/uploads/2015/05/Literature-Review-Summary-Defining-the-Marginalized.pdf (accessed May 2025).

WMO (2024) Record carbon emissions highlight urgency of Global Greenhouse Gas Watch. See https://wmo.int/media/news/record-carbon-emissions-highlight-urgency-of-global-greenhouse-gas-watch (accessed May 2025).

1 Tourism in a Warming World: Who Emits and Who Pays the Price?

Ya-Yen Sun, Futu Faturay and Wanru Zhou

This chapter examines the critical role of climate justice in global tourism by focusing on two core issues: emissions contribution and tourism demand redistribution. Drawing on the latest global data, we map the distributional inequalities in emissions and climate impacts, detailing a country's contribution to tourism emissions and the corresponding shifts in travel demand. We found that high-income countries currently generate two thirds of tourism-related greenhouse gas emissions and are expected to attract even more tourists by 2050 due to favourable weather conditions. In contrast, low-income and climate-vulnerable destinations, especially Small Island Developing States, face a high risk of economic instability as tourism demand decreases and climate impacts escalate, despite contributing less than 1% of the global tourism footprint. The chapter concludes with policy recommendations that emphasise targeted climate finance, transparent emissions reporting and international support to ensure that tourism contributes to equitable and sustainable development in a rapidly changing climate.

Introduction

Climate justice in global tourism hinges on two central considerations that have received significant attention from international policy frameworks and multilateral institutions. The first pillar is fair action on emissions mitigation. A relatively small group of high-income countries and affluent markets drive a large share of tourism activities, and consequently, they are responsible for a disproportionate amount of tourism-related greenhouse gas emissions (GHGs). The Paris Agreement laid the groundwork by establishing the principle of 'common but differentiated responsibilities', emphasising that countries with higher emissions should take the lead in reducing their carbon footprints

(UNFCCC, 2021). This mandate was further reinforced by the Glasgow Declaration on Climate Action in Tourism at COP26 (UN Tourism, 2021), which called for the tourism sector to commit to halving its emissions by 2030 and achieving net zero by 2050. Such commitments are crucial for curbing high-emission activities, especially long-haul flights and luxury travel, that are incompatible with global climate targets. Simultaneously, many countries with relatively low levels of tourism activity are among the most vulnerable to climate change, facing threats such as rising sea levels, extreme weather events and ecosystem degradation.

The second pillar is fair action on adaptation and resilience. Many destinations that suffer the most severe consequences of climate change – particularly small island developing states and countries in the Global South – are heavily dependent on inbound tourism for economic growth and job creation (Scott *et al.*, 2019). However, these regions often lack the financial resources and institutional capacity to safeguard their communities against escalating climate hazards. In response, the Loss and Damage Fund was officially established at COP27 in 2022 to provide financial assistance to vulnerable developing countries facing irreversible climate-related disasters, marking a historic breakthrough after decades of advocacy (UNFCCC, 2022). At COP28, nations agreed on the operational framework of the fund, including its governance and funding mechanisms, ensuring that financial support can be mobilised to help communities rebuild and enhance resilience against future climate impacts (UNFCCC, 2023). Specific actions supporting tourism were further emphasised in the Baku Declaration on Enhanced Climate Action in Tourism, endorsed at COP29 (2024). The declaration explicitly integrated tourism into national climate agendas by calling for innovative financing mechanisms and capacity-building measures to support adaptation in the most at-risk regions.

Beneath these two pillars lies the fundamental questions of who has contributed the most to tourism-related carbon emissions and whose destinations and tourism sectors bear the greatest climate costs. Despite its economic benefits, tourism remains a carbon-intensive sector, contributing significantly to global greenhouse gas emissions. Recent estimates indicate that tourism accounted for approximately 8.8% of global emissions before the COVID-19 pandemic, and emissions from tourism expanded at approximately 3.5% per year, 2.5 times the growth rate of the global economy (Sun *et al.*, 2024). As the global economy recovers from the pandemic, travel demand is rebounding rapidly; in 2024, aviation traffic has already surpassed pre-pandemic levels by nearly 4% (IATA, 2025). This surge is largely driven by travellers from highly developed economies such as the United States, Germany, the United Kingdom, Australia and Canada (UN Tourism, 2025), further highlighting the discrepancy in recovery capacities and their related contributions to climate change within the tourism sector.

At the same time, climate change is reshaping tourism demand worldwide (Grillakis *et al.*, 2016). Warming temperatures are extending peak tourism seasons in traditionally cooler regions, creating new opportunities for destinations at higher latitudes or in alpine areas. However, these potential benefits are counterbalanced by significant risks. Rising sea levels are eroding beaches, and challenges such as water scarcity, biodiversity loss and extreme weather events, such as hurricanes, wildfires and floods, are disrupting local ecosystems and undermining tourism infrastructure. Consequently, the redistribution of tourism demand is a double-edged sword: while some destinations may benefit from extended seasons and increased appeal, others risk a decline in attractiveness or may even become uninhabitable (Zhou *et al.*, 2024). This shifting landscape has direct implications for tourism sectors, their employees and local communities, as the benefits of tourism become concentrated in areas less vulnerable to climate hazards, while high-risk destinations struggle to maintain their economic viability.

Distribution inequality refers to '(in)equitable distribution of benefits and burdens across different groups and generations from the impacts and actions related to climate change' (Rastegar & Becken, 2024: 8). In this chapter, we examine two key dimensions of distributional inequality in tourism and climate justice. First, we analyse the unequal distribution of tourism-related greenhouse gas emissions among individual countries. Drawing on a global tourism carbon emissions analysis of 175 countries (Sun *et al.*, 2024), we map out the magnitude of tourism carbon emissions each country produced during the pre-COVID period. Second, we explore how climate change is reshaping tourism demand unevenly across destinations. Using the data from the first meta-analysis on how climate change affects both domestic and inbound tourism demand (Zhou *et al.*, 2024), we discuss tourism demand redistribution patterns, providing insights into which countries are likely to benefit and which are poised to lose as global travel demand responds to a changing climate.

By mapping major tourism emissions producers and the destinations that bear the climate costs, our analysis reveals a vicious cycle: countries, mainly in the Global North, that generate the majority of tourism-related carbon emissions may experience a net gain in tourism due to a warming climate and will continue to emit more. Meanwhile, nations in the Global South, despite contributing significantly less to tourism emissions, are likely to suffer losses in tourism revenue. The chapter concludes with recommendations for international climate agreements, such as those negotiated at COP meetings, warning that, without timely intervention, disparities in emissions production and demand redistribution will likely worsen under the post-pandemic tourism boom.

Unequal Distribution: How Countries Contribute to Global Tourism Carbon Emissions

Economic development of a country significantly influences many facets of tourism development and later their contribution to global GHGs. From a demand side, travel is income elastic and a luxury good (Smeral, 2010). Rising income status not only implies people demand more travel but also different forms of travel. When countries rise from developing to developed status, the travel format residents engage will evolve from primary domestic travel, then to short-haul international trips and eventually to long-haul inter-continental journeys. The evolution of trips also implies different goods and services are being consumed. Visitors from and in low-income countries consume more unprocessed food and road transport with limited commercial hospitality services, demonstrating that, for this income group, travel mostly involves the bare necessities and produce limited emissions per journey (Lenzen et al., 2018). In contrast, visitors from and within high-income countries demand a high proportion of transport (especially air travel), shopping, accommodation and restaurant services, reflecting their expectations for convenience, comfort and enjoyment, resulting in a higher carbon footprint.

Besides the type of goods and service consumed, income level critically determines how much people can afford and are willing to spending on travel. In 2019, global tourism expenditure reached US$6.0 trillion. If dissecting this value by income status, around 60% of spending (61.7%) was contributed by residents of high-income countries, followed by upper–middle-income countries (30.5%) and lower–middle-income countries (7.5%) (Figure 1.1a). Travel spending from low-income countries contributed significantly less (0.4%). Tourists from high-income countries also exhibited the highest per capita tourism expenditure, averaging US$2523 per year (Figure 1.1b). This figure is five times higher than the per capita tourism expenditure of upper–middle-income countries and nearly 90 times greater than that of low-income travellers, amounting to US$30 per capita per year. It is important to note that the share of outbound travel (black bar in Figure 1.1b)

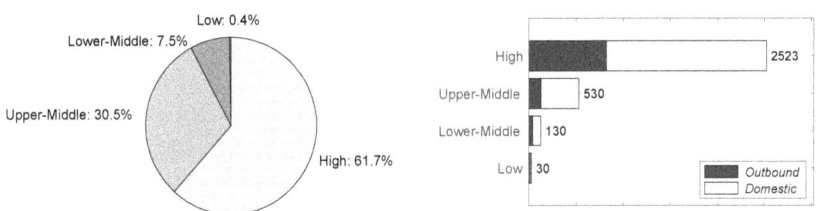

Figure 1.1 (a) Tourism expenditure share (left) and (b) per capita tourism expenditure (right) for four income groups in 2019
Source: Authors.

increases along with a country's economic status. Demand for overseas travel and flight services is strongly cointegrated with income, as people become wealthier, they are likely to travel more internationally.

While sorting tourism expenditure is relatively straightforward, allocating tourism emissions to countries is far more complex. This complexity arises from the involvement of both domestic and foreign producers serving both domestic residents and international visitors in each host destination, making it challenging to assign carbon emissions to individual countries. For example, which country should be held responsible for the emissions generated when a Japanese visitor consumes French-made bottled water in China? The literature recommends addressing emissions responsibility from two perspectives: the consumption perspective (residence-based accounting, RBA) and the destination perspective (destination-based accounting, DBA) (Lenzen *et al.*, 2018). RBA attributes emissions to the traveller's country of origin, covering both domestic and outbound tourism, thereby reflecting the carbon footprint generated by residents' travel spending. DBA, on the other hand, assigns emissions to the destination country where tourism-related activities occur, encompassing both domestic and inbound tourism, thus capturing the carbon footprint associated with tourism expenditures at the destination.

In this chapter, RBA data is used to illustrate each country's contribution to the global tourism carbon footprint. RBA data serves as a proxy for measuring the extent to which residents of individual countries consume travel, both domestically and internationally, and how that consumption contributes to global tourism emissions. This approach aligns with the argument that the consumer is the key actor who triggers demand, enjoys the benefits and privileges of travel and therefore bears responsibility for the emissions generated in the production process (Peters & Hertwich, 2008).

The RBA data discussed below draws from a global tourism carbon emissions study that analysed tourism carbon footprint for 175 countries between 2009 and 2019 (Sun *et al.*, 2024), providing the first portrayal of how residents' travel footprint differs across countries.

Emissions by traveller's country

During the pre-COVID period, global tourism consumption peaked in 2019, generating a total carbon footprint of 5.2 Gt CO_2-e and accounting for 8.8% of global GHGs. Demand was found to be highly saturated in certain destinations and enjoyed by privileged groups. The RBA reveals that the distribution of emissions is highly concentrated in high-income countries, which accounted for 54.5% of total emissions; 1.7 times higher than the share of upper–middle-income countries (Figure 1.2a). Lower–middle-income countries contributed 13.3%, while low-income countries generated less than 1% of total emissions. This imbalance underscores the uneven distribution of global tourism-related emissions.

Seven high-income countries, including the US, Germany, the UK, Japan, Canada, France and Italy, were the largest contributors, collectively accounting for nearly 40% of total global tourism emissions. The US led with a 19.1% share in 2019, while the other six countries contributed nearly equal shares.

Among upper–middle-income countries, China dominated, contributing 14.5% of global tourism-related emissions, ranking second only to the US. This is primarily driven by China's large population and rapid economic development, including significant growth in its tourism sector. Other populous countries such as Mexico, Russia and Indonesia accounted for 2.9%, 2.2% and 2.1% of global tourism emissions, respectively.

In the lower–middle-income group, India was the dominant emitter, contributing 5.7% of global tourism emissions in 2019. Other countries such as Egypt, Vietnam and the Philippines also ranked among the top emitters in this category, with contributions ranging from 0.5% to 1.2% of the total.

Low-income countries collectively accounted for just 0.9% of global tourism-related emissions, with individual country contributions typically below 0.1%. The largest emitters in this category were primarily African countries, including Ethiopia, Sudan, Uganda, Mozambique and the Democratic Republic of Congo.

Per capita emissions

Per capita carbon emissions, calculated as the total carbon footprint divided by the population, provide a proxy for assessing an individual's contribution to climate impact. In tourism, per capita footprints vary substantially across income groups (see Figure 1.2b). In 2019, high-income countries averaged 2.2 tons CO_2-e per capita, dramatically higher than the figures for upper–middle-income (0.6 tons CO_2-e), lower–middle-income (0.2 tons CO_2-e) and low-income countries (0.1 tons CO_2-e), representing a difference of approximately 20 times between high- and low-income nations.

Among high-income countries, Brunei Darussalam, Singapore, Iceland, Kuwait and Norway exhibited the highest per capita tourism emissions in 2019. These nations' economic prosperity and strong international travel tendencies contributed to disproportionately high emissions per resident.

In upper–middle-income countries, Malaysia stood out as the only country in this category with per capita tourism emissions exceeding 2 tons CO_2-e. Other countries, such as Mexico, Moldova and Bulgaria,

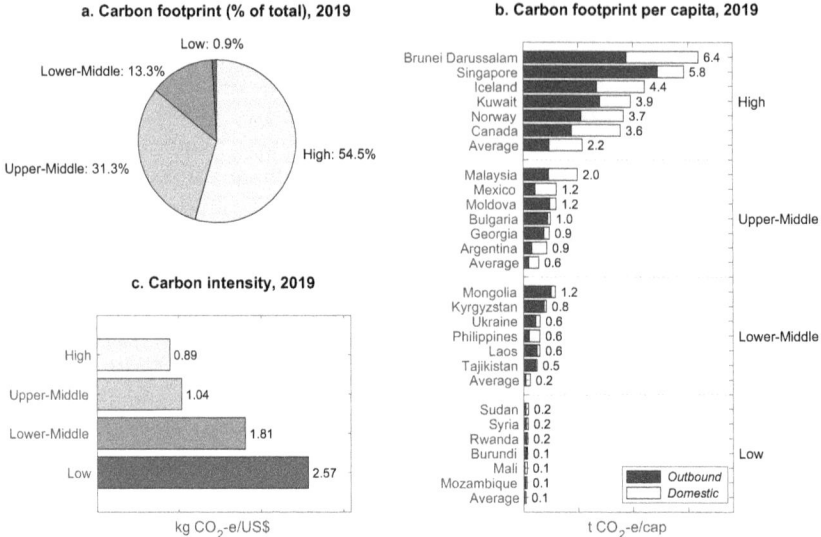

Figure 1.2 (a) Carbon footprint share (top left), (b) per capita tourism carbon footprint (right) for four income groups in 2019 and (c) carbon intensity (bottom left)
Source: Authors.

had per capita carbon emissions ranging between 1 and 2 tons CO_2-e, while the majority of nations in this group remained below 1 ton CO_2-e.

In the lower–middle-income category, Mongolia was the only country with per capita tourism emissions above 1 ton CO_2-e, whereas the remaining countries in this group had per capita emissions below this threshold. All countries in low-income group recorded per capita tourism emissions below 0.2 tons CO_2-e in 2019, highlighting the stark inequality in tourism-related emissions across different income levels.

Emissions intensity

Carbon intensity, defined as emissions per dollar spent, serves as a proxy for assessing technological advancement and the adoption of low-carbon practices in tourism-related industries. To lower energy intake per unit of operation and subsequently to lower emission levels, innovation in technology, including technological and operational improvements, plays a critical role.

The technological capacity of tourism firms, however, is clearly associated with the economic status of countries where the advanced economies have the privilege of accessing capital, manpower, innovation ideas and technology transfer more than developing countries (Gössling, 2011). In addition, the economic structure of a society determines how a product is obtained, processed and transferred via the domestic and international supply chains, which then determines their emissions contents. When countries embrace international trade and free-trade agreements, destinations are more likely to benefit from these 'green imports' (such as electric vehicles, recycled materials and energy efficient aircrafts) and reduce the carbon emissions from tourism (Sun, 2019). Developed countries, due to their higher economic status, typically have greater access to energy-efficient goods and services compared to developing countries (Yu *et al.*, 2022). This advantage allows their tourism sectors to adopt energy-efficient equipment, improve operational efficiency and reduce emissions more effectively.

Tourism emissions intensity therefore shows a clear correlation with income levels. In 2019, low-income countries recorded a tourism carbon intensity of 2.57 kg CO_2-e/US\$, followed by lower–middle-income countries at 1.81 kg CO_2-e/US\$, upper–middle-income countries at 1.04 kg CO_2-e/US\$ and high-income countries at 0.89 kg CO_2-e/US\$ (Figure 1.2c). Although the purchasing power of a US dollar is generally higher in lower-income countries, the data supports the claim that every dollar spent on tourism-related activities in high-income countries generates 65% fewer emissions compared to spending in low-income countries. This lower carbon intensity in wealthier nations can be attributed to better infrastructure, more fuel-efficient transportation and increased investment in sustainable tourism initiatives.

Unequal Distribution: How Climate Change Impacts Tourism Across Countries

Carbon emissions act like a blanket to trap solar heat, resulting in global warming. The year 2024 is the first year with an average temperature clearly surpassing 1.5°C above the pre-industrial baseline (Copernicus Climate Change Service, 2025). This means that the 2030 temperature threshold for significant emission reductions set by the Paris Agreement has been exceeded. Unprecedented global warming has led to variability in weather conditions (e.g. temperature, precipitation) and a high incidence of extreme weather events worldwide (WMO, 2024). These climatic factors have been identified as affecting the attractiveness of destinations to tourists, leading to a shift in tourism demand (Scott, 2021; Zhou *et al.*, 2024).

Research on how climate change impacts tourism demand is primarily generated from econometric modelling to measure the demand sensitivity to meteorological conditions. The general view is that hotter regions will see fewer tourists, while cooler regions will benefit from increased tourism demand due to global warming (Amelung *et al.*, 2007; Hamilton *et al.*, 2005; Matei *et al.*, 2023). However, the shifting pattern is far more complex, compounded by a variety of climate risks and region-specific changes (Gössling *et al.*, 2012). For example, despite the fact that Mauritius and Thailand are hot all year round, elevated temperatures still attract tourists to Thailand but discourage them from Mauritius (Chen *et al.*, 2017; Wannapan & Chaiboonsri, 2017). In addition to temperature, visitors have unique responses to each of the weather conditions (humidity, rainfall and sunshine duration) making it more challenging to generalise the travel demand dynamics (Bae & Nam, 2020; Fauzel, 2020).

The interplay between weather and tourism demand is complex, and current research is only beginning to reveal these dynamics. Knowledge remains fragmented, with most information coming from case studies and news reports. Here, we review findings from the only meta-analysis reported to date on the dynamics between climate factors and tourism demand. Zhou *et al.* (2024) analysed 34 econometric studies comprising 290 climate elasticity estimates to synthesise travel demand elasticity at the global level across six climate types. In this study, global warming elasticity of tourism demand is defined as 'the percentage change in tourism demand corresponding to a 1% change in the respective warming indicators (i.e., temperature, relative humidity, rainfall, and sunshine)' (Zhou *et al.*, 2024: 1770).

The analysis found that, holding all else constant, a 1% increase in the warming indicators is associated with a decrease in travel demand of 0.22% in tropical regions, 0.12% in arid regions and 0.26% in Mediterranean regions, while travel demand increases by 0.13% in subtropical regions, 0.11% in marine west coast regions and 0.14% in continental destinations. These findings confirm a global shift: climate change has redistributed visitors from tropical, arid and Mediterranean destinations to humid subtropical, marine west coast and continental areas.

A forecast of tourism demand changes by 2050 further outlines the likely benefits and costs incurred by destination countries. Based on the business-as-usual scenarios from the World Bank's Climate Change Knowledge Portal, tourism demand changes by income group are forecasted as follows: low-income countries (−0.33%), lower–middle-income countries (−0.42%), upper–middle-income countries (−0.05%) and high-income countries (0.66%). The general conclusion is that the lower the economic status of the host country, the more likely its community will experience reduced tourism demand due to climate change.

In terms of individual destinations (Figure 1.3), some countries will be worse off (referred to as the Loss group), while others will benefit (referred to as the Gain group). Of the 193 countries modelled, the Loss group comprises 127 countries (66%) that are expected to suffer a decline in visitors and tourism receipts by 2050 due to climate change. Among these, low-income and lower–middle-income countries make up half of the Loss group. It is important to note that all 37 Small

Figure 1.3 Global map of tourism demand changes by 2050. Dark grey represents countries with tourism demand losses (the Loss group) and light grey represents countries with tourism demand growth (the Gain group). The icons (square and circle) identify the hotspot countries for each of the two groups which will experience larger percentage changes in travel demand

Island Developing States (SIDS) modelled are predicted to face tourism losses ranging from –0.45% to –1.03% per 1% change in the respective warming indicators.

By contrast, 66 nations (33% of the 193 countries) fall into the Gain group, with tourism demand expected to increase due to changing climate patterns. Three-quarters of these nations are high- and upper–middle-income countries. These include some of the most developed economies with the highest per capita incomes, such as the US, Germany, the UK, Luxembourg and Australia. China, one of the leading countries in the upper–middle-income category, is also expected to welcome more tourists as global temperatures continue to rise.

Discussion

Tourism contributes to climate change and is also impacted by it, reflecting the interplay of two unequal distribution patterns discussed above. Figure 1.4 displays these inequalities by mapping countries into Loss and Gain groups across four income categories. The size of the bubbles represents tourism carbon emissions from residents' travel; larger bubbles indicate a larger tourism footprint in 2019.

A significant disparity between high-income and low-income countries is observed. The figure suggests an uneven distribution of climate change

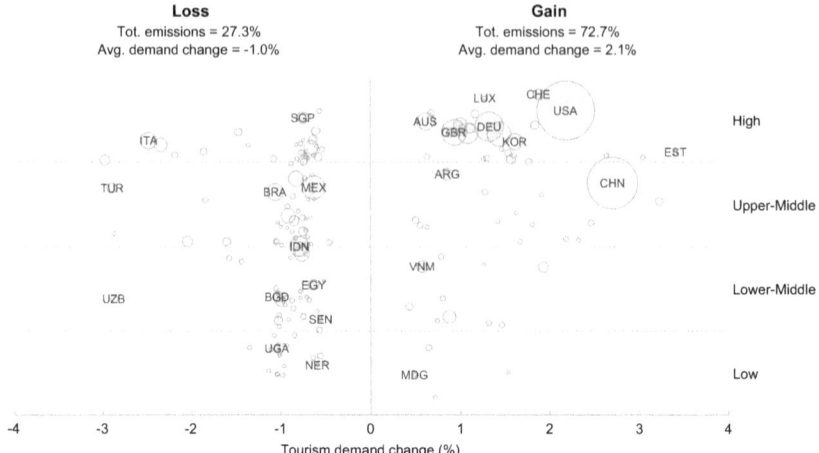

Figure 1.4 Climate change impacts on tourism demand by country and income group in 2050. The X-axis represents tourism demand changes in 2050 due to temperature increases. The Y-axis represents the 2019 Gross National Income (GNI) per capita, adjusted to fit the graph. The bubble size represents tourism carbon emissions in 2019
Source: Authors.

impacts on tourism across income levels. Most 'Loss' countries, those projected to lose tourism demand by 2050, are lower-income or developing nations, often located in hotter climates, while many 'Gain' countries, expected to see increased demand, are higher-income nations in cooler, higher-latitude regions. In other words, countries that have contributed the most to global emissions (typically wealthy, northern hemisphere nations such as the US, China, the UK and Germany) are projected to gain tourists (Figure 1.4, right). These are the countries whose residents travel the most, generating the largest emissions, while their local tourism industries are expected to benefit from climate change. In contrast, countries that have contributed relatively little to global emissions, often poorer Southern Hemisphere nations, are projected to lose tourists and experience reduced tourism revenue (see Figure 1.4, left). Climate injustice in tourism is evident: the Gain group produces around 73% of tourism emissions but is projected to benefit from shifting travel patterns with a 2% net increase, whereas the Loss group, which contributes around 27% of emissions, faces an additional 1% decline in visitor numbers (see Figure 1.4).

The magnitude of tourism demand change also varies widely by country. Some destinations may see only modest changes, while others could experience substantial surges (exceeding 2%) in tourism demand. For example, net increases in tourism in prominent destinations such as the US (2.2%) and China (2.7%) could generate higher commercial value than the entire tourism sector of some smaller countries. It is important to note that the figure represents national demand changes, implying that substantial regional variations are likely due to intra-regional substitution. For instance, projections for Europe under high-warming scenarios indicate that southern coastal regions (e.g. parts of Greece, Spain and Italy) could see tourist numbers fall by 5~10% or more, while cooler northern regions (e.g. parts of the UK or Scandinavia) might experience increases of a similar magnitude (+5~15%+) (Matei et al., 2023). SIDS are especially at risk, as many rely heavily on climate-sensitive tourism, and tourism can comprise a large share of their GDP, as seen in countries such as Saint Lucia, Seychelles and Cape Verde. These countries could suffer significant economic losses if visitor numbers drop, whereas higher-latitude countries may enjoy notable gains.

Overall, tourism demand is less likely to vanish than to relocate. Tourists are expected to choose different destinations rather than stop travelling altogether. Analysis has indicated that, even with regional losses, total tourism in Europe might still grow slightly under climate change as travellers shift northward or to other seasons (Matei et al., 2023). This suggests that, while global tourism revenues and emissions could remain high in the future, they may become concentrated in different areas. Economically, tourism income may flow away from vulnerable low-income economies, exacerbating their challenges, and toward higher-income countries that can attract redirected travel. In particular, lower–middle-income and low-income countries tend to face the highest vulnerability under climate change. A global tourism vulnerability index, which examines tourism assets, operating costs, exposure to extreme weather and social unrest and adaptive capacities, supports these findings, indicating that countries in Africa, the Middle East, South Asia and SIDS face the highest vulnerabilities (Scott et al., 2019). These countries are also those most likely to experience a reduction in tourism demand. Given that a higher proportion of women youth and low-income workers are employed in these tourism sectors (Sun et al., 2022), these workers face an increased risk of reduced working hours, lower pay and layoffs under a deteriorating tourism sector, leading to further income inequality in the society and social unrest in local communities.

Conclusion

Travel is a luxury good, income elastic and carbon intensive. These characteristics create significant disparities in the global tourism industry, raising critical concerns about climate justice. Based on the information discussed earlier, we summarised these disparities in three areas: spending patterns, emissions contributions and the differential impact of climate change on tourism.

- Expenditure and emissions: High-income countries account for 66% of global tourism expenditure and 55% of the tourism carbon

footprint, whereas low-income countries contribute less than 0.5% of global tourism spending and under 1% of emissions.
- Per capita footprint: The average per capita tourism footprint in high-income countries is approximately 2.2 tonnes CO_2-e per annum, compared to just 0.1 tonnes in low-income countries.
- Climate impact on demand: By 2050, half of high-income countries are projected to welcome more tourists due to favourable climatic conditions, while three quarters of lower–middle and low-income countries, including all SIDS, are expected to face reduced tourism demand.

These patterns imply that increasing tourism and carbon emissions will intensify distributional inequality, with high-income countries continuing to travel, emit more and receive more tourists, while low-income destinations bear most of the climate cost amid reduced tourism demand.

Breaking this vicious cycle requires robust global policy interventions, with emissions mitigation as the top priority. The tourism sector's emissions are increasing at about 3.5% per annum, highlighting the urgent need for effective measures. Encouraging more countries to commit to the Glasgow Declaration on Climate Action in Tourism and Baku Declaration would compel governments to publicly pledge emission reductions and reorient their pro-growth tourism strategies. However, a uniform short-term reduction, such as a uniform 5% cut in tourism demand across all countries to cut emissions, is not an equitable approach. The IPCC (2014) suggests that a 'fair contribution' to emissions mitigation must consider equality, responsibility, capability and cost effectiveness. Consequently, historical and cumulative tourism emissions by country should be factored into international climate agreement, placing a greater mitigation responsibility on high-income countries with larger populations and higher per capita emissions.

Assigning priority and responsibility for mitigating tourism emissions also requires transparent reporting and regular review of the sector's climate footprint. In 2024, the United Nations World Tourism Organisation (UN Tourism) approved the Statistical Framework for Measuring the Sustainability of Tourism (MST), an internationally agreed reference for assessing the economic, social and environmental aspects of tourism, including carbon emissions. This framework enables countries to measure and report tourism-related emissions consistently and harmoniously. The resulting sectoral assessments can be used both to monitor industry progress, especially in high-emitting sectors such as aviation and cruise lines, and to inform international negotiations on emissions mitigation and support for vulnerable countries.

Furthermore, it is imperative to prioritise financial support for vulnerable destinations. Low-income and climate-vulnerable countries, particularly SIDS, must have access to international climate finance mechanisms such as the Adaptation Fund, Green Climate Fund and the COP28 Loss and Damage Fund. Simplified application procedures, targeted capacity building and dedicated funding windows for tourism adaptation should be instituted to ensure that SIDS and Least Developed Countries receive timely support. Given these nations face heightened environmental risks and financial burdens, such as increased insurance premiums, rising debt from climate disasters and fragile economies, access to funding is often challenging (Kikstra et al., 2022). Climate finance mechanisms must prioritise communities most affected by climate-related tourism losses, such as those experiencing hurricane damage to island resorts or drought affecting safari tourism.

It is important to note that funding should enhance both mitigation and adaptation capacities. Reducing the disparity in carbon intensity is essential for harmonising mitigation efforts and ensuring a more sustainable future for global tourism. Currently, low-income countries produce 70% more emissions per dollar of tourism spending compared to their high-income counterparts. This high emissions-per-dollar output in developing economies reflects their limited capacity to deliver low-emission, low-impact tourism services. Targeted interventions to reduce carbon intensity in low-income countries will help drive overall emission reductions.

A major challenge facing the global Loss and Damage Fund is the insufficient financial support from industrialised countries. By 2023, contributions from pledged nations amounted to only about US$700 million – less than 0.2% of the estimated annual requirement (Pill & Hammersley, 2024). The tourism sector can help address this issue by implementing a levy on international aviation. Aviation accounts for over 20% of the global tourism carbon footprint, yet its usage is highly unequal; only 2–4% of the global population flew internationally in 2018, with the top 1% of most frequent flyers responsible for more than 50% of commercial aviation emissions (Gössling & Humpe, 2020). A proposed carbon tax, US$6 on economy-class long-haul flights and US$62 on business class, could potentially raise US$8–10 billion per year (Minninger, 2024). Such taxes would internalise the environmental costs of travel while providing critical funding for vulnerable destinations. Revenues should be channelled to the Loss and Damage Fund and earmarked for rapid response and recovery in tourism communities affected by extreme weather events or slow-onset climate impacts, ensuring that dedicated funds are available when disasters strike.

These recommendations advocate a cohesive approach that requires wealthy nations and the tourism industry to lead in reducing emissions, directs resources to those most at risk and establishes institutions that integrate tourism into global climate solutions. As tourism is poised for rapid growth in the post-COVID era, along with its associated carbon emissions, without these policy interventions, tourism risks becoming a driver for further climate injustice.

References

Amelung, B., Nicholls, S. and Viner, D. (2007) Implications of global climate change for tourism flows and seasonality. *Journal of Travel Research* 45 (3), 285–296. https://doi.org/10.1177/0047287506295937.

Bae, J. and Nam, S. (2020) An analysis of the effect of climate indicators on tourism demand: A case study of Jeju Island. *Journal of Policy Research in Tourism, Leisure and Events* 12 (2), 185–196. https://doi.org/10.1080/19407963.2019.1585363.

Chen, C.-M., Lin, Y.-C., Li, E.Y. and Liu, C.-C. (2017) Weather uncertainty effect on tourism demand. *Tourism Economics: The Business and Finance of Tourism and Recreation* 23 (2), 469–474. https://doi.org/10.5367/te.2015.0513.

COP29 (2024) COP29 declaration on enhanced climate action in tourism. COP29. See https://cop29.az/en/pages/cop29-declaration-on-enhanced-climate-action-in-tourism (accessed March 2025).

Copernicus Climate Change Service (2025) Global Climate Highlights 2024. See https://climate.copernicus.eu/global-climate-highlights-2024 (accessed March 2025).

Fauzel, S. (2020) The impact of changes in temperature and precipitation on tourists arrival: An ARDL analysis for the case of a SIDS. *Current Issues in Tourism* 23 (19), 2353–2359. https://doi.org/10.1080/13683500.2019.1639639.

Gössling, S. (2011) *Carbon Management in Tourism*. Routledge.

Gössling, S. and Humpe, A. (2020) The global scale, distribution and growth of aviation: Implications for climate change. *Glob Environ Change* 65, 102194. https://doi.org/10.1016/j.gloenvcha.2020.102194.

Gössling, S., Scott, D., Hall, C.M., Ceron, J.-P. and Dubois, G. (2012) Consumer behaviour and demand response of tourists to climate change. *Annals of Tourism Research* 39 (1), 36–58. https://doi.org/10.1016/j.annals.2011.11.002.

Grillakis, M.G., Koutroulis, A.G., Seiradakis, K.D. and Tsanis, I.K. (2016) Implications of 2°C global warming in European summer tourism. *Climate Services* 1, 30–38. https://doi.org/https://doi.org/10.1016/j.cliser.2016.01.002.

Hamilton, J.M., Maddison, D.J. and Tol, R.S.J. (2005) Effects of climate change on international tourism. *Climate Research* 29 (3), 245–254. https://doi.org/10.3354/cr029245.

IATA (2025) Global air passenger demand reaches record high in 2024. IATA. See https://www.iata.org/en/pressroom/2025-releases/2025-01-30-01/#:~:text=Total%20full%2Dyear%20traffic%20in,record%20for%20full%2Dyear%20traffic (accessed March 2025).

IPCC (2014) *Climate Change 2014: Mitigation of Climate Change. Contribution of Working Group III to the Fifth Assessment Report of the Intergovernmental Panel on Climate Change*. C. U. Press.

Kikstra, J.S., Nicholls, Z.R.J., Smith, C.J., Lewis, J., Lamboll, R.D., Byers, E., Sandstad, M., Meinshausen, M., Gidden, M.J., Rogelj, J., Kriegler, E., Peters, G.P., Fuglestvedt, J.S., Skeie, R.B., Samset, B.H., Wienpahl, L., van Vuuren, D.P., van der Wijst, K.I., Al Khourdajie, A....Riahi, K. (2022) The IPCC Sixth Assessment Report WGIII climate assessment of mitigation pathways: from emissions to global temperatures. *Geosci. Model Dev.* 15 (24), 9075–9109. https://doi.org/10.5194/gmd-15-9075-2022.

Lenzen, M., Sun, Y.-Y., Faturay, F., Ting, Y.-P., Geschke, A. and Malik, A. (2018) The carbon footprint of global tourism. *Nature Climate Change* 8 (6), 522–528. https://doi.org/10.1038/s41558-018-0141-x.

Matei, A., García-León, D., Dosio, A., Silva, F., Barranco, R. and Ciscar, J.-C. (2023) *Regional Impact of Climate Change on European Tourism Demand*. Publications Office of the European Union, Luxembourg. https://doi.org/10.2760/899611.

Minninger, S. (2024) The Loss and Damage Fund: Costs and benefits for the tourism sector. Tourism Watch. See https://www.tourism-watch.de/en/article/the-loss-and-damage-fund-costs-and-benefits-for-the-tourism-sector/#:~:text=International%20Taxation,ten%20billion%20USD%20per%20year (accessed March 2025).

Peters, G.P. and Hertwich, E.G. (2008) Post-Kyoto greenhouse gas inventories: Production versus consumption. *Climatic Change* 86 (1–2), 51–66. https://doi.org/10.1007/s10584-007-9280-1.

Pill, M. and Hammersley, G. (2024) *A Climate Loss and Damage Fund that Works*. Lowy Institute.

Rastegar, R. and Becken, S. (2024) Embedding justice into climate policy and practice relevant to tourism. *Journal of Sustainable Tourism* 33 (10), 2011–2028. https://doi.org/10.1080/09669582.2024.2377720.

Scott, D. (2021) Sustainable tourism and the grand challenge of climate change. *Sustainability* 13 (4), 1966. https://www.mdpi.com/2071-1050/13/4/1966.

Scott, D., Hall, C.M. and Gössling, S. (2019) Global tourism vulnerability to climate change. *Annals of Tourism Research* 77, 49–61. https://doi.org/10.1016/j.annals.2019.05.007.

Smeral, E. (2010) Impacts of the world recession and economic crisis on tourism: Forecasts and potential risks. *Journal of Travel Research* 49 (1), 31–38. https://doi.org/10.1177/0047287509353192.

Sun, Y.-Y. (2019) Global value chains and national tourism carbon competitiveness. *Journal of Travel Research* 58 (5), 808–823. https://doi.org/10.1177/0047287518781072.

Sun, Y.-Y., Li, M., Lenzen, M., Malik, A. and Pomponi, F. (2022) Tourism, job vulnerability and income inequality during the COVID-19 pandemic: A global perspective. *Annals of Tourism Research Empirical Insights* 3 (1), 100046. https://doi.org/10.1016/j.annale.2022.100046.

Sun, Y.-Y., Faturay, F., Lenzen, M., Gössling, S. and Higham, J. (2024) Drivers of global tourism carbon emissions. *Nat Commun* 15 (1), 10384. https://doi.org/10.1038/s41467-024-54582-7.

UN Tourism (2021) The Glasgow Declaration on climate action in tourism. UN Tourism. See https://www.unwto.org/the-glasgow-declaration-on-climate-action-in-tourism (accesssed March 2025).

UN Tourism (2025) Tourism statistics. See https://www.unwto.org/tourism-statistics/key-tourism-statistics (accessed March 2025).

UNFCCC (2021) The explainer: The Paris Agreement. UNFCCC. See https://unfccc.int/news/the-explainer-the-paris-agreement#:~:text=Are%20all%20countries%20treated%20the,including%20investing%20in%20adaptation%20measures (accessed March 2025).

UNFCCC (2022) COP27 reaches breakthrough agreement on new "Loss and Damage" Fund for vulnerable countries. UNFCCC. See https://unfccc.int/news/cop27-reaches-breakthrough-agreement-on-new-loss-and-damage-fund-for-vulnerable-countries (accessed March 2025).

UNFCCC (2023) COP28 agreement signals "beginning of the end" of the fossil fuel era. UNFCCC. See https://unfccc.int/news/cop28-agreement-signals-beginning-of-the-end-of-the-fossil-fuel-era (accessed March 2025).

Wannapan, S. and Chaiboonsri, C. (2017) Sustainable international tourism demand in Thailand: The case of Chinese tourists. *Aktual'ni problemy ekonomiky* 191, 163.

WMO (2024) *State of the Global Climate 2023*. World Meteorological Organization. https://doi.org/https://library.wmo.int/idurl/4/68835.

Yu, Y., Yamaguchi, K. and Kittner, N. (2022) How do imports and exports affect green productivity? New evidence from partially linear functional-coefficient models. *J Environ Manage* 308, 114422. https://doi.org/10.1016/j.jenvman.2021.114422.

Zhou, W., Faturay, F., Driml, S. and Sun, Y.-Y. (2024) Meta-analysis of the climate change-tourism demand relationship. *Journal of Sustainable Tourism* 32 (9), 1762–1783. https://doi.org/10.1080/09669582.2024.2354882.

2 Climate (In)Justice in Tourism: Meanings and Perspectives from Industry and Academic Leaders

Shenshen He, Bobbie Chew Bigby, Freya Higgins-Desbiolles and Bryan Grimwood

This chapter explores tourism's complex role in global climate (in)justice through a thematic analysis of interviews with 17 tourism professionals from across sectors and geographies. Findings reveal how tourism contributes to climate injustice through disproportionate carbon emissions, economic inequality, labour exploitation and exclusionary governance practices. Participants highlighted how the benefits of tourism are concentrated among wealthy actors, while resource-constrained communities absorb the environmental and social costs, including those associated with climate change. Drawing on distributive, procedural and recognition justice frameworks, the chapter illustrates how climate justice is understood – and often limited – in tourism discourse and practice. While some participants acknowledged deeper structural and colonial dynamics, such critiques remain marginal. The chapter calls for a shift away from high-carbon, growth-driven tourism toward regenerative, accountable models that centre equity and local participation. In doing so, it contributes to a growing body of scholarship urging critical reflection on the legitimacy and future of tourism in an era of accelerating climate crisis.

Introduction

Tourism is closely connected to global climate change and plays a significant role in exacerbating climate injustice. The tourism industry's carbon emissions, especially from aviation, cruise industries and long-distance travel, have disproportionately contributed to global climate change. According to Hickel (2020), high-income countries in the Global

North have contributed the most to carbon emissions, while low-income countries in the Global South are the most vulnerable to the negative impacts of climate change. Moreover, the economic benefits of tourism are unevenly distributed, with multinational companies dominating the industry and local economies often failing to benefit (Cheer *et al.*, 2021). This is especially the case in the Global South, where resource-strained communities suffer the costs of environmental damage and social conflict, as well as significant labour exploitation (Cheer *et al.*, 2021). These issues highlight the structural injustices in tourism and the need for a climate justice approach that focuses on resource strained communities and promotes more equitable tourism development amidst ongoing climate impacts (Bigby *et al.*, 2024).

The purpose of this chapter is to contextualise and illuminate meanings of climate (in)justice in tourism. Recognising that 'climate justice' has long been under-explored and under-prioritised in the context of tourism (Tourism Panel on Climate Change, 2023), the chapter draws specifically on a thematic analysis of interviews with 17 tourism professionals and experts from across the globe, including individuals working in the tourism industry, government, academia and community-based tourism initiatives. These interviews provide diverse and grounded understandings of how climate (in)justice is interpreted by those directly engaged in the tourism sector. By exploring and interpreting these diverse perspectives, this chapter contributes to linking the discussion of global climate justice to specific practices of the tourism industry. Although climate justice has received increasing attention within broader discourses of sustainable development and climate change, practical explorations of how climate (in)justice is understood in the context of tourism are rather limited. Understanding these meanings is important because how tourism professionals interpret climate justice can influence how they understand responsibilities, set priorities and take action. This chapter therefore provides insights into ethical and structural dimensions of climate change as experienced and understood by the representatives of the tourism system and highlights the importance of integrating justice considerations into future climate and tourism policy.

Literature Review

There is a wealth of research that explores the relationship between tourism and climate change. Current literature identifies several key themes, including tourism's significant contribution to global carbon emissions, the vulnerability of tourism destinations to climate impacts, governance issues related to climate change mitigation and tourism industry responses to climatic challenges that foster resilience and adaptative capacity (Becken & Kaur, 2022). For example, tourism activities contributed approximately 8 to 11% of global carbon emissions in 2019, with air travel and cruises contributing the most (Bigby *et al.*, 2024). Meanwhile, small island developing countries and less developed coastal regions face threats such as sea level rise, extreme weather and ecosystem degradation (Fanning & Hickel, 2023). These risks of global climate change are threatening ecological and cultural livelihoods, as well as the core assets and resources of many tourism destination communities (Bigby *et al.*, 2024). In addition, existing research also highlights many complex issues facing tourism climate governance, such as how emission reduction actions can be effectively implemented (Becken, 2014) and how the resilience of destinations facing climate impacts can be enhanced (Higgins-Desbiolles *et al.*, 2022).

Climate justice and injustice have been recently introduced to the tourism and climate change literature as a set of frameworks for engaging with the interconnected challenges and repercussions of tourism's relationship with climate change. Climate justice emphasises that climate change is not only a scientific or technological issue, but also a moral and social justice issue (Sultana, 2021). The central focus of climate justice is placed on the disproportionate impact climate change has on different groups and how these can be corrected in a fair and equitable manner (Sultana, 2021). The theoretical repertoire of climate justice consists of distributive justice (the fair distribution of resources, risks and responsibilities) (Hickel, 2020), procedural justice (ensuring that vulnerable groups are included in decision-making processes associated with planning mitigation and adaptation strategies) (Sultana, 2021) and 'climate coloniality' (the historical exploitation of resources in the vulnerable regions by developed countries) (Fanning & Hickel, 2023). It is noteworthy that Boluk *et al.* (2019) criticised the UN Sustainable Development Goals for ignoring such historical, systematic inequalities and called on tourism researchers to deeply integrate climate justice perspectives to address the fundamental imbalances.

To an extent, tourism scholars have begun to apply a climate justice orientation in tourism studies, particularly in terms of evaluating how tourism activities exacerbate climate-related social and ecological injustices. Recent literature points to a focus on three main issues: the inequitable distribution of responsibility for tourism's carbon emissions, the often-unfair distribution of economic, social and other benefits from tourism (Bellato *et al.*, 2024) and the frequent marginalisation of Indigenous peoples and local communities in tourism governance (Whyte, 2010). Bigby *et al.* (2024), for instance, observe that, while rich tourists' frequent luxury tourism consumption leads to high carbon emissions, the negative consequences of climate change are mainly experienced by low-income regions and communities. Moreover, as Rastegar and Ruhanen (2023) explain, the lack of local participation and transparency in the tourism decision-making process exacerbates the dispossessions and displacements of Indigenous communities from critical governance processes.

Other relevant research has centred on Indigenous rights, regenerative tourism and a values-based model of tourism governance (Becken & Kaur, 2022). Curtin and Bird (2022: 461), for example, emphasised the importance of Indigenous agency and the practices of 'hosting, connection, and sharing' to achieve authentic reconciliation and cultural protection, challenging the traditional tourism model centred on commodification. Bellato *et al.* (2024) further elaborated on the concept of regenerative tourism, emphasising that tourism should go beyond the purpose of economic development and become a positive driving force for ecological restoration and community empowerment, abandoning past models of resource plunder implied by the tourism's extractive characteristics. In a similar vein, Becken and Kaur (2022) proposed that tourism governance should be clearly oriented towards local ecological, cultural and wellbeing values, rather than outdated models focused strictly on economic growth. For example, Aotearoa New Zealand integrated the concept of regenerative tourism into its national budget to promote intergenerational equity and ecological sustainability (Budget New Zealand, 2023). These discussions are examples that point to instances where tourism practice and research are shifting from the traditional growth model to new directions aligned more closely with concepts of justice and regeneration.

In this chapter, we contribute to the emerging literature on tourism and climate justice by illuminating how tourism professionals and experts understand climate (in)justice. Although the existing literature has engaged the climate justice concept and its importance for the tourism industry, the literature lacks a specific analysis of the actual understandings and responses of tourism professionals or practitioners to these issues. This study bridges the gap between theoretical research and practical reality by conducting in-depth interviews with tourism industry, academic and community leaders to analyse their specific understanding and practical experience of distributive justice (allocation of responsibilities and benefits of emission reduction) (Hickel, 2020), procedural justice (participation and transparency) (Sultana, 2021) and recognition justice (the value of Indigenous peoples and local knowledge). This practice-based research approach not only enhances the application scope of climate justice theory but also provides potential policy guidance and practical supporting evidence to help promote a more equitable and sustainable transformation of the tourism industry (Bellato *et al.*, 2024).

Methodology

This chapter builds on the report *Climate Justice in Tourism: An Introductory Guide* (Bigby *et al.*, 2024). As a 'white paper' contribution, the report explored the implications of climate justice in tourism for an industry-oriented audience (Bigby *et al.*, 2024). For this chapter, we analysed the 17 semi-structured interviews informing the report and which engaged experts representing various sectors, including business, government, academia, the non-profit sector, NGOs and community-based organisations and covered a wide geographic scope. The online interviews followed a flexible structure to explore questions concerning the role of tourism in climate (in)justice, existing inequities and strategies for resilience and adaptation. Interview participants shared their insights on climate (in)justice in tourism and best practices in promoting climate justice through open-ended discussions. All interviews were audio recorded and transcribed, with each interview lasting between 30 and 80 minutes.

To analyse the interviews, we used a thematic approach (Braun & Clarke, 2006) that allowed us to illuminate meanings of climate (in)justice in tourism and opportunities to promote climate justice in the industry. Congruent with Braun and Clarke (2006), our thematic analysis followed an inductive rather than deductive approach. Specifically, instead of trying to identify and construct themes into an existing theoretical framework, our analysis remained close to the data itself to allow themes to emerge (Potwarka *et al.*, 2016).

Our thematic analysis consisted of six phases (Braun & Clarke, 2006). The first involved familiarising ourselves with the interview transcripts. Members of our research team thoroughly read and re-read the transcripts to ensure a deep understanding of the content. Initial interpretations and reflections were noted during this stage. The second phase involved generating initial codes and more focused codes. We approached this process by systematically reviewing the 17 interviews to identify key patterns, recurring terms and significant features related to our research questions. We used open coding to highlight terms or phrases that directly appeared in the participants' transcripts. We ensured that the codes we identified reflected the data, including words or phrases that participants themselves used (van den Hoonaard, 2015). Additionally, during this phase we noted potential relationships between codes and how particular transcript excerpts fitted into broader categories. Our focused coding process involved identifying codes and potential relationships that were particularly relevant to the research questions and phenomena under study (Creswell & Poth, 2018). Here, we reviewed transcripts and open codes, removed the redundancies and started grouping codes into more useful, focused ones (Braun & Clarke, 2006). The third phase involved organising codes into broader themes by identifying patterns and connections (Braun & Clarke, 2006). The next phase involved reviewing themes. Our research team members reviewed the emergent themes and subthemes, comparing them against the coded extracts and the entire dataset to ensure that they reflected the data (Braun & Clarke, 2006). We also discussed the extent to which the themes reflected the details of the participants' responses, and where needed, we combined or revised the themes to match more closely the interview data (Braun & Clarke, 2006). The fifth phase involved defining themes in plain terms. Finally, the last phase involved writing up and substantiating the thematic interpretations by incorporating verbatim participant quotations into our descriptions of the themes and subthemes (Braun & Clarke, 2006).

The subsequent section presents the results of our thematic analysis, highlighting key themes and interpretations that emerged within the interview data. Focus is placed on illustrating how tourism professionals and experts conceptualise climate (in)justice. This includes their perceptions of justice in the distribution of climate responsibilities and benefits, recognition of marginalised voices and participation in climate-related decision-making processes. Prior to proceeding, we should note that the research reported herein followed ethical guidelines approved by the University of Waterloo's Research Ethics Board. Additionally, we recognise that our perspectives as researchers engaged with climate justice, social justice and sustainability in tourism, and community-based research, unavoidably shape our approach to interpreting the findings.

Analysis: How Climate (In)Justice is Understood among Tourism Experts and Practitioners

The role of tourism in global climate (in)justice is complex and contradictory. Participants in this research expressed that tourism both contributes to and is negatively impacted by climate change. Specifically, they described how tourism relies on the natural environment and stable climatic conditions but also serves as a major global emitter of carbon emissions. Additionally, the participants emphasised that, while the tourism industry contributed significantly to global carbon emissions, it often held a disproportionately limited accountability in global climate governance. For example, although aviation and cruise industries are major emitters, effective regulatory measures and sufficient emission reduction targets remain inadequate, which are illustrated by the extremely low global adoption rate (below 5%) of Sustainable Aviation Fuels (Geoffrey Lipman, personal communication, 6 December 2023). Furthermore, economic benefits generated by tourism frequently flowed disproportionately to multinational corporations, marginalising local communities and preventing equitable responsibility and benefit sharing (Megan Morikawa, personal communication, 9 January 2024). Participants also expressed concerns about the tourism industry's limited accountability for emissions, inequitable distribution of economic benefits, poor labour conditions and exclusion of local and historically marginalised voices from decision-making processes.

Consistent with the aims in this chapter, themes emerging from our analysis centred on illuminating what climate (in)justice means to tourism experts. In the following subsections, we describe four central themes: the disproportionate carbon emissions associated with tourism, the inequitable distribution of tourism's costs and benefits, the exploitation of labour across tourism's global supply chains and the lack of transparency, accountability and local involvement in tourism governance. Together, these themes provide insight into how tourism professionals understand the ethical and structural aspects of climate (in)justice in the tourism sector, and how these understandings may influence more ethical and sustainable responses to climate change.

Disproportionate carbon emissions in tourism

As noted above, tourism is a major source of global carbon emissions, but the responsibility for these is inequitably distributed. Participants reiterated these points, emphasising that most global tourism carbon emissions come from a few wealthy, high-income countries, while low-income countries and communities were more vulnerable to negative climate impacts. In more concrete terms, this imbalance manifests in travellers from wealthy countries continuing to experience the perks of high-carbon travel, while low-income countries and resource-constrained communities experience an increasing number of extreme weather events, rising sea levels and ecosystem damage caused by climate change. The most climate vulnerable and resource-constrained communities thus bear the brunt of climate change. As James Higham, Professor of Sustainable Tourism at Griffith University (Australia), noted:

> Tourism emissions from global tourism [is] increasing very, very rapidly... driven by demand, and that demand is being driven by a very small number of countries, and a very small proportion of wealthy individuals within those countries... 50% of the population don't fly at all in a given year. Whereas 70% of all flights are consumed by only 15% of the population. And so we have these enormous inequities. And growth in air travel has been driven by that 15%. (James Higham, personal communication, 10 January 2024)

James Higham's perspective indicated that, although a large percentage of the global population did not participate in air travel, a small number of high-income individuals from certain countries still consumed most of the carbon budget. The continuance of this high-carbon travel model made the Global South the biggest victim of climate change. Ayako Ezaki, the co-founder of TrainingAid, an international tourism training and capacity-building organisation, emphasised a similar point based on her experience. She stated, 'So many of the smaller economies have 0.00001% of the emissions compared to larger countries like Japan' (Ayako Ezaki, personal communication, 6 December 2023).

This disparity highlights the disproportionate distribution of both emissions' responsibility and climate impacts, with countries contributing the lowest historically to carbon emissions bearing the highest environmental costs. This imbalance in climate justice was further discussed in the Tourism Panel on Climate Change (TPCC) Stocktake, which stressed the disproportionate effects on developing countries, including Small Island Developing States (SIDS), which suffered the negative environmental impacts despite their minimal contribution to global emissions (Tourism Panel on Climate Change, 2023).

Interview participants generally viewed the aviation and cruise industries as key sources of unjust carbon emissions in tourism. As stated by Daniela Subtil of Stay Grounded, a Europe-based global NGO focused on reducing aviation and promoting climate justice, aviation is one of the most carbon intensive forms of transportation with providing significant challenges in meaningful decarbonisation (Daniela Subtil, personal communication, 6 December 2023). She emphasised that current reduction efforts in aviation are insufficient, as the industry continues to operate at highly unsustainable emission levels without substantial change. According to the TPCC Stocktake (2023), although sustainable aviation fuel (SAF) offered the potential to reduce emissions by 50% or more, its large-scale production faces obstacles such as high financial costs and the need for further research and development (Tourism Panel on Climate Change, 2023). Likewise, Geoffrey Lipman of SUNx Malta, an EU-based NGO dedicated to promoting climate-friendly travel globally, highlighted that:

> Sustainable aviation fuel can only account for 5% of our flying. So really, we're making a big thing about it. But that's a lot of marketing bullshit. And we think there should be a guarantee of sustainable aviation fuel for LDCs (Least Developed Countries) and SIDS (Small Island Developing States). (Geoffrey Lipman, personal communication, 6 December 2023)

As the above discussion suggests, while tourism's carbon emissions are disproportionately driven by wealthy countries and high-income individuals, low-income countries and vulnerable communities suffer the most severe climate impacts. Although aviation is a primary source of emissions, the minimal use of SAF and the industry's seemingly superficial commitments to carbon neutrality in many instances further exacerbate global climate injustice.

Inequitable distribution of tourism's costs and benefits

In this section, three key ideas emerged from the interviews regarding the disproportionate distribution of tourism's costs and benefits. These include: (1) the concentration of tourism's economic benefits in multinational corporations and high-end markets, (2) the financial leakage that limits local economic gains and (3) the disproportionate distribution of resources, particularly in communities facing water scarcity and other challenges.

First, although tourism is often promoted as a tool for economic growth, interview participants suggested that its benefits were often concentrated among multinational corporations, large chain brands and high-end markets, with local communities and small businesses frequently receiving minimal returns. Instead of economic gains, local communities tend to suffer much of the environmental and social costs of tourism. This disproportionate distribution of economic benefits has, in many cases, amplified economic inequality and failed to promote social equity (Bennike & Nielsen, 2024). As Ayako Ezaki noted, 'we might be creating economic opportunities, but at the same time, they might not be fairly paying the people who make those things possible' (Ayako Ezaki, personal communication, 6 December 2023).

Second, financial leakage, where a significant portion of tourism income 'leaks' out of local economies and into the coffers of external investors and international chains, remains a continued issue in many tourism destinations, particularly those relying on international tourists (Dwyer & Forsyth, 2006). Megan Morikawa, Global Director of Sustainability at Iberostar Group, a Spanish-based luxury hospitality chain, emphasised that this phenomenon was especially severe in the Caribbean region, where profits from luxury resorts, often owned by international companies, are sent back to investors' home countries (Megan Morikawa, personal communication, 9 January 2024). According to Megan Morikawa, 'Jamaica has historically suffered from leakage, financial leakage, and they're struggling to figure out the financial model for how they turn a resilience centre into broad scale for good' (Megan Morikawa, personal communication, 9 January 2024). This point highlights the challenges that local economies experience in securing and sustaining the economic benefits provided by tourism that all too often are not shared by the community but instead leak out to non-local tourism stakeholders.

Financial leakage not only reduces the positive impacts of tourism on the local economy but also makes many tourism-dependent communities more vulnerable to economic shocks, particularly in areas where tourism is the primary source of income. In such communities, changes in tourist arrivals that are caused by natural disasters, political unrest or economic recessions, can result in unexpected decreases in earnings, leaving local residents without alternative livelihoods or financial security. Such crises are likely to occur more frequently as a result of climate change.

Third, the growth of tourism is often accompanied by the inequitable distribution of local resources, especially in areas facing water scarcity, land shortages and unstable housing markets. In many dry areas, the high water consumption of the tourism industry leads to competition for resources with local communities. Shivya Nath, a travel writer and sustainability consultant based in India, whose work focuses on responsible tourism and community-centred climate action, illustrated this in stating:

> I was recently in Morocco, in Marrakech, where there's been a prolonged drought. In places like that, water consumption is really high, for example, swimming pools use a lot of water. Communities already feel water is scarce, and this has created tension. Locals are naming tourism as the reason for water shortages, while people in the tourism industry, in turn, blame local communities for using water in public baths, where baths are a part of their cultural tradition and have existed for centuries. (Shivya Nath, personal communication, 18 January 2024)

In this account, Shivya Nath illustrates a classic tension in several destinations: to promote tourism development, governments often prioritise infrastructure investment for tourists at the expense of supporting the basic needs of local communities. For example, large-scale tourism infrastructure construction in coastal areas not only exacerbates ecological damage but can also result in the forced expropriation of land and forced relocation of local communities. This development model reflects an inequity in decision-making priorities within the tourism industry, where governments tend to satisfy the needs of tourists rather than the long-term wellbeing of local communities. As Christina Gale, Sustainable Tourism Manager at the South Pacific Tourism Organisation (SPTO), based in Fiji and working across Pacific Island nations, stated:

> Sea level rise has exacerbated coastal inundation and its impacts on nearby communities, particularly in areas where tourism infrastructure is concentrated. These communities are now increasingly affected by strong wave action and flooding linked to both climate change and the placement of tourism developments along the coast. (Christina Gale, personal communication, 6 December 2023)

As the above quotations suggest, while tourism can drive economic growth, the economic benefits are extremely disproportionately distributed, with local communities and small businesses often failing to receive a fair return. At the same time, tourism's intensive resource consumption continues to impact many aspects of communities, ranging from inadequate housing access to inequitable infrastructure investment, as governments often prioritise tourists over local residents.

The exploitation of labour in tourism's global supply chains

The tourism industry relies on a wide range of global supply chains, from aviation and hotels to handicraft production and other services, with many labourers in low-wage, unstable jobs lacking social security and long-term stability. While the tourism industry generates many jobs globally, interview participants generally agreed that this growth often comes at the expense of exploited labour, particularly in developing countries and in Global South tourism destinations. The following quotation from Daniela Subtil illustrates this phenomenon:

> [The] most extreme example is aviation or air traffic... The tourism industry, impoverishes people, the way we see it right now... mass tourism but also it is not only mass tourism, is really based on precarious work. And if we are just actually getting more people that are in poverty that work in the tourism industry, so people that are employed, but still poor, and are just more vulnerable to the climate crisis. (Daniela Subtil, personal communication, 6 December 2023)

The seasonal nature of the tourism industry forces many tourism industry employees to suffer long overtime hours during peak seasons, while facing layoffs during the off season, making it hard for them to achieve long-term financial stability. In a joint interview, Stroma Cole and Angela Kalisch, sustainability scholars affiliated with the University of Westminster and the UK-based organisation, Equality in Tourism, highlighted the exploitative conditions facing many women workers in the tourism sector. They stated: 'Mostly they're women, paid, you know, very little, the working conditions, they work under... and look at the cruise industry, for example... working conditions under deck that needs to be addressed... sexual harassment, gender-based violence, at work, and so on' (Stroma Cole & Angela Kalisch, personal communication, 11 December 2023). This labour exploitation is most obvious in sectors such as cruise lines, accommodation and food services, where female workers generally received lower wages and work under challenging conditions.

Furthermore, while women occupy many positions in the tourism industry, interview participants observed that their salaries are generally lower than those of men in equivalent positions. Additionally, women are more often restricted to low-paid positions with few opportunities for promotion, such as housekeepers, receptionists and restaurant waitresses. Workplace harassment and gender discrimination in tourism settings are also issues that interview participants identified. Specifically, they indicated that women are more vulnerable to these injustices while the tourism industry often lacks effective complaint and protection mechanisms, which adds additional layers of potential harassment and vulnerability upon women's tourism work experience. Indeed, many women chose to remain silent in the face of injustice so as not to lose their jobs. As Daniela Subtil pointed out

> BIPOC [Black, Indigenous, People of Colour] people, because they are... really the hidden force of this tourism industry... gender justice, I think this is such an important dimension, how women are, of course, being a very marginalized group, how their work is so precarious and how it's linked to sexual exploitation and sex trafficking and all of those very disturbing activities... BIPOC people and migrants are

just so exploited... really living the worst working conditions in the tourism industry... this billion multi-billion dollar industry is just possible because of all of that exploitation. (Daniela Subtil, personal communication, 6 December 2023)

Tourism's global supply chains rely on low-wage, unstable labour, particularly in the cruise, accommodation and aviation industries, where workers often faced exploitation. Although supporting the entire industry, these grassroots workers rarely receive fair compensation or labour protection, suffering a disproportionate cost of tourism's growth.

The lack of transparency, accountability and local involvement in tourism

Three key subthemes emerged from our analysis that relate to the role of tourism governance in climate justice: (1) the lack of transparency and accountability in tourism governance, (2) the exclusion of local communities from decision-making processes and (3) the prioritisation of tourism infrastructure over local needs. Each of these issues exacerbates the inequities faced by local communities and contributes to the ongoing climate (in)justice within the industry.

First, for participants in this study, the governance structure of the tourism industry generally lacked transparency. Governments and large enterprises were often not subject to public oversight in decision-making processes, resulting in many tourism development projects that failed to fully consider the interests of local communities. Additionally, the tourism industry's weak accountability mechanisms allowed companies to avoid responsibility for environmental and social damage, while local residents were excluded from tourism policymaking and had difficulty applying a substantive influence on the direction of industry development. Participants generally believed that the tourism industry's governance mechanisms had serious deficiencies in terms of information disclosure and transparency in decision making. Companies could take advantage of this gap to engage in 'greenwashing' and while continuing to claim compliance with sustainability standards in the absence of third-party audit. As Stroma Cole and Angela Kalisch pointed out:

> I believe corporate accountability and enforcement of that accountability and transparency is really important... In order to have a social license, you need to be able to tell people what you're doing for it, and it's still not happening... there's lots of greenwashing going on... where there is reporting possibly it's kind of tokenistic, then it's not independently verified. And there is no transparency. (Stroma Cole & Angela Kalisch, personal communication, 11 December 2023)

Geoffrey Lipman pointed out that action by the tourism industry may be illusory: 'I think the establishment culture promises to meet all of the targets, and to do things in a, in a just way, and then does a lot of marketing hype to allow it to continue business as usual...' (Geoffrey Lipman, personal communication, 6 December 2023). This reflected how large travel companies often create the illusion of fulfilling social responsibilities, while in reality, they maintained the same profit-driven practices, using marketing to mask their unchanged operations.

Second, the role of local communities in tourism governance mechanisms was often marginalised, which was another major issue repeatedly mentioned by the interviewees. At many tourism destinations, policy formulation and project planning were often dominated by government authorities, industry associations or external investors, while residents, who were directly affected, were marginalised. Very often, residents did not realise the impact of tourism projects until they were implemented, and they usually did not have the opportunity to participate in policy discussions or express their needs at an early planning stage. Some participants expressed that many tourism policies were formulated without considering the realities of local communities, making it difficult to implement these policies and ensure that they truly benefit the community. As Shivya Nath stated:

> I think that's where it starts, because a lot of the policies are crafted such that they don't even reach the people on the ground, who are working on a grassroots level, or they make no sense to them. (Shivya Nath, personal communication, 18 January 2024)

Moreover, the tourism governance mechanism often imposed additional barriers on economically unstable communities, making it difficult for them to truly participate in tourism development. The economic barriers mentioned by Shivya Nath further exacerbated the marginalisation of local communities, leaving them not only without a voice in decision-making process but also struggling to gain fair opportunities in the tourism economy through entrepreneurship. She stated: 'You need to have a certain amount of financial access to be able to even start a tourism initiative' (Shivya Nath, personal communication, 18 January 2024).

Third, tourism development and environmental restoration initiatives often prioritised enhancement of tourism infrastructure and ecological conservation over the needs of local communities, which exacerbated community dissatisfaction and economic vulnerability. Some tourism initiatives might even result in the forced relocation or the loss of the land on which they lived for generations, as Judy Kepher Gona expressed it:

> When restoration happens, the communities are not considered in that restoration... And if there are opportunities for reinvestment and restoration, they will focus on how to save the wildlife and how to restore the land in protected areas. And very little will go out to take care of the communities that are also ravaged the pastoralists who lose their livestock

during drought because of climate change, and they lose pasture and more of space, with the erosion and loss of vegetation during excess rains. But the work that is done in community compared to the investments that are made to keep protected areas going and safe for tourists. When we say we want to keep places safe for tourists, we forget that these places… are first and foremost [the] homes of people. (Judy Kepher Gona, personal communication, 6 December 2023)

As the above quotations suggested, lack of transparency in tourism governance and the absence of local communities from decision making led to unfair development patterns. Economic barriers and expanding tourism infrastructure at the expense of community-driven infrastructure further marginalised these communities, preventing them from benefiting fairly from the industry's growth.

Discussion

Through an analysis of interviews with 17 tourism experts from different regions and institutions, this chapter has illuminated the complex roles that tourism plays in climate (in)justice and further explored the limitations within the tourism industry regarding its understanding of justice issues. The interviews illustrate that, while participants generally recognise the dual role of the tourism industry in climate justice, most perspectives in tourism remain focused on distributional injustices, such as carbon emissions responsibility, economic benefits and resource allocation, with limited reflection on deeper structural injustices.

For example, in terms of distributive justice (the fair distribution of resources, risks and responsibilities) (Hickel, 2020), several participants highlighted the severe asymmetry between carbon emissions in the tourism industry and the groups most affected by them. James Higham (personal communication, 10 January 2024) pointed out that 70% of aviation carbon emissions came from 15% of wealthy travellers, which obviously contrasts with the long-standing climate disasters faced by countries in the Global South (Hickel, 2020). Meanwhile, despite the aviation industry's commitments to using sustainable fuels, Geoffrey Lipman (personal communication, 6 December 2023) criticised the global application rate of SAFs remaining below 5%, far short of emission reduction targets. These findings illustrate both the shifting of emission reduction responsibilities and how developed countries in global climate governance sometimes use technical discourse to cover their continued dominant position.

The unfair distribution of economic benefits is a structural issue of 'financial leakage' (Dwyer & Forsyth, 2006), which exacerbates climate (in)justice. Megan Morikawa (personal communication, 9 January 2024) used the Caribbean region as an example to point out that a large portion of tourism profits flow to international markets, leaving local communities unable to receive substantial benefits. This financial leakage not only weakens local economic resilience but also makes it harder for these communities to build adaptive capacity to climate change, as they lack the resources to invest in necessary climate mitigation and adaptation measures. Ayako Ezaki (personal communication, 6 December 2023) further noted that, although tourism produced many jobs, these jobs have not improved the economic resilience of locals but have instead frequently made poverty and inequality worse. This pattern of transnational corporate dominance and community marginalisation, as emphasised by Cheer *et al.* (2021), is an extension of institutional inequality within the tourism capitalist system.

In addition, the exploitation of labour, especially in the cruise industry, deepens the challenges faced by local communities. Stroma Cole and Angela Kalisch (personal communication, 11 December 2023) highlighted cases of exploitation of women and migrant workers in the cruise industry, further showing gender and class harassment within the global tourism supply chain and demonstrating the absence of labour justice. These labour injustices are linked to climate justice because they reflect the systemic inequalities that prevent affected communities from participating in decision-making processes and addressing the broader environmental and social challenges caused by climate change.

Finally, the lack of transparency and local community participation in tourism governance was one of the key factors leading to climate (in)justice. Participants generally agreed that the decision-making process of many tourism projects and policies lacks transparency, and local communities were not effectively involved. Shivya Nath (personal communication, 18 January 2024) mentioned that many policies were designed without taking full consideration of the actual needs and interests of local communities, which led to difficulties in achieving the expected results during policy implementation and even exacerbated the dissatisfaction of local communities. In their academic analysis, Rastegar *et al.* (2023) emphasised that, although many tourism projects claim to focus on sustainability and environmental protection, in practice, external investors and industry associations often dominate decision making, while the voices of local communities were frequently marginalised. Tuck and Yang (2012) argued that a decolonial approach necessitates the recovery of land and local knowledge and the real participation of local residents and Indigenous peoples as not only a proper process of governance, but also an important way to recover from colonial violence and ecological damage. Therefore, increasing governance transparency and local participation is fundamental to promoting the realisation of more holistic forms of climate justice in the tourism industry.

However, what is more notable is that awareness of procedural justice (participation and transparency) (Sultana, 2021) and structural justice is underrepresented in most interviews. While participants, such as Shivya

Nath and Christina Gale, mentioned the absence of local communities in tourism decision making, only a few, such as Daniela Subtil and Judy Kepher Gona, further contextualised tourism issues within broader political, economic and climate crisis frameworks. Daniela Subtil pointed out that the high-carbon nature of tourism should not be simply viewed as an 'industry issue', but rather integrated into a systemic critique of the entire consumerist and growth logic. Judy Kepher Gona goes further to question the existing financial systems that ignore the value of land as an asset for securing funding, emphasising how 'structural neglect' removes communities' equal access to development opportunities.

As the concept of 'climate coloniality' (i.e. the historical exploitation of resources in resource-constrained regions by developed countries) makes clear, the climate crisis is not only a matter of carbon emissions but is also tied to the continuation of oppressive colonial structures (Fanning & Hickel, 2023). Tourism, as an industry rooted in colonial dynamics and expansionism, often continues to operate on the extraction of nature and labour. Therefore, when discussing climate justice, we must not only focus on who emits how much and who is affected, but also ask: Who sets the rules for development? Who holds the power to speak? And who is systematically excluded?

What deserves further reflection is that tourism, as a typically non-compulsory, recreational and high-emission consumption activity, must be re-evaluated for its legitimacy in the current climate emergency (Palm, 2025). Tourism is not essential for survival; it often comes with a high carbon footprint and high resource consumption, yet it is usually packaged or promoted as a symbol of rights or development. This logic reflects a privileged structure in which only those with strong economic capabilities and high mobility can continue to enjoy this right, while the Global South and future generations suffer the environmental costs.

Given these concerns, it is important to recognise that climate change is only one dimension of the broader 'polycrisis' in which tourism operates (Rastegar *et al.*, 2023). Economic, political, social and ecological crises are interconnected and mutually reinforcing, continuously expanding the injustices within the global tourism system. Therefore, a narrow focus on the climate change dimension alone is insufficient to fully demonstrate the structural challenges and underlying causes faced by the modern tourism industry within the broader context of these crises. We live in an era of multiple interconnected global crises, including breaking climate thresholds, the gradual breakdown of ecosystems, social unrest and growing financial inequalities. Against this background, continuing to tolerate the uncontrolled expansion of the tourism industry would seem to require ignoring its existence as a 'harmful luxury' and shifting its negative impacts onto resource-constrained regions and future generations (Smith, 2019). As a high-carbon, non-essential consumption activity, the legitimacy of tourism can no longer be defended only on the grounds of economic contributions or cultural exchange alone (Palm, 2025). Instead, this consumption-driven practice should be thoroughly re-evaluated within the framework of global climate justice, reassessing whether its social and environmental costs are justified and who is suffering these costs while others obtain the subsequent power and privileges.

The findings from the 17 interviews illustrate that, while a few interviewees, such as Daniela Subtil, clearly pointed out a systematic critique of the high-carbon nature of tourism, this critique remains marginal in the broader tourism industry. This reflects a deeper structural blind spot, with few within the tourism industry truly challenging a fundamental question: why do we continue to prioritise protecting this non-essential, high-emission consumption activity when the global carbon budget is nearly exhausted? The absence of such critical perspectives contributes to the systemic inertia that prevents the necessary structural changes in the industry.

Ultimately, the findings from this research illustrate the complex role of tourism in climate (in)justice and highlight inequalities in areas such as carbon emissions, the distribution of economic benefits, labour exploitation and transparency in governance (Fanning & Hickel, 2023). It seems clear that the tourism industry should take more equitable measures to reduce carbon emissions, ensure the fair distribution of economic benefits, improve labour conditions and enhance the participation and transparency of local communities in governance (Bigby *et al.*, 2024).

Conclusion

Based on a thematic analysis of 17 interviews with tourism experts, this chapter has explored the tourism industry's complex roles in global climate justice as both a contributor to and a casualty of climate change. As our analysis illustrates, interview participants' understanding of climate (in)justice reveals deep-rooted injustices in the tourism industry, including disproportionate carbon emissions in tourism globally, inequitable distribution in tourism, labour exploitation and a lack of transparency and local involvement in tourism governance. While tourism may provide economic opportunities in many regions, participants indicated that its growth is often driven by a small number of high-income countries, while environmental and social costs are absorbed primarily by resource-constrained countries and environmentally vulnerable communities. Furthermore, financial leakage, the industry's frequent avoidance of trade unions and accountability, and the absence of local communities in tourism governance, makes it difficult for residents to benefit.

Importantly, interview participants emphasised the need for improving distributive justice around areas such as carbon emissions responsibility, economic benefit allocation and resource distribution. However, as implied in our analysis, the tourism industry has a tendency to overlook questions of procedural justice and structural justice. In this context, as Fanning and Hickel (2023) point out, 'climate coloniality' is deeply grounded in the historical logic of colonialism, which continues to shape the disproportionate extraction of resources and labour in the tourism industry. This framework of 'climate coloniality' illuminates how the tourism industry's practices, rooted in both its historical colonial past and its contemporary operations, contribute to the continuation of structural injustices, particularly in how resources are extracted from historically marginalised communities (Fanning & Hickel, 2023). The ongoing marginalisation of these communities in decision-making processes further exacerbates their vulnerability to climate change impacts. The interviews also support Sultana's (2021) analysis on the consequences of an absence of structural justice: the existing tourism governance model marginalises local communities and Indigenous knowledge systems due to a lack of transparency and local participation.

The research reported in this chapter draws attention to a general lack of critical thinking within the tourism system regarding tourism as high carbon emitting 'discretionary consumption' (Palm, 2025). As the global climate has already crossed the 1.5°C threshold and increasing numbers of ecosystems are nearing collapse, traditional arguments linking tourism to economic benefits or cultural exchange are no longer sufficient to support the validity of tourism as a luxury consumption activity (Copernicus Climate Change Service, 2025). But only a few interviewees consistently expressed these concerns, indicating a broader lack of awareness for critical structural evaluation of the tourism industry.

Acknowledgements

The authors would like to acknowledge the Travel Foundation for its support in commissioning and assisting the development of the original report on climate justice and tourism that is cited in this chapter. The interviews under focus in this chapter all emerged from this research process that was initiated by the Travel Foundation. The authors also wish to acknowledge and thank Mr Jeremy Smith, co-author of the Climate Justice and Tourism report, for all of his insights, contributions, energy and time that he dedicated to this research process and report writing.

References

Becken, S. (2014) Water equity – contrasting tourism water use with that of the local community. *Water Resources and Industry* 7–8, 9–22. https://doi.org/10.1016/j.wri.2014.09.002.
Becken, S. and Kaur, J. (2022) Anchoring 'tourism value' within a regenerative tourism paradigm – a government perspective. *Journal of Sustainable Tourism* 30 (1), 52–68. https://doi.org/10.1080/09669582.2021.1990305.

Bellato, L., Frantzeskaki, N. and Nygaard, C. (2024) Towards a regenerative shift in tourism: applying a regenerative conceptual framework toward swimmable urban rivers. *Tourism Geographies* 26 (8), 1361–1380. https://doi.org/10.1080/14616688.2024.2358306.

Bennike, R.B. and Nielsen, M.R. (2024) Frontier tourism development and inequality in the Nepal Himalaya. *Journal of Sustainable Tourism* 32 (4), 773–794. https://doi.org/10.1080/09669582.2023.2174129.

Bigby, B.C., Smith, J. and Higgins-Desbiolles, F. (2024) Climate justice in tourism: An introductory guide (Travel Foundation). See https://www.thetravelfoundation.org.uk/climatejustice/ (accessed May 2025).

Boluk, K.A., Cavaliere, C.T. and Higgins-Desbiolles, F. (2019) A critical framework for interrogating the United Nations Sustainable Development Goals 2030 Agenda in tourism. *Journal of Sustainable Tourism* 27 (7), 847–864. https://doi.org/10.1080/09669582.2019.1619748.

Braun, V. and Clarke, V. (2006) Using thematic analysis in psychology. *Qualitative Research in Psychology* 3 (2), 77–101. https://doi.org/10.1191/1478088706qp063oa.

Budget New Zealand (2023) Support for today- Building for tomorrow. See https://www.treasury.govt.nz/sites/default/files/2023-05/b23-wellbeing-budget.pdf (accessed April 2025).

Cheer, J.M., Lapointe, D., Mostafanezhad, M. and Jamal, T. (2021) Global tourism in crisis: Conceptual frameworks for research and practice. *Journal of Tourism Futures* 7 (3), 278–294. https://doi.org/10.1108/jtf-09-2021-227.

Copernicus Climate Change Service. (2025) Global Climate Highlights 2024. See https://climate.copernicus.eu/global-climate-highlights-2024 (accessed May 2025).

Creswell, J.W. and Poth, C.N. (2018) *Qualitative Inquiry and Research Design: Choosing Among Five Approaches*. 4th edn, SAGE Publications Inc.

Curtin, N. and Bird, S. (2022) 'We are reconciliators': When Indigenous tourism begins with agency. *Journal of Sustainable Tourism* 30 (2–3), 461–481. https://doi.org/10.1080/09669582.2021.1903908.

Dwyer, L. and Forsyth, P. (2006) *International Handbook on the Economics of Tourism*. Edward Elgar.

Fanning, A.L. and Hickel, J. (2023) Compensation for atmospheric appropriation. *Nature Sustainability* 6 (9), 1077–1086. https://doi.org/10.1038/s41893-023-01130-8.

Hickel, J. (2020) Quantifying national responsibility for climate breakdown: An equality-based attribution approach for carbon dioxide emissions in excess of the planetary boundary. *The Lancet Planetary Health* 4 (9), e399–e404. https://doi.org/10.1016/s2542-5196(20)30196-0.

Higgins-Desbiolles, F., Blanchard, L.-A. and Urbain, Y. (2022) Peace through tourism: Critical reflections on the intersections between peace, justice, sustainable development and tourism. *Journal of Sustainable Tourism* 30 (2–3), 335–351. https://doi.org/10.1080/09669582.2021.1952420.

Palm, M. (2025) Rethinking 'discretionary' travel: The impact of night and evening shift work on social exclusion and mobilities of care. *Travel Behaviour and Society* 40, 101030. https://doi.org/10.1016/j.tbs.2025.101030.

Potwarka, L.R., Tepylo, H., Fortune, D. and Mair, H. (2016) Launching off but falling fast: Experiences of becoming more physically active in response to the Vancouver 2010 Olympic Winter Games. *Event Management* 20 (3), 297–312. https://doi.org/10.3727/152599516x14640225219155.

Rastegar, R. and Ruhanen, L. (2023) Climate change and tourism transition: From cosmopolitan to local Justice. *Annals of Tourism Research* 100, 103565. https://doi.org/10.1016/j.annals.2023.103565.

Rastegar, R., Higgins-Desbiolles, F. and Ruhanen, L. (2023) Tourism, global crises and justice: Rethinking, redefining and reorienting tourism futures. *Journal of Sustainable Tourism* 31 (12), 2613–2627. https://doi.org/10.1080/09669582.2023.2219037.

Smith, O. (2019) Luxury, tourism and harm: A deviant leisure perspective. In T. Raymen and O. Smith (eds) *Palgrave Studies in Crime, Media and Culture* (pp. 305–323). Palgrave Macmillan. https://doi.org/10.1007/978-3-030-17736-2_14.

Sultana, F. (2021) Critical climate justice. *The Geographical Journal* 188 (1), 118–124. https://doi.org/10.1111/geoj.12417.

Tourism Panel on Climate Change (2023) Tourism and Climate Change Stocktake 2023. See https://tpcc.info/ (accessed May 2025).

Tuck, E. and Yang, W. (2012) Decolonization is not a metaphor. *Decolonization* 1 (1), 1–40.

Van den Hoonaard, D.K. (2015) *Qualitative Research in Action: A Canadian Primer*. Oxford University Press Canada.

Whyte, K.P. (2010) An environmental justice framework for Indigenous tourism. *Environmental Philosophy* 7 (2), 75–92. https://www.jstor.org/stable/26168043.

3 Towards a Just Transition in Aviation Climate Policy: Implications for Tourism

Gerben Broekema, Vishal Babajee, Ramón Fisac García and Daniel Scott

After decades of steady international tourism growth – boosting economies, reducing poverty and lowering global inequality – ambitious aviation climate policies now risk reversing these gains. A just transition to a low-carbon economy must be fair and inclusive, protecting workers, destinations and vulnerable groups impacted by climate action. This chapter uses the COVID-19 pandemic's travel bans as a case example to assess the distributional effects of sharp drops in international tourism, examining impacts on three country groups during the 2021 summer holiday season. Findings suggest that decreased international tourism increases between-country inequality, with severe consequences for tourism-dependent countries. In contrast, tourism-outbound nations may benefit from regional travel and domestic spending, but also face rising challenges like labour shortages, accommodation pressures and overtourism. Tourism generates environmental and social externalities – pollution, habitat loss, cultural commodification and inequality – which must be addressed in European Union (EU) policy. A review of existing and proposed EU aviation climate measures reveals a likely reduction in long-haul travel demand and affordability, echoing the pandemic's uneven impacts and risking a divergence in tourism prosperity. Policymakers and non-state actors are urged to adopt a broader view of justice when shaping sustainable air travel policies to ensure equity in the low-carbon transition.

Introduction

Larger networks, higher frequency and lower cost have made long distance travel significantly more accessible over the past decades. This has benefited travellers, as well as countries, businesses and workers in the aviation sector, its supply chain and sectors reliant on the movement of people and goods by air. Global air connectivity has grown at a rate of 4.7% in terms of passengers between 1990 and 2019 vs. a global economic growth of 3.0% over the same period (IEA, 2020). While there are many nuances regarding the impact of increased travel, recent research shows that international tourism reduces poverty and decreases inequality (Lagos & Wang, 2022). Other research supports the important role of tourism in economic development in emerging economies, with co-benefits for the rest of the economy of developing tourism infrastructure (Khan *et al.*, 2020). The relationship between tourism growth and economic development is typically not a one-way causal relationship but is proven to be positively linked for a set of 123 countries in their development between 1995 and 2019 (Cárdenas-García *et al.*, 2024). At the same time, there are also diverse social and environmental concerns about international tourism (Rastegar & Becken, 2024). One particular discrepancy is the carbon footprint.

Different strategies exist to address the challenge of climate change and the balance of greenhouse gases (GHG) emissions. Net zero CO_2 emissions, also referred to as carbon neutrality, are achieved when *anthropogenic* CO_2 emissions are balanced globally by anthropogenic CO_2 removals over a specified period (IPCC, 2018). Removals can take place through offset mechanisms (like carbon capture, reforestation or purchasing carbon credits). The goal is not to eliminate all emissions, but to ensure that any residual emissions are balanced by equivalent removals, achieving a 'net' impact of zero on the climate. Real zero, on the other hand, means eliminating all GHG emissions without relying on offsets. Therefore, the mechanisms are focused on the elimination of fossil fuels, the development of renewable energy use and clean technologies. While real zero represents a more ambitious goal and stronger climate impact, it presents technical barriers to change rapidly a system that is built on fossil fuels. These rapid changes can generate social externalities in vulnerable populations that need to be analysed with care.

Global warming will affect citizens across the world but is expected to increase global inequality with poorer countries being affected worse than rich countries (Diffenbaugh & Burke, 2019; IPCC, 2022).

Awareness of the need to address carbon emissions is widely shared throughout the international community (195 country signatories to the Paris Climate Agreement), including among industry and business associations. As a response, the aviation sector has brought forward a decarbonisation strategy to pursue a net zero target by 2050. This roadmap leans heavily on the transition to Sustainable Aviation Fuels (SAF) and hydrogen, in addition to technology improvements such as more fuel-efficient aircraft and electric aircraft. Complementing these measures is the Carbon Offsetting and Reduction Scheme for International Aviation (CORSIA), adopted by the International Civil Aviation Organization (ICAO), which aims to stabilise international aviation emissions at 2020 levels by 2030. With SAF being considerably more expensive than regular jet fuel (3 to 4 times), air fares are expected to become more expensive as the blending rate of SAF is increased (regulatory or voluntary), potentially reducing demand and affecting tourism-dependent countries (WTTC, 2023).

Environmental interest groups and other observers challenge the sector's decarbonisation strategy in terms of its ambition in terms of timelines and feasibility (Scott & Gössling, 2021). Academic research confirms that there are challenges and downsides in terms of negative spill-over effects on other sectors in achieving the sector's ambition, stating that policies lowering demand expansion may be unavoidable (Bardon & Massol, 2023; Gössling *et al.*, 2015; Scott & Gössling, 2021). In recent years, various countries and regions have implemented or proposed regulations and measures to limit aviation flights to reduce their environmental impact. These include aviation-related taxes, movement caps to curb noise and GHG emissions and behavioural changes such as '*flygskam*' (Swedish for 'flight shame') (Baran, 2019). There is increasing pressure to introduce policy measures that directly or indirectly manage demand (reduce growth or reshape inbound market segments).

The challenge is that in defining what climate change mitigation measures to take, there is little consideration of the (spatial pattern of) short- and mid-term consequences (e.g. the consequences on employment or inequality in destination countries). A 2023 International Monetary Fund (IMF) Working Paper points to the importance of including this lens:

> Our qualitative and quantitative scenario analysis suggests that the mid-transition – a period during which fossil-fuel and low-carbon energy systems co-exist and transform at a rapid pace – could have profound stability and resilience implications for global trade and the international financial system. (Espagne *et al.*, 2023: 2)

Similar concerns have been raised specific to international tourism flows (Gössling *et al.*, 2015; Pentelow & Scott, 2011).

In addition, the intra-generational goal of sustainability refers to the challenge of ensuring equitable resource distribution and environmental protection within the present generation. It focuses on addressing disparities between different regions, communities and socioeconomic groups to promote fairness in access to resources, economic opportunities and environmental quality. The inter-generational goal of sustainability concerns the responsibility of the current generation to preserve natural and economic resources for future generations. It is rooted in the principle of sustainable development, which aims to meet present needs without compromising the ability of future generations to meet theirs (World Commission on Environment and Development, 1987). The achievement of both goals presents a dilemma in the short and the long term as the consecution of one of these goals with a business-as-usual approach might endanger the consecution of the other. Policymaking needs to address this challenge through forward-thinking governance, long-term planning and a commitment to sustainable innovation and climate justice.

A transition that has distributional effects worsening inequality and negatively impacting vulnerable communities and/or groups of workers stands at odds with the concept of a just climate transition, in which reducing inequality and leaving no one behind are key elements. In order for the transition of the aviation sector to be just, it is important to assess how (potential) policy measures affect countries and citizens in different geographies.

While there are many studies assessing impacts of changing patterns of trade induced by climate change, there are few studies assessing the impact of structural changes in the pattern of tourism. Assessing the impact of less travel, either as a result of measures directly affecting demand or indirectly through price effects, is challenging. Over the past decades, air travel and international tourism have grown steadily, interrupted only by a few, often short, periods of decline because of economic downturns. Also, in terms of price, air travel has become more and more affordable a rate of 1.7% per year with airfare prices decreasing between 2014 and 2025 by 44% in real terms (IATA, 2024). At the costs of air travel have declined, the average trip distance has increased (Gössling et al., 2018; Rosselló-Nadal & Santana-Gallego, 2024).

The decline in international air travel during the COVID-19 pandemic provides insight into the spatial patterns of reduced mobility and the geography of its wider socioeconomic consequences. Governments swiftly implemented international travel restrictions as a primary and highly impactful measure to curb the pandemic. These restrictions led to a complete halt in international leisure air travel and tourism, triggering the most significant economic disruption worldwide in decades.

The chapter assesses the high-level effects of reduced international air travel on three distinct groups of countries in the 2020–2022 period, exploring the differences in impact on several key economic indicators over time. From this, insights are obtained regarding to what extent current and proposed aviation climate policies can be considered just, with several implications for policymakers in tourism-outbound and tourism-inbound countries.

The Spatial Pattern of Impact of COVID-19 Travel Restrictions

Impact of travel restrictions on international tourism

In 2020, international travel restrictions were the first and one of the most impactful measures taken by countries to limit the spread of the COVID-19 virus. While physical business meetings and social interactions could partly be substituted by digital interactions, travel restrictions meant a full collapse of international leisure air travel and tourism, causing the most severe disruption of the global tourism economy since World War II. The World Tourism Organization (UNWTO) (2023) estimated global international arrivals declined 72% in 2020 (versus 2019) and 69% in 2021. The World Travel & Tourism Council (WTTC, 2020) estimated global tourism job losses to be around 174 million in 2020. With an over-representation of women, youth and low-skilled labour in tourism, as well as a high proportion of micro-, small- and medium-sized enterprises impacted, the COVID-19 pandemic hit these demographic and social groups the hardest, worsening gender- and within-country income inequality (ILO, 2021).

Even if the pandemic had a terrible impact in all communities of the world, the impacts were magnified in regions. Some of the most affected countries were those that can only be reached by aircraft or ship, and that were not part of a regional travel policy, such as the Schengen-travel restrictions adopted by the European Union (EU). Countries such as the Bahamas, Maldives and Fiji were completely closed for international travel. The immediate impact was felt across the entire tourism industry, where activity completely stopped within a fortnight. With the prolongation of travel restrictions, the multiplier effect of reduced consumer spending and business closures further deteriorated the situation, resulting in economic and social upheavals in tourism-dependent countries, despite significant government intervention. However, even within Europe, key tourism destination countries such as Spain, Greece and Portugal were affected strongly by travel restrictions. Spain saw its expenditure by foreign tourists drop from EUR 82 billion in 2019 to EUR 21 billion (−74%) in 2020 and EUR 35 billion in 2021, while tourism employment dropped from 2.7 million jobs only to 2.2 million (−17%) in 2020 and 2.3 million in 2021 largely thanks to government interventions (INE, 2022).

In sharp contrast, in the main outbound source markets for leisure air travel, such as continental Europe, the impact of COVID-19 travel bans was completely the opposite to that of Spain and the Small Island Developing States (SIDS) described above. Holidays were spent domestically or in neighbouring countries. In the Netherlands, domestic holiday travels in July–September quarter in 2020 were up by 27% from 6.0 million to 7.6 million compared to the same quarter in 2020 (CBS, 2020). And those not taking a holiday spent their holiday budget on local services and products. Many vacationers enjoyed taking domestic holidays for a change and inhabitants of popular tourism hotspots such as Paris, London and Amsterdam appreciated the decrease in international tourists.

Pandemic impacts on tourism economies

To examine the difference in economic impact for different types of tourism economies, three categories of countries were selected:

- **Small Island Developing States (SIDS):** Small Island Developing states with a large tourism sector completely dependent on international air connectivity (20+% of jobs), with income levels well below European averages and very tight travel restrictions in the initial two years of COVID-19. Examples of these countries are Fiji, Mauritius and Seychelles.
- **Tourism-inbound destinations (TID):** Mainly Mediterranean countries with large tourism sectors (~10 to 15% of jobs), but also sizeable domestic tourism markets, and part of the EU and Schengen area (resulting in limited travel restrictions in 2021). Examples of these countries are Spain, Portugal and Greece.
- **Tourism-outbound countries (TOC):** Mainly high-income North-European countries characterised by a large outbound tourism market and a relatively small dependency on tourism (~5% of jobs). Examples of these countries are Germany, Netherlands and Denmark.

The key metric used for assessing the economic impact is the change in Gross Domestic Product (GDP) representing the level of economic activity. The GDP dynamics alone, however, does not provide a full picture of the impact on national economies, as government interventions to maintain jobs and avoid bankruptcies moderated GDP impact in some countries. Such interventions were funded by increased borrowing by the State. However, this rise in debt cannot be deemed sustainable. For this reason, also the government debt as percentage of GDP is analysed. The analysis includes the change in 2020 and 2021 vs. 2019, as this provides a view on the duration of the impact. The pandemic impact on credit rating of a country also had important implications for borrowing costs.

The analyses reveal very different levels of tourism economic impact during the COVID-19 travel bans (Figure 3.1):

- **Small Island Developing States (SIDS):** Economies shrunk to levels 10–20% below 2019 and government debt levels increased by 20–40% (as % of GDP).
- **Tourism-inbound destinations (TID):** Economies contracted by around 10% in 2020, with a small contraction in 2021. Government debt levels increased by an average of 20%, but with higher levels in some countries (e.g. Greece 41% in 2020).
- **Tourism-outbound countries (TOC):** In Northern Europe, by 2021 countries had already fully recovered from the less than 5% GDP contractions that occurred in 2020, with minor increases in government debt ratios.

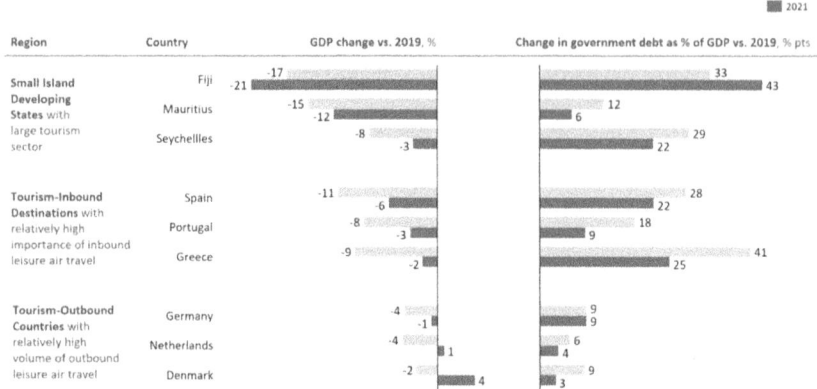

Figure 3.1 GDP and government debt as per cent to GDP development in 2020 and 2021 vs. 2019
Source: Analysis by authors based on data from World Bank (2023).

Additional insight into the differential impacts of pandemic travel disruptions can be obtained by looking at the variation in GDP growth in the third quarter (summer holiday season) of 2021 vs. 2019 and the balance of tourist nights of each country. Figure 3.2 shows a clear correlation between tourism balance and GDP growth. Prosperous countries (represented by the size of bubble) with a net tourism-outbound market faired relatively well in the summer of 2021, likely due to increase domestic travel and limited impact of reduced inbound tourism. The increase in prices (e.g. for holiday accommodation) might partly explain the high GDP growth. At the other end of the spectrum, net tourism-inbound countries with generally lower GDP per capita

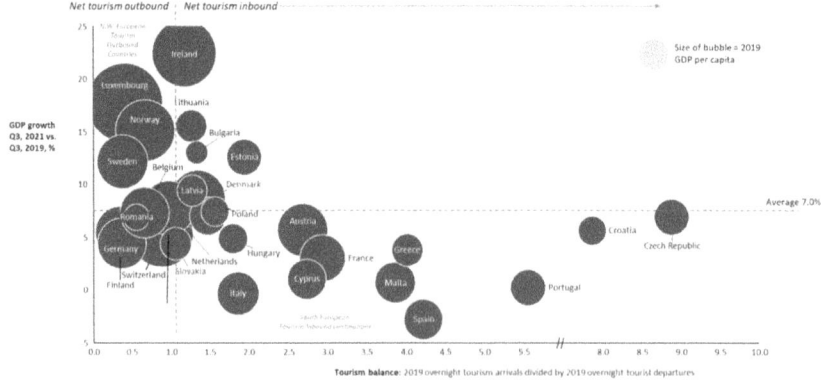

Figure 3.2 Economic growth vs tourism balance pre-COVID to COVID
Source: Analysis by authors using data from Eurostat (2025) and UNWTO (2025).

levels performed relatively poorly. Croatia's relatively mild economic impact compared to other countries with a strong inbound-tourism ratio can be explained by its non-Schengen status attracting travellers from the United States and other countries that could not travel into the Schengen area in 2021.

Both at a global level as well as at an intra-EU level, there are clear indications that a change in travel volume asymmetrically affects countries with tourism-dependent economies, with Small Island Developing States being affected significantly and limited impacts in North European countries. The results of this analysis provide an important analogue for anticipated carbon disrupted travel patterns.

Assessing Current Aviation Climate Policies in Europe

In recent years, several countries and regions have implemented or proposed regulations and measures to limit aviation to reduce its environmental impact, including aviation-related taxes and movement caps to curb noise and greenhouse gas (GHG) emissions (e.g. Netherlands Environmental Assessment Agency, 2024). In parallel with these mechanisms, social responses (behavioural changes, civic pressures and protests) have emerged in recent years, such as the 'flight shame' movement, that captures the growing awareness and guilt associated with the environmental impact of air travel. Different measures aim to limit growth or even reduce air travel, such as introducing or increasing aviation-related taxes or noise/emission-based airport charges, including aviation emissions in carbon emissions markets, or applying emission caps or restricted airport gate slots.

European policy measures and mechanisms

In Europe, several climate policies are already in place or are being developed or considered. These measures can be grouped into five main categories:

(1) Introducing/increasing aviation-related taxes or noise/emission-based airport charges: national aviation taxes (either flat rate or distance dependent such as the UK Air Passenger Duty), jet-fuel excise duties and value added taxes (VAT).
(2) Including aviation emissions in carbon emissions markets: European Trading System (ETS) where aviation's free allowances will be reduced by 100% towards 2026, potentially including the so-called non-carbon dioxide (CO_2) climate effect of aviation (the result of the net warming effect of emitting CO_2 at altitude).[1]
(3) Applying caps: movement caps to curb noise and GHG emissions, carbon emission or outright bans such as those applied in France for

certain short-haul flights (flights of one hour or less where high-speed rail options are available).
(4) Targeting behavioural change: creating consumer awareness of environmental impact, encouraging travellers to avoid taking the plane and extending travel stays.
(5) Initiatives for the use of cleaner fuel and setting of environmental standards: mandates for blending Sustainable Aviation Fuel, development of hydrogen-aviation, emission-intensity (CO_2 per km or NOx per Landing- and Take-Off cycle).

Impact of Policy Measures

Intra-European air travel will see large differences in the impact of fuel- and emission-based costs. Countries in the centre of Europe, including high income countries like Germany, the BeNeLux-countries (Belgium, Netherlands and Luxembourg) and France would be the least affected. On the other hand, countries or regions at the extremities of the EU, such as Malta, Crete and Finland, or the Canary Islands, will experience much greater cost increases due to higher average flight distances as well as a likely reduction of (the growth of) air travel. By 2030, the cost of a 2000 km flight could rise by EUR 50–150 depending on factors like the ETS price, SAF blending, kerosene tax and VAT applicability to intra-EU flights. Except for Finland, Sweden, Norway and Ireland, these countries have a higher dependency on tourism and a lower level of prosperity than centrally located countries and will be impacted by changes in travel costs and/or see changes in travel mode away from air travel.

Long-haul flights between the EU and non-EU countries are not covered (yet) by the ETS but by the less impactful CORSIA system. However, factors like Sustainable Aviation Fuel (SAF)-blending, pressure on higher quality offset projects, potential inclusion of emissions in ETS resulting in the full ETS-price of CO_2 being applicable rather than the much lower carbon cost of ICAO-certified carbon projects, and distance-dependent aviation taxes could make long-haul flights to countries such as Mauritius more expensive by EUR 150–250 by 2030. This would represent a 20–30% price increase (Figure 3.3), which would significantly impact demand for international leisure travel given the strong price elasticity of demand.

In addition, the cost increases outlined in Figure 3.3 means that citizens of non-EU countries are exposed to EU aviation policies and taxes aimed at achieving the EU's aviation carbon reduction targets. The demand growth from these countries, because of higher levels of income and a much stronger growing population, cannot be accommodated. National aviation taxes are paid by non-EU citizens with receipts flowing into the budget of EU countries. Tourism-dependent and lower income

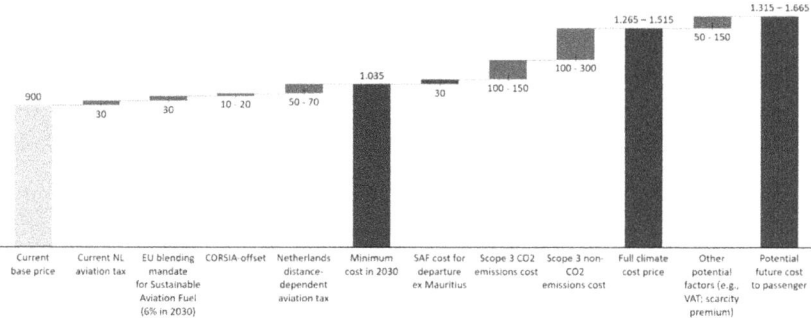

Figure 3.3 Potential increase in a long-haul ticket price with full implementation of aviation climate policy measures. Example: price composition for an Amsterdam–Mauritius ticket in EUR
Source: Analysis by authors using data on emissions from Atmosfair (2025).

countries cannot afford to introduce such taxes in their own country as it would hurt their economy and own citizens too much. Citizens of remote islands have few to no alternatives to flying when visiting family and friends in Europe, leading to increased social isolation. This asymmetry of higher air travel prices can be considered socially regressive.

Movement caps, or in the case of the Netherlands, potentially a reduction in annual flights at Amsterdam Schiphol Airport, and behavioural changes will further reduce air travel demand and/or will push air fares up as a result of reduced supply (Netherlands Environmental Assessment Agency, 2024). Leisure markets are expected to be among the first routes to be affected.

Table 3.1 provides a systematic overview of policy measures already implemented and/or being considered at EU or a national level to address carbon emissions, noise or other negative externalities of air travel. While each measure has a clear well-intended primary effect, each measure also has unintentional secondary effects that have distributional effects between countries in terms of economy, accessibility or other socioeconomic factors.

A Zero Emissions Tourism (ZET) scenario developed by Peeters and Papp (2024) highlights the need for a fairer distribution of airlift capacity to address inequities in global tourism policies. Peeters and Papp (2024) argue that carbon neutrality can be achieved while enabling global tourism to remain inclusive and equitable through the redistribution of aviation resources like SAF and airport slots, allowing regions with few viable alternatives to air travel to maintain essential tourism access while significantly reducing emissions.

Applying the dynamic described in the previous section, it becomes clear that the current and planned EU climate- and environment-related aviation policies will have a negative economic and social impact. This

Table 3.1 Intended and unintended effects of climate change and other environmental impact measures reductions

Strategy	Action	Intended primary effect	Unintended secondary effects
Raising the cost of air travel	1. Increase in taxes (national aviation taxes; EU Energy Tax Directive)	Reduced demand (growth)	Reduced (growth of) inbound visitors for tourism-dependent countries
	2. Including aviation emissions in carbon emissions markets and higher cost of carbon	Higher prices leading to reduced demand	Reduced accessibility of air travel for lower-income individuals
Reducing (growth of) air travel (growth caps; degrowth)	3. Including caps through regulation	Decrease in airline capacity	Scarcity premiums of airlines; waste due to less efficient use of capacity; reduced economies of scale (resulting in higher airport charges)
	4. Reducing flight volume through affecting behaviours (e.g. flight shaming)	Decrease in air travel demand	Increase in consumption and associated carbon footprint in other sectors resulting from changes in spending patterns
Reducing the impact of aviation	5. Developing new fuels (SAF), improving efficiency, creating new technologies	Decrease in CO_2 emissions	Opportunity cost of land/green energy use for other purposes
Compensating the impacts of aviation	6. Developing projects (e.g. for sustainable energy) to compensate carbon footprint generated while also contributing to other SDGs	Absorption of CO_2 Resources for the transition in developing countries Improvement on other SDGs	Risk of under compensation due to poor quality of projects Potentially undesirable social consequences of certain carbon reduction projects

is especially the case for remote, often less prosperous and more climate vulnerable, countries. More prosperous European tourism-outbound countries, with lower climate vulnerability, at the same time will benefit both from proceeds from aviation-related taxation, sales of ETS rights and a higher share of budget being spent locally instead of in tourism-dependent countries along the Mediterranean and beyond (comparable to the dynamic during the COVID-19 pandemic). This asymmetrical impact, characterised by widening disparities in prosperity and marked inter-country inequality, falls short of the principles of a just transition.

Conclusion and Implications for Aviation Climate Policy Development

Among the intra-generational challenges, the reduction of global inequality is of utmost importance for nations across the globe, regardless of their level of development. However, some countries and regions at the early and medium stages of the economic integration process seem to be constrained by the combination of four factors:

- *Geographical isolation*: these islands and archipelagos are normally far away from mainland or the metropolis, with limited connectivity that is mainly ensured by air.
- *Economic dependence*: they rely on increased trade and tourism because of historical reasons (colonisation processes and dependence on the metropolis, dependence on a bigger economy) or geographical reasons (beauty of the natural landscape, isolation and peace to escape from urban contexts).
- *Exposure to environmental hazards*: they are expected to be disproportionally affected by climate change (exposure to natural catastrophes and having limited mitigation and adaptation capacity).
- *Economic vulnerability*: they have limited resources due to their insularity and they can be negatively impacted by key mitigation measures adopted by their bigger commercial partners (which could reverse the trend of reducing inequality).

These countries/regions are numerous and there is not a single typology. Among them, we can identify peripheral European regions (e.g. The Antilles, Departements d'Outremer, Açores and Canary Islands), non-EU leisure travel destinations (e.g. Maldives and Mauritius) and Small Island Developing States (SIDS) that base their economies on the exchanges with the European Union.

At the same time, typical tourism-outbound markets are already today short of labour supply and lack space (and public support) for accommodation of more tourism locally. Therefore, the challenge is how to consider all environmental and social impacts that condition the development of these economies and how to design policies so that these factors are integrated (Robins *et al.*, 2018). For policymakers, the following questions offer salient considerations to assess the socioeconomic implications of aviation and tourism climate policies and the balance between sustainability and equity:

- How can we define a just transition that incorporates international equity, in addition to existing national-level interpretations of the concept (a combination of the intra-generational justice towards the disadvantaged collectives and the inter-generational justice towards the future generations)?
- What are the socioeconomic implications of environmental policies taken in Europe for those regions that are dependent on European trade and tourism? And how feasible and beneficial is it for tourism-outbound markets to accommodate more tourism demand locally?

- How can we measure and compare the effects in both the inter-generational and intra-generational perspectives and to build a triple bottom line?

Some policies on climate change focus on the desirable long-term effects (e.g. the reduction of CO_2 concentration in the atmosphere and the salient impacts on ecosystems and society), but do not fully consider the short- and mid-term consequences (e.g. the consequences on employment or wealth creation). It is a complex equation where the prevalence of positive or negative effects in the balance is difficult to calculate and make informed trade-offs. Nonetheless, climate justice demands that we avoid policies that disproportionately affect the most vulnerable.

Addressing these inequities is critical to ensuring a just transition in the aviation industry and global tourism (Becken & Scott, 2023) and a research priority highlighted by Scott (2024). As we witnessed during the COVID-19 pandemic, the multiplier effect of reduced consumer spending and employment and business closures further deteriorated the situation, resulting in economic and social upheavals in tourism-dependent countries, despite significant government intervention.

It is of critical importance to explore and define solutions that allow for the same or even higher global carbon reductions, while improving rather than worsening inequality and positively contributing to sustainable development. A truly just transition demands climate policies that harmonise sustainability with economic equity, ensuring that global carbon reduction efforts uplift rather than marginalise the most vulnerable regions.

Note

(1) Aviation's non-CO_2 climate effects stem from emissions at high altitudes, including nitrogen oxides (NO_x), water vapour and contrails, which can alter atmospheric chemistry and form heat-trapping clouds. These effects significantly amplify aviation's total climate impact, making it 2–4 times greater than CO_2 alone.

References

Atmosfair (2025) Emissions calculation. See https://www.atmosfair.de/en/standards/emissions_calculation/ (accessed April 2025).

Baran, M. (2019) How 'flygskam,' or 'flight shame,' could change the way we all travel. *AFAR*, 25 June. See https://www.afar.com/magazine/how-flygskam-or-flight-shame-could-change-the-way-we-all-travel (accessed April 2025).

Bardon, P. and Massol, O. (2023) Decarbonizing aviation with sustainable aviation fuels: Myths and realities of the roadmaps to net zero by 2050. IFP Energies nouvelles. See https://ifp.hal.science/hal-04374158/file/Decarbonizing%20Aviation%20with%20sustainable%20Aviation%20Fuels%20Myths%20and%20Realities%20of%20the%20Road%20Maps%20to%20Next%20Zero%20by%202050.pdf (accessed February 2025).

Becken, S. and Scott, D. (2023) Tourism and climate change: Global stocktake 2023. Tourism Panel on Climate Change. See https://institutetourism.com/wp-content/uploads/2023/12/TPCC_Stocktake-2023-Report_For-distribution.pdf (accessed April 2025).

Cárdenas-García, P.J., Brida, J.G. and Segarra, V. (2024) Modeling the link between tourism and economic development: evidence from homogeneous panels of countries., *Humanities and Social Sciences Communications* 11 (308), 1–12. https://doi.org/10.1057/s41599-024-02826-8.

Centraal Bureau voor de Statistiek (CBS) (2020) Nederlanders in zomer 2020 vaker op vakantie in eigen land. See https://www.cbs.nl/nl-nl/nieuws/2020/52/nederlanders-in-zomer-2020-vaker-op-vakantie-in-eigen-land (accessed February 2025).

Diffenbaugh, N.S. and Burke, M. (2019) Global warming has increased global economic inequality. *Proceedings of the National Academy of Sciences* 116 (20), 9808–9813. https://doi.org/10.1073/pnas.1816020116.

Espagne, E., Oman, W., Mercure, J., Svartzman, R., Volz, U., Pollitt, H., Semieniuk,G. and Campiglio, E. (2023) Cross-Border Risks of a Global Economy in Mid-Transition. IMF Working Paper No. 2023/184. See https://www.imf.org/en/Publications/WP/Issues/2023/09/08/Cross-Border-Risks-of-a-Global-Economy-in-Mid-Transition-538950 (accessed October 2025).

Eurostat (2025) Real GDP per capita. See https://ec.europa.eu/eurostat/databrowser/view/sdg_08_10/default/table (accessed April 2025).

Gössling, S., Scott, D. and Hall, C.M. (2015) Inter-market variability in CO_2 emission-intensity in tourism: Insights for energy policy and carbon management. *Tourism Management* 46 (1), 203–212. https://doi.org/10.1016/j.tourman.2014.06.021.

Gössling, S., Scott, D. and Hall, C.M. (2018) Global trends in length of stay: Implications for destination management and climate change. *Journal of Sustainable Tourism* 26 (12), 2087–2101. https://doi.org/10.1080/09669582.2018.1529771.

IATA (2024) Strengthened profitability expected in 2025 even as supply chain issues persist. International Air Transport Association. See https://www.iata.org/en/pressroom/2024-releases/2024-12-10-01/ (accessed February 2025).

IEA (2020) World air passenger traffic evolution, 1980-2020, International Energy Agency, Paris. See https://www.iea.org/data-and-statistics/charts/world-air-passenger-traffic-evolution-1980-2020 (accessed July 2024).

ILO (2021) An uneven and gender-unequal COVID-19 recovery: Update on gender and employment trends 2021. International Labour Organization. See https://www.ilo.org/publications/uneven-and-gender-unequal-covid-19-recovery-update-gender-and-employment (accessed July 2024).

INE (2022) Tourism Satellite Account of Spain: Main Aggregates. Year 2021. Instituto Nacional de Estadística. See https://www.ine.es/en/prensa/cst_2021_en.pdf (accessed October 2024).

IPCC (2018) Glossary. Special report: Global warming of 1.5°C. Intergovernmental Panel on Climate Change. See https://www.ipcc.ch/sr15/chapter/glossary/ (accessed April 2025).

IPCC (2022) Summary for policymakers. In H.-O. Pörtner, D.C. Roberts, E.S. Poloczanska, K. Mintenbeck, M. Tignor, A. Alegría, M. Craig, S. Langsdorf, S. Löschke, V. Möller and A. Okem (eds) *Climate Change 2022: Impacts, Adaptation, and Vulnerability. Contribution of Working Group II to the Sixth Assessment Report of the Intergovernmental Panel on Climate Change* (pp. 3–33). Cambridge University Press.

Khan, A., Bibi, S., Lorenzo, A., Lyu, J. and Babar, Z.U. (2020) Tourism and development in developing economies: A policy implication perspective. *Sustainability* 12 (4), 1618. https://doi.org/10.3390/su12041618.

Lagos, K. and Wang, Y. (2022) International tourism and poverty alleviation: cross-country evidence using panel quantile fixed effects approach. *Journal of Travel Research* 62 (6), 1347–1371. https://doi.org/10.1177/00472875221119978.

Netherlands Environmental Assessment Agency (PBL) (2024) *What are just and feasible climate targets for the Netherlands?* Seehttps://www.pbl.nl/system/files/document/2024-06-pbl-2024-what-are-just-and-feasible-climate-targets-for-the-Netherlands_5594.pdf (accessed July 2024).

Peeters, P. and Papp, B. (2024) Pathway to zero emissions in global tourism: Opportunities, challenges, and implications. *Journal of Sustainable Tourism* 32 (9), 1784–1810. https://doi.org/10.1080/09669582.2024.2367513.

Pentelow, L. and Scott, D. (2011) Aviation's inclusion in international climate policy regimes: Implications for the Caribbean tourism industry. *Journal of Air Traffic Management* 17, 199–205. https://doi.org/10.1016/j.jairtraman.2010.12.010.

Rastegar, R. and Becken, S. (2024) Embedding justice into climate policy and practice relevant to tourism. *Journal of Sustainable Tourism* 33 (10), 2011–2028. https://doi.org/10.1080/09669582.2024.2377720.

Robins, N., Brunsting, V. and Wood, D. (2018) Investing in a just transition: Why investors need to integrate a social dimension into their climate strategies and how they could take action. See https://www.lse.ac.uk/granthaminstitute/wp-content/uploads/2018/06/Robins-et-al_Investing-in-a-Just-Transition.pdf (accessed July 2024).

Rosselló-Nadal, J. and Santana-Gallego, M. (2024) Toward a smaller world. The distance puzzle and international border for tourism, *Journal of Transport Geography* 115, 103809. https://doi.org/10.1016/j.jtrangeo.2024.103809.

Scott, D. (2024) Tourism and the climate crisis. *Journal of Sustainable Tourism* 32 (9), 1709–1724. https://doi.org/10.1080/09669582.2024.2391911.

Scott, D. and Gössling, S. (2021) Destination net-zero: What does the international energy agency roadmap mean for tourism? *Journal of Sustainable Tourism* 30 (1), 14–31. https://doi.org/10.1080/09669582.2021.1962890.

UNWTO (2023) Impact assessment of the COVID-19 outbreak on international tourism. United Nations World Tourism Organization. See https://www.unwto.org/impact-assessment-of-the-covid-19-outbreak-on-international-tourism (accessed April 2025).

UNWTO (2025) UN Tourism Data Dashboard. See https://www.unwto.org/tourism-data/global-and-regional-tourism-performance (accessed April 2025).

World Bank (2023) *World Development Indicators*. See https://databank.worldbank.org/source/world-development-indicators (accessed April 2025).

World Commission on Environment and Development (1987) *Our Common Future*. Oxford University Press.

WTTC (2020) 174m travel & tourism jobs could be lost due to COVID-19 and travel restrictions. World Travel & Tourism Council. See https://wttc.org/news-article/174m-travel-and-tourism-jobs-could-be-lost-due-to-covid-19-and-travel-restrictions (accessed July 2024).

WTTC (2023) Sustainable aviation fuel: Opportunities and implications for tourism destinations. World Travel and Tourism Council. See https://researchhub.wttc.org/product/sustainable-aviation-fuel-opportunities-implications-tourism-destinations (accessed April 2025).

4 Stay Grounded: Re-Imagining Tourism for a Grounded Climate-Just Future

Angel Sulub and Daniela Subtil Fialho

This chapter presents a case study of Stay Grounded (SG) as an important example of a social movement action to promote climate justice, authored by two of its coordinators. SG is a global network fighting for a radical reduction of air traffic and a just mobility system. Aviation represents the pinnacle of climate injustice and SG argues for the degrowth of air traffic as the only possible pathway compatible with climate, ecological and social justice. Since much of aviation-induced demand is coupled with air-traffic-dependent tourism, the struggle against airport expansion is also a fight against a predatory tourism model. To illustrate, we offer a case study from the Mayan people of Mexico, where struggles against airport expansion, touristification and associated extractivism are closely linked. The 'development' model employed contributes to ethnocidal processes that destroy Indigenous identity and lifeways.

As a result of this analysis, we argue for rapid reduction of air traffic and more just mobilities in order to respond to the threat of climate and social collapse brought on by climate change. This chapter contributes to discussions on a just transition to more climate and socially just futures.

Disclaimer: This chapter took inspiration in the work of the SG Network. However, the views represented are of the responsibility of the authors and should not be attributed to SG.

Introduction

The aviation industry, a cornerstone of global capitalism and hyper-mobility, is increasingly scrutinised for its significant and undeniable role in exacerbating the climate crisis and, as a consequence, social inequalities. This chapter embarks on a comprehensive exploration of these interconnected issues through the presentation of a case study of the work of the Stay Grounded (SG) Network.

This global network was formally established in 2016 and evolved out of decentralised coordinated actions against airport expansion worldwide. The foundational common understanding was that these were not isolated local struggles fighting the direct impacts of air traffic in their territories, but rather a part of a global struggle against the growth of a global industry in part responsible for destroying our climate. Nowadays SG has more than 200 member organisations worldwide, including climate justice groups, non-governmental organisations (NGOs), trade unions, initiatives fostering alternatives to aviation like night trains, organisations supporting communities resisting offsets and other false solutions and anti-touristification groups.

In the second section, we establish the context and perspectives from which we write this communication, introducing the scope of SG's work. We highlight the stark contrast between the urgent need to address the climate crisis and the aviation industry's relentless growth trajectory. Here, we examine the false solutions proposed by the aviation sector, including carbon offsetting, technological improvements and so-called Sustainable Aviation Fuels (SAF), which fail to mitigate the industry's environmental impact. After delving into SG's mission to combat aviation's disproportionate contribution to global heating, we investigate the symbiotic relationship between the aviation and tourism industries, illustrating how mass tourism has flourished alongside the expansion of air traffic. This has contributed to skyrocketing levels of greenhouse gas emissions (GHG) and socioenvironmental conflicts, underscoring the importance and urgency of the resistance to airport expansion. We propose the degrowth of the aviation industry embedded in a framework of global climate justice as the only way forward to effectively respond to the emergency of the climate crisis.

Focusing on the touristification process – the 'process of transformation of a place into a tourist space' (Hernández & de la Calle Vaquero, 2023) – in Southeastern Mexico, the third section reveals the profound transformation of Mayan territories due to the expansion of airport infrastructure and the associated tourism development, emphasising the colonial and extractivist foundations of these processes. We showcase how local resistance movements challenge these exploitative practices and advocate for the rights of Indigenous communities through an anti-capitalist and anti-colonial lens. In this context we present *Permanecer en la Tierra*, SG's regional network for Latin America & the Caribbean (LAC).

Finally, in the fourth section, we make the case for a just transition away from air-traffic related tourism, underscoring the urgent need for a paradigm shift that prioritises decarbonisation, climate justice and equity in tourism practices. As we navigate the complexities of climate collapse and social inequities, we call for a radical and steep reduction of the aviation and tourism industries to foster a more just and sustainable future for all. By being co-authored by an SG coordinator from the Global North and an SG coordinator from the Global South, this chapter suggests the diversities of knowledges and experiences we need to draw on and models the kinds of collaborations and solidarities required between the peoples of the Global North and the Global South in order to build climate just futures.

Context

Grappling with climate emergency: *Green* means grounded

Opposing directions: Incipient climate collapse and aviation growth

The scientific consensus is unequivocal on the ongoing failure to address the climate crisis. Physical climate science and continual direct measurements are showing that amplifying feedbacks are occurring for ice sheet loss in Antarctica and Greenland (Jacob, 2023) and for the Atlantic Meridional Ocean Circulation (AMOC) (van Westen *et al.*, 2024). Climate catastrophes are increasing in number and intensity; it is now evident that we have entered an era of great danger, even to casual observers. Apocalyptic flooding in Libya killing more than 11,000 people (Al Jazeera, 2023), one third of the country being underwater in Pakistan (Unicef, 2023), the 'worst drought in a century' occurring in southern Africa (Al Jazeera, 2024) and the Philippines experiencing six typhoons in a month (Tandon, 2024) are just some examples among many. In the last decade, there has been one 'hottest year on record' after another (NASA Global Climate Change, n.d.).

In 2024, the world experienced its first year with a temperature anomaly above the 1.5°C Paris Agreement threshold (Copernicus, 2024), but science will not consider that target to be exceeded until it has occurred as a 20-year average (Bevacqua *et al.*, 2025). The impacts are already intolerably intense worldwide. The smallest fraction of a degree of temperature rise matters. While adapting to impacts is surely needed, what is more vital is quickly ending CO_2 emissions – or any adaptations will fail. Fossil fuels have to be kept in the ground and a rapid and just transition into a decarbonised society is mandatory.

Despite the incipient climate collapse, the aviation industry continues on the only trajectory available under capitalism: growth. Between 2013 and 2019, emissions from passenger aircraft increased by 33% (Graver *et al.*, 2020). As of 2020, aviation is responsible for approximately 4% of observed human-induced global heating (Klöwer *et al.*, 2021), which includes CO_2 and non-CO_2 emissions, the latter being responsible for two thirds of aviation's total climate impact (SG, 2022c).

In 2024, the industry celebrated a new all-time high in air traffic and plans to double it between 2024 and 2043 (IATA, 2024). This leaves the aviation industry at odds with the climate emergency, because, in addition to being one of the most carbon-intensive activities (Klöwer *et al.*, 2021), aviation is a hard-to-abate industry (Urban *et al.*, 2024). In other words, decarbonising aviation remains a mirage and green flying an illusion.

Aviation's response to the climate crisis: False solutions and greenwashing

International aviation, which accounts for around 65% (as of 2016) of the aviation industry's CO_2 emissions (ICAO, 2016), was left out of the 'ambiguous' Paris Agreement (Schleussner *et al.*, 2024). The responsibility to tackle international aviation emissions has been delegated to the International Civil Aviation Organization (ICAO) (Delbecq *et al.*, 2023). The ICAO has the *Carbon Offsetting and Reduction Scheme for International Aviation* (CORSIA) as one of its main strategies to supposedly reduce CO_2 emissions from air transport. Offsetting combined with unproven, unfair, high-cost and risky carbon capture technologies (SG, 2024), should after all be responsible for eliminating 19% of CO_2 emissions to help the aviation industry to meet its pledge of achieving 'net zero' emissions (NZE) by 2050, according to the International Aviation Transport Association (IATA) (IATA, n.d.).

Adopted in 2016, CORSIA has been widely discredited and described as a false solution to the climate crisis (Biofuelwatch, 2019; Transport & Environment (T&E), 2019). The agreement is not legally binding and will only be mandatory from 2027 onwards. The market-based scheme only includes international flights (with some exemptions) and does not cover non-CO_2 emissions. Under CORSIA, the carbon budget for aviation to stay within 1.5°C would be spent by 2030 and only around 5% of the total climate impact (or 14% of CO_2 emissions) of aviation may be covered in 2030 (SG, 2023c) (see Figure 4.1).

Carbon offsetting does not reduce emissions as the avoided emissions achieved by offsetting projects are neutralised by the emissions the offsets are bought for. Offsetting is fundamentally a license to pollute. Many projects have been proved fraudulent due to dubious accounting criteria.

Carbon offsetting is also unfair and reproduces neocolonial dynamics, since it justifies high emissions from a wealthy minority, while grabbing resources essential to the majority (SG, 2023c). The Global Atlas of Environmental Justice (EJAtlas, n.d.), under the label of 'biodiversity conservation conflicts' and the Map of False Solutions to the Climate Crisis in Latin America and the Caribbean (PLACJC & DCJ, 2024) document such socioenvironmental conflicts.

With CORSIA's reputation crumbling and public awareness about the climate harm of air transport increasing, the aviation industry looks to technological fixes to continue 'business as usual'. Technological developments and the so-called SAF – also known as alternative fuels

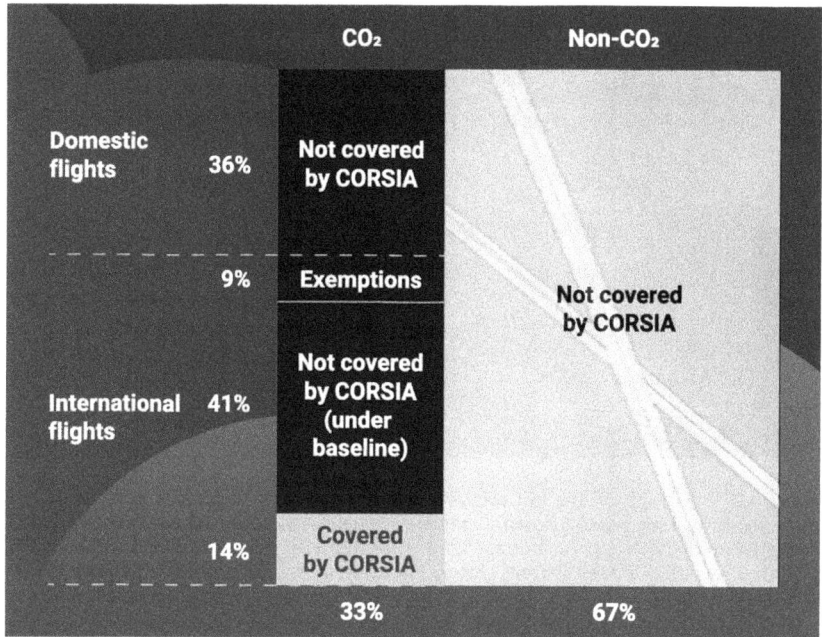

Figure 4.1 Amount of CO_2 emissions and type of flights that CORSIA may cover (SG, 2023c)

or fossil fuel substitutes – are the other main strategies for emissions' reduction of the aviation industry. Yet, none of these alleged solutions addresses air-traffic growth nor tackles aviation's climate impact.

According to IATA, technological developments will be responsible for 13% of the mix of actions the industry takes to achieve NZE by 2050 (IATA, n.d.). Technological improvements in aircraft indeed lead to fuel efficiency gains and therefore could reduce emissions from aviation. The problem is that historically air-traffic growth has outweighed reduction of emissions through efficiency gains and the trend remains consistent – a phenomenon known as the 'rebound effect' (SG, 2021a) (see Figure 4.2).

The estimated long-term fuel efficiency gains of 1.3% per year (Fleming & Lépinay, 2019) will be insufficient to balance out emissions from the projected air-traffic growth at an annual average rate of 3.8% until 2043 (IATA, 2024). Airlines upgrading the class of seats and flying further or faster also cancel out efficiency gains (SG, 2021a).

The hype over the development of electric and hydrogen aircraft has also obscured the serious limitations those technologies present. The only electric airplanes likely to be certified this decade will remain small and only suitable for short-haul flights, which could be replaced by much more energy-efficient public transport on the ground. Technical

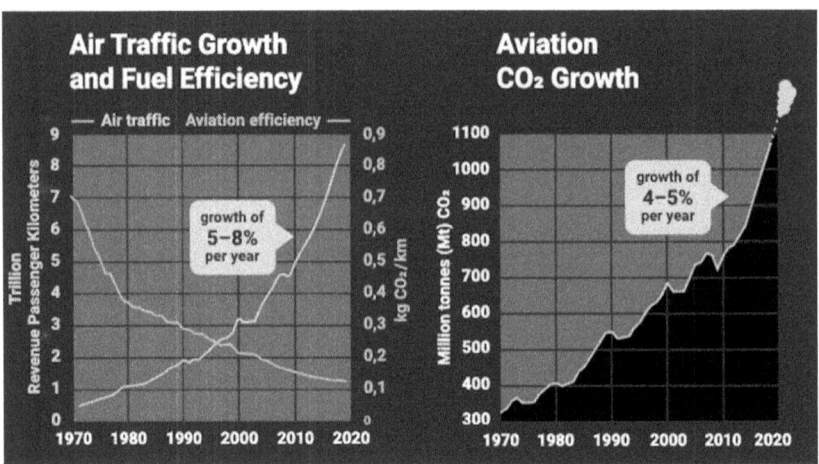

Figure 4.2 The graph on the left shows how, between 1970 and 2020, efficiency improvements have slowed down over time (downward curve), whereas the rate of air-traffic growth has increased over time. The graph on the right shows sustained growth of aviation CO_2 emissions, closely linked to air-traffic growth and despite efficiency gains (SG, 2021a)

challenges persist as 25–30 kg of batteries are needed to equal 1 kg of fuel and batteries are too heavy to replace most jet fuel and combustion engines (SG, 2021b).

Aircraft propelled by hydrogen too will remain restricted to short-haul flights until 2050. Moreover, hydrogen flights will not be 'zero' emissions as long as hydrogen is mostly produced from non-renewable energy. Even if produced from renewables, hydrogen-powered aircraft will still emit NOx and generate contrail cirrus that have a higher climate impact than CO_2 today (SG, 2021c). Powering aircraft with green hydrogen would skyrocket the demand for renewable electricity resources and, in the current geopolitical world system, likely reinforce practices of greenwashing neocolonialism. Under this dynamic – already underway and documented – the demand for huge amounts of renewable energy to maintain the consumerist lifestyle in the Global North and of global elites continues and intensifies the plundering of resources in the Global South, while concealing this process with labels of green policies and economic development (Amouzai & Haddioui, 2023).

SAF emerges as the panacea for aviation's CO_2 emissions and disproportionate climate impact. IATA estimates that 65% of the emissions reduction to achieve NZE by 2050 comes from the use of SAF (McCausland, 2024). These fossil fuel substitutes are liquid hydrocarbon fuels that can be used with existing aircraft in place of jet fuel produced from fossil fuels. They can be broadly categorised into two varieties: biofuels produced from biomass sources and synthetic electro-fuels (e-fuels) produced using electricity. The latter, also known as 'Synfuels' or Power-to-Liquid (PtL) fuels, are still decades away from production at industrial scale, need a huge amount of renewable energy and would leave NOx and contrail cirrus issues unaddressed (SG, 2021e).

Biofuel production, and particularly the production using hydrotreated vegetable oils (HVO), is the form of SAF currently demonstrating commercial viability. The industry projects that, by 2030, the share of biofuels will 'very likely' be 6.5% (i.e. nearly 30 billion litres) of the global jet fuel use (ATAG, 2021). But in 2023, biofuel production fell short of meeting intermediary production goals and did not surpass 600 million litres, representing only 0.2% of global jet fuel use (IATA, 2024). The promise of aviation biofuel scale up remains to be fulfilled. Even in the most optimistic projections of aviation biofuel use, it is not expected that such fuel will represent a large percentage of total fuel consumption over the next few decades, due to the substantial growth in air traffic and fuel consumption. Biofuels can still produce CO_2 emissions and will not completely eliminate non-CO_2 emissions (SG, 2021d). In fact, biofuels can cause more GHG emissions than the fossil fuels they are meant to replace (see Figure 4.3).

Replacing fossil jet fuel, even if partially, with biofuels poses other serious social and environmental risks. Biofuel production can use various sources of biomass as agricultural crops (first generation biofuels) and industrial, agricultural, cultural, municipal or household waste (second generation biofuels), e.g. vegetable oils (soy, palm, corn or rapeseed/canola oils), used cooking oil, animal fats including tallow, lard, fat residues from slaughterhouses, corn husks, forest resources or food waste (SG, 2021d; Biofuelwatch, 2025).

Since a very limited quantity of 'sustainable waste' is available globally (T&E, 2025), the run for feedstocks will increase demand for

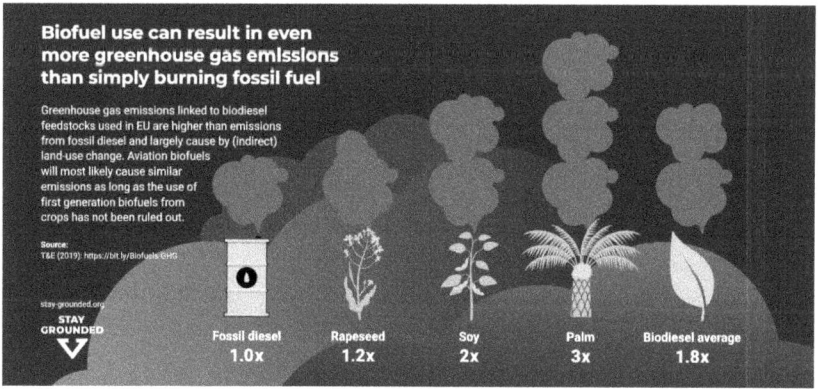

Figure 4.3 Illustration showing how (indirect) land-use changes associated with biofuel production can lead to higher GHG emissions than using fossil fuels (SG, 2021d)

crop-based biomass in the form of 'waste' and/or 'byproduct' as well as prompt a new market for industrial livestock farming. This phenomenon is being increasingly described as 'fat grab' (Biofuelwatch, 2025), and it is poised to drive deforestation and biodiversity losses, impact the health of communities and grab more land and resources in Global South territories. Rising food costs and land conflicts are very likely outcomes (SG, 2021d). Conflicts over the expansion of aviation biofuel production have been already documented in Brazil (Souza, 2023) and Paraguay (HEÑÓI et al., 2022).

The aforementioned illusory solutions to handle aviation emissions are not in line with the reality of our climate emergency, because they are embedded in the misleading goal of reaching NZE by 2050. NZE, or 'carbon neutrality', is achieved when any remaining anthropogenic CO_2 emissions are balanced globally by anthropogenic CO_2 removals (IPCC, 2022). The net zero CO_2 emissions goal is central to nearly every climate strategy promulgated by industry and governments. This concept is designed to allow some 'hard to abate' emissions like aviation to continue as long as equivalent quantities of CO_2 are removed from the atmosphere, either via natural carbon cycle processes, by increasing biological or geochemical sinks of CO_2 or by using industrial processes to capture CO_2, whose feasibility at scale remains to be proven.

But NZE by 2050 does not mean the cumulative emissions of aviation will stay within a 1.5°C CO_2 budget (Figure 4.4).

In fact, the aviation industry plans to continue to emit large amounts of CO_2 until 2050. To stay within the CO_2 budget to meet the 1.5°C target, the industry would have to start drastically cutting its emissions now. Otherwise, the budget will be exceeded around 2030. Also, the

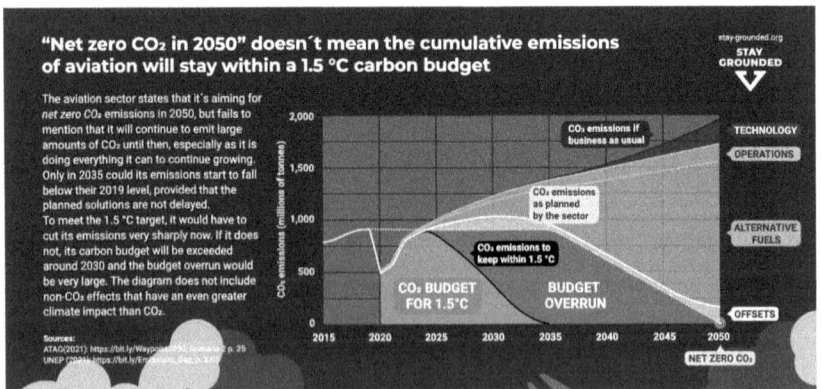

Figure 4.4 Graphic shows how current growth rate of air traffic and corresponding emissions will, before 2035, overrun the CO_2 budget for aviation to stay below the 1.5°C threshold temperature increase. If the planned 'solutions' are not delayed, only in 2035 could aviation emissions start to fall below 2019 levels (SG, 2022a)

NZE strategy of aviation only accounts for CO_2 emissions reduction, leaving non-CO_2 effects – that have an even greater climate impact than CO_2 – unchecked (SG, 2022a).

The aviation industry's reliance on false solutions and greenwashing campaigns to justify its growth are then a sort of denialism of the inevitable: air-traffic reduction. As the next section will demonstrate, reducing air traffic is unavoidable and a key pillar of climate justice in tourism.

Stay Grounded: Air-traffic reduction is also a matter of justice

While the climate crisis disproportionally and mostly impacts the world's poorest, only a small minority of the world's population flies at all. In 2018, only 2 to 4% of the world's population flew internationally, while about 80% had never flown at all and only 1% was responsible for half of all aviation emissions (Gössling & Humpe, 2020). According to the authors, this small minority is also the wealthiest one. Importantly, the wealthiest top 10 % use 75% of air transport energy.

Also in 2018, 90% of aviation emissions came from high-income or upper–middle-income countries (Graver *et al.*, 2019). The unfairness of aviation is blatant; the total contribution of aviation to observed human-induced global heating (4% as of 2020) was higher than the contributions of some entire countries (e.g. India, Canada) or even continents (e.g. Africa, South America) (Ritchie, 2019).

But the injustice of aviation goes beyond emissions and the climate crisis. To have (more) air traffic, (more) airports are needed and the construction of new airport infrastructure or expansion of existing ones has caused severe social and ecological damage on the ground. The Map of Airport-related Injustice and Resistance (Figure 4.5) is a collection of those stories of destruction, loss and resistance (EJAtlas, 2025). A collection that is poised to grow given that hundreds of new airports or airport expansions are planned to fuel the skyrocketing growth of the aviation industry (SG, 2018).

As airports represent an essential infrastructure for the globalised capitalist economy – needed for the trade of goods, business travel, tourism, as well as the deportation of unwanted 'travellers' (Herrero, 2019) – they open the door for countless associated 'development' projects that have similar destructive impacts on local communities. More often than not, such projects impact low-income communities and even leave them more exposed to the catastrophic impacts of looming climate breakdown.

Communities neighbouring airports and under flight paths are exposed to dangerous levels of noise and air pollution on a daily basis (SG, 2023b). Land grabbing, evictions, deforestation, loss of agricultural land, sea reclamation, erasure of wetlands and mangroves, plunder of water and other essential goods and resources and human rights violations are

Figure 4.5 The Map of Airport-related Injustice and Resistance documents socioenvironmental conflicts resulting from airport expansion projects worldwide. The dots in the map represent documented conflicts (EJAtlas, 2025, used with permission)

among the preconditions and/or consequences of airport construction and maintenance. Families evicted from their homes and robbed of their farmland are often left with poor or no financial compensation. The economic hardships resulting from those devastating processes are immense. In addition, the destruction of natural barriers to climate extremes, e.g. mangroves, leaves communities further exposed to climate catastrophe, increasing their socioeconomic vulnerability (EJAtlas, 2025).

Addressing the root causes of the climate, ecological and global injustices has therefore been at the centre of SG's mission since its foundation, focusing on what we consider the pinnacle of climate injustice: aviation. Resisting airport expansion is understood as being a piece of the puzzle in the fight against the injustices brought by global capitalism, a system that prioritises wealth accumulation that deepens social inequality to the detriment of people's wellbeing and dignity, ecological preservation and climate stability. Tourism acts as a vital component of this capitalist system and its focus on wealth extraction. The next section considers how the growth of aviation is inextricably tied to the ever-present growth goals of tourism.

Aviation and tourism: Growing side by side

Of the more than 100 cases of airport-related conflicts documented in the Map of Airport-Related Injustices and Resistance (EJAtlas, 2025), 30% are connected to passenger-oriented airport projects aiming to expand tourism. An increasing number of airport projects, mostly in the

Global South, are 'aerotropolis', where new airports are integrated in aviation-dependent commercial and industrial areas comprising tourism resorts, hotels, shopping malls and complementary facilities.

Airports indeed play a vital role in opening new destinations and markets for the tourism sector, often inviting predatory and violent models of touristification, whose harm goes far beyond GHG emissions (Devine & Ojeda, 2017). An expanded account of how airport infrastructure expansion and touristification processes are connected – and their impact tremendous – will be presented in the following section, exploring the example of Southeastern Mexico.

The advent of mass tourism is inextricably tied to the rise of commercial aviation (Duval, 2013). Since the 1960s the constant growth of the tourism industry, and in particular of international leisure travel, has accompanied the sustained increase of air traffic. Improved international air connectivity has opened up opportunities for tourism development and the touristic potential of certain regions and cities has prompted new international routes and destinations (Moreno et al., 2015). Illustratively, the share of international tourists travelling by air has increased from 35% in 1980 (Papatheodorou, 2021) to 59% in 2019 (UNWTO, 2020). Over a longer timespan, we witnessed a growth from 25 million international arrivals in 1950 (Akdağ & Oter, 2011) to 1460 million in 2019 (UNWTO, 2020), representing a more than 58-fold increase.

Following the liberalisation of air transport since the 1980s, the aviation industry, and consequently the number of air passengers, grew exponentially, in which the introduction of low-cost carriers in the aviation market played a major role, particularly at the beginning of this century. Exploiting opportunities in (international) leisure travel (Moreno et al., 2015), the 'low-cost revolution' supported frequent flying, promoted a tourism boom and sparked intense competition among airlines; aviation grew faster than ever before (ICAO, n.d.).

The increasing climate impact of tourism raises alarm bells as aviation is one of the fastest-growing sources of GHG emissions. Tourism is recognised as one of the economic sectors growing at the fastest pace globally and the number of tourists travelling by air rises continuously. As of 2019, according to Sun et al. (2024), tourism emissions represented 8.8% of total global GHG emissions and between 2009 and 2019, the sector's emissions increased by 1.5 Gigatonne (Gt) CO_2-equivalent, increasing at a 3.5% annual rate, more than double the rate of the global economy. Aviation accounts for 52% of the direct emissions of global tourism emissions and is the tourism sub-sector with the biggest emissions growth (Sun et al., 2024). The tourism industry, by fuelling air-traffic growth, points in the opposite direction of what is needed to address the climate emergency. A sudden change of course is mandatory. In the next section, the vital pathway of degrowth in aviation is analysed.

Degrow aviation for climate justice

The steep reduction of aviation is a vital measure at our disposal to meet the emergency of the climate crisis. The solutions that the aviation industry presents to decarbonise air transport are, after all, not solutions. Apart from being ill equipped to tackle current emissions, let alone projected ones, and stop aviation's contribution to global heating, they are profoundly unjust (Haßler *et al.*, 2019).

From offsetting and SAF production to the production of astronomical amounts of renewable energy, the burden of 'decarbonising' aviation is poised to fall on marginalised communities, mostly located in the Global South. If the aviation industry's plans materialise, people that have probably never set foot on an airplane will see their lands, water, forests and other essential resources taken away to implement aviation's false solutions, while being hit the worst by the looming climate breakdown.

Moreover, technological developments in aircraft, SAF production and CO_2 removal technologies require massive financial investment, almost entirely relying on public money. Instead of spending taxpayer's money on 'decarbonising' a transport mode that serves a global elite, public investments should go into meeting the essential needs of the majority: adequate food, housing, public transportation, sustainable energy and mobility, education and a clean environment. This logic should also prompt questioning as to why states continue to subsidise the aviation industry, in the forms of tax exemptions or bailouts, while public transport continues to deteriorate (Haßler *et al.*, 2019).

As the world must phase out fossil fuels and transition into a society powered by clean renewable energy to avoid the worst of the climate crisis and reverse it, the fair distribution of resources and the principle of sufficiency need to be at the heart of this process. There is only so much renewable energy that can be produced globally without further jeopardising ecological balance, as there is a limit to the amount of land available to serve as natural carbon sinks to balance out GHG emissions. For instance, if all jet fuel used in 2019 was to be replaced by e-fuels, that would require two and half times the renewable electricity available globally in that year (SG, 2021e). The application of renewable energy should prioritise economic sectors, utilities and other activities directly responding to the majority's essential needs, instead of air transport. Such prioritisation should fairly differentiate between frivolous flights (e.g. holiday, military, business and private jet flights) – the majority of the flights – and migrants flying sporadically to see their families in faraway locations, or flights for medical emergencies (SG, 2022b).

Degrowing aviation then implies a radical transformation of the global economic system, overcoming the capitalist society organised around profit to prioritise people's needs, ecological wellbeing and global justice. Following the principle of climate justice, the degrowth of aviation must start by cutting flights in the countries that are most responsible for the climate crisis and progressively reduce air traffic globally. Historically low-emitter countries and/or territories with a great level of dependence on air traffic for socioeconomic stability should be given more time to downscale aviation.

The degrowth of aviation must translate into a just transition for all whose livelihoods depend on the aviation and interdependent industries like tourism. Among other measures, this includes retraining workers of those industries and securing them jobs in other sectors and supporting entire communities and regions to repurpose their economic models (Stay Grounded & PCS, 2021).

As countries, corporations and wealthy citizens with the greatest historical responsibility, mostly from the Global North, must take the lead in rapidly reducing emissions, they also have the obligation to pay their ecological, climate and colonial debts. Reparations in the form of financial transfers, technology transfers, patent waivers, debt cancellation as well as policies positively responding to the displacement, damage and losses caused by climate catastrophe must be put in place (SG, 2023a). Those are part of the necessary steps to respond to the climate emergency while addressing historical and systemic power imbalances and discrimination and having marginalised groups and communities at the centre of these transformational processes and transitions. This too is crucial to degrow aviation to achieve climate justice.

The next section offers a case study to illustrate these points. It provides an historical analysis of how the development of aviation and tourism in Quintana Roo, Mexico, has further marginalised Indigenous and peasant communities. This case study reinforces demands for justice and reparations and highlights how the struggles to degrow aviation and tourism are complementary.

An Account of the Touristification Process in Southeastern Mexico: Quintana Roo

The violent transformation of the life of the Mayan peoples: Militarisation and tourism as a continuation of the colonial model

The transformation of the Mayan territory in recent decades is immeasurable. Populations that until the mid-20th century lived in autonomy are now dependent on the capitalist tourism model that has taken away their territory and enslaved them in the hotels of Cancun. Little remains of those Mayan peoples from the centre of the Federal Territory of Quintana Roo,[1] with their own economic, administrative, political, commercial and spiritual ways of life.

The Mayan peoples were stripped of their autonomy by a military campaign led by the then President Porfirio Díaz in order to exterminate the people who called themselves Mayas Masewales, an autonomous nation that coexisted with the nascent Mexican state and resisted the colonisation of Mexico.

Mayan self-sufficiency and autonomy were the product of a direct and sacred relationship with the land and autonomous ways of life. The Mayan *milpa* was (and is) the centre of life. This is an ancestral food cultivation system characterised by polyculture, with the central axis being the planting of corn, around which the rituality of the people revolves. Moreover, meliponiculture (a traditional practice of beekeeping with native stingless bees), poultry farming and other local economic practices have had a community and solidarity-based meaning and impact.

The violence experienced with the militarisation during the time of Díaz is the immediate antecedent of the new campaign of militarisation and touristification that has been dismantling the *milpa*. This new wave deploys political and administrative measures to encourage and protect the arrival of investments in energy, tourism, airports, industry and real estate.

The army is in charge of the construction and administration of communication and infrastructure works such as the already operational Felipe Ángeles International Airport in Mexico City. The military are the builders of three sections of the *Tren Maya* (Mayan Train, in English), and they were also in charge of the construction of the military base and Tulum International Airport. They are also the builders of hotels and tourist infrastructure (Carrillo, 2023). Mexico is not alone in this dynamic; similar situations of hyper-militarised territories are found in the Tamil region in Sri Lanka, where the military also runs hotels, restaurants and other tourist services (Teknomers, 2022).

The presence of the army on beaches and in tourist centres such as Cancun and the Riviera Maya is now part of everyday life. In 2021, the creation of the Quintana Roo Tourist Battalion with more than 1400 agents gave rise to a new landscape on the beaches of the Mexican Caribbean, where the sun, sea and sand are accompanied by tourists lying on sun loungers and the army with powerful weapons walking among them (Cid & Breña, 2023). Likewise, members of the Navy who walk along Playa del Carmen's Fifth Avenue are now in demand by visitors for souvenir photographs (Mexía, 2022).

Cancun: The birth of the monster with a thousand heads

Fifty-five years ago, a monster was born on the Yucatan peninsula, its name: Cancun.

The Mexican government and the world's largest financial institutions created the first *Centro Integralmente Planeado* (in English, Integrally Planned Centre) in Mexico. To this end, they built Cancun International Airport and the road infrastructure that would allow air and land communication to this new tourist centre. The emergence of tourist development hubs has generated a large-scale induced migration to these regions, with an overwhelming and excessive growth of cities such as Cancun and Playa del Carmen and with all the social inequalities that this entails. The latter has the highest annual growth rate in Mexico and Latin America at 7.9% (Rodríguez, 2024).

From 1970 to 2020, the population of Quintana Roo grew by 2000%, going from 88,200 inhabitants to more than 1,857,900 (INEGI, n.d.). Such rapid development and migration have also brought social tensions and problems. The establishment of organised crime has brought generalised violence. Cancun has become a paradise for crime, extortion (Hernández Ruiz, 2022), human trafficking, sex tourism and femicide (Aragón Falomir, 2024).

Already with hundreds of hotels and millions of tourists visiting every year, the growth trajectory persists. In 2024, more than 30 million arrived at the Cancun International Airport (Vázquez, 2025), with a total of 20 million visitors who stayed overnight in hotels. According to the Tourism Promotion Council of the state of Quintana Roo, Cancun receives nearly 50% of Mexico's international tourism. In 2025, the state of Quintana Roo expects to receive 22 million tourists who will stay overnight in hotels and accommodation centres – 10% more than in 2024. It is important to note that, in 2024, 5300 new hotel rooms were built, bringing the total to 134,000 in the state of Quintana Roo (Macronews, 2025).

However, when considering the welfare of the Mayan peoples, Cancun presents a model of failed 'development'. Nevertheless, it has been successful for the purposes of its architects, including the Inter-American Development Bank (IDB), the multinational corporations that have settled there and, of course, the Mexican federal government, driven by growth metrics and profit goals. While economic indicators reflect the 'success' of the tourist industry, the profound changes in the last five decades in the Mayan territory point towards a nightmarish process: a violent transition from food self-sufficiency and autonomy to almost total dependence on tourism.

Impacts of touristification on the territories

The tourist boom in Cancun has been accompanied by a huge wave of economic, social, cultural and environmental impacts. The natural devastation is plain to see: devastation and fragmentation of the forest, erosion of beaches, disappearance of mangroves and coastal dunes, danger of extinction of many species of flora and fauna, destruction of reefs, massive arrival of sargassum seaweed, salinisation of aquifers, to name only a few.

On the social front and closely linked with the environmental devastation, Mayan peoples were forced to leave the *milpa*, rural life and their ancestral ways of working to become employees in the tourism industry. The jobs sustaining the industry – builders, waiters, bartenders, chambermaids and others – were and still are the ones available, offering outrageously precarious labour conditions. This is a precarity that does not allow a dignified life for those workers who live in overcrowded conditions on the outskirts of the cities with no guarantee of basic public services, including access to water, good housing and easy transport. This transformation of the life of the Mayan communities cannot be understood without recalling the fast-paced urbanisation in Cancun, spreading through the state of Quintana Roo. The physical and symbolic space of the Mayan communities was completely reconfigured.

In Quintana Roo, there are 2180 rural localities and 27 urban ones. In other words, 98% are rural villages and 2% urban areas. However, in the 2% of urban areas lives 90% of the total population, some 800,000 people (INEGI, 2020). Urban areas such as Cancun and Playa del Carmen are home to the largest number of Mayan people who are in these cities as labour. Despite their numbers, they are the invisible, the despised, the ones who give the city a bad image, unless they have a uniform that identifies them as a servant of tourism. Then they are useful.

Projects such as the Mayan Train aim to replicate the Cancun and Riviera Maya model in rural regions with exceptional natural wonders. It will create at least 19 cities with a tourist vocation. The environmental, social, cultural and economic impact is overlooked; not to mention the increase in crime and drug trafficking. While tourism development generates immeasurable economic activity, this industry mainly benefits the transnational corporations that invest in this model of development: large hotels, energy companies, agribusiness and real estate companies.

Resistance: The pathway to *Permanecer en la Tierra* (Stay Grounded)

In December 2023, a group of Tulum residents held a peaceful demonstration to protest and denounce the arbitrariness of the construction of the new international airport and military airbase in their community. The banners read: 'ecocidal airport, illegal airport, military airport' (Figure 4.6). It was at that time that a new airport was inaugurated that seeks to intensify tourism in the central region of Quintana Roo, south of Cancun.

This demonstration showed the resistance that exists in the region; the discontent of many people who have seen how their territory is gradually being transformed and taken away from them.

Permanecer en la Tierra, the Latin American and Caribbean regional network that forms part of the global network Stay Grounded, also participated in this action as part of the work it carries out to make visible the impacts of aviation and airport expansion in the world, particularly in tourist areas where aviation plays a fundamental role for the growth of the sector.

Renovations have been announced at three more airports in the region, in Chetumal, Cozumel and Mérida. Similar trends are observed

Figure 4.6 Residents in Tulum protest the new international airport and military airbase in Tulum. Photo credit: Permanecer en la Tierra (used with permission)

in the rest of the country, as Mexico prepares to host another high-emitting FIFA World Cup in 2026, together with Canada and the United States (Gobierno de México, 2025). This will be another year of airport expansion and air-traffic growth for further touristification.

Permanecer en la Tierra is promoting the campaign 'Turismo Mata' (in English, 'tourism kills') in 2025, to make visible the impacts of the tourism industry in the territories where it develops, as well as the link between aviation and tourism. This campaign has been echoed by organisations, groups and individuals – in Mexico, El Salvador, Dominican Republic and the Canary Islands – who have seen how touristification threatens local life.

But the resistance and the struggle to defend the territory in Mayan villages has a long history. This struggle has been present for centuries and it is what has allowed local communities to continue living and preserving their culture. Likewise, the touristification of the Mayan territory has also been met with resistance.

In 1969, the inhabitants of the Mayan village of Chumpón, in the municipality of Felipe Carrillo Puerto and near Tulum, opposed the construction of what is now the most important federal highway in Quintana Roo, Highway 307. The construction of this highway was the trigger for the tourist infrastructure in Quintana Roo. Federal troops protected construction workers, as the Mayans' discontent at the presence of Mexicans in their territory was well known[2] (Congresso Nacional Indígena, 2022).

Touristification created 'Cancun', a project conceived and designed in the offices of international organisations together with the Mexican government planned to generate economic growth for the benefit of the government and these corporations. That was the beginning of the great plundering of the beaches, the lagoons, the *cenotes* (sacred sites for the Mayan) which were turned into merchandise, into products for a model of massive urban growth fuelled by the tourism industry. This process has exploited Mayan labour and used the Mayan name, culture and past for the enrichment of the large corporations. These corporations profit from the accommodation of tourists, their mobility, the construction of infrastructure for their use and the services they demand.

The intensification of mass tourism is intended to expand to the entire peninsula with infrastructure projects that include trains, airports, cruise ports and a series of works, aiming to reorder the territory, with the most affected being the native peoples. These projects contribute enormously to the climate crisis and show the unsustainability of the global developmentalist and consumerist system, of which tourism is implicated.

But as the wave of capitalist projects intensifies, so too does the resistance and the collective construction of other ways of living, and the network *Permanecer en la Tierra* is putting its efforts into this. Its existence and work reinforce the global aspect of the SG network: '*an airport rejected somewhere is a rejection of airports anywhere*'. The regional network also exposes how aviation and the associated tourism industry follow a neocolonial model which, in turn, underpins capitalist expansion worldwide. This reality underpins the commitment to anti-capitalist and anti-colonial approaches of this struggle and in the shared demands for climate and social justice. Only by overcoming the current hegemonic systems can new worlds of justice and dignity be brought into being and thriving.

The Case for a Just Transition Away From Air-Traffic-Related Tourism

In a world ravaged by climate catastrophe, phasing out fossil fuels as quickly as possible is imperative. We should start by eliminating frivolous carbon-intensive activities while implementing a just transition into decarbonised societies that prioritise meeting people's essential needs over wealth accumulation. In this scenario, we must aim at ending air-traffic-related tourism, while taking into account major inequalities defining the tourism sector and addressing climate injustices.

Enormous inequalities characterise the tourism industry. This ranges from who gets to travel and enjoy the perks of tourism activities and who is responsible for most of GHG tourism emissions, to which countries dominate most of the tourism activity and those which are the most economically dependent on tourism. In order to implement a rapid transition away from the air-traffic-dependent tourism model, we must consider such inequalities and gaps.

In 2016, 80% of all international arrivals took place within the same region. Intraregional international travel in Europe represented 50% of all intraregional international travel worldwide, of which 39% international arrivals were by air (UNWTO & International Transport Forum, 2019). Between 2009 and 2019, visitors (including both domestic and international travels) from USA, China and India alone made up 39% of tourism global emissions that, combined with the visitors of the following 17 countries (most from the Global North and wealthy nations) with the highest emissions, contributed 75% of tourism global emissions (Sun *et al.*, 2024). Those emissions are closely linked with the use of air transport for hypermobile and carbon intensive lifestyles. In 2023, 42% of international tourism was concentrated in ten counties: France, Spain, the US, Italy, Turkey, Mexico, the UK, Germany, Greece and Austria (inbound tourism) (UNWTO, 2024; WP Travel, n.d.).

On the one hand, those statistics emphasise the responsibility of countries with high climate, ecological and colonial debts – with some exceptions – for driving tourism global emissions. On the other hand, these statistics also clearly suggest that initiating the process of phasing out air-traffic-dependent tourism and degrowing tourism itself should happen in those countries (Sun *et al.*, 2024). Apart from the aforementioned debts, a majority of those are either Global North countries, advanced, strong and/or fast-paced growing economies or have a degree of economic diversity that puts them in an advantageous position to implement structural economic transformations.

In contrast, the visitors of the 155 countries with low tourism emissions contributed to merely 25% of the sector's global emissions. This staggering inequality is reinforced by the fact that some of those countries are low-lying islands and low-income countries with very little responsibility for the climate crisis but are disproportionately and negatively impacted by it (Sun *et al.*, 2024). Moreover, among those 155 countries, are countries whose economies are extremely dependent on international tourism, e.g. the Maldives, Bahamas or Fiji (Navarro-Drazic & Lorenzo, 2021; Ortega & Ribeiro, 2024).

As low emitters and with low responsibility for the climate crisis, countries like the Maldives, Bahamas and Fiji should be given more time to transform their economies away from the air-traffic-dependent tourism model. Those countries should also be paid reparations that would not only cover for loss and damages caused by the climate crisis, but also compensate and support communities whose livelihoods are currently dependent on tourism that is air-transport-dependent.

In addition, the tourism industry is dominated by multinational corporations (Stabler *et al.*, 2009) that, in taking advantage of and pushing for policies of market deregulation, have accumulated capital at the expense of local communities and ecosystems where tourism development takes place. The wealth of those corporations should be used to pay their share in reparation payments to communities impacted by tourism and for funding a global just transition.

The steepest reduction of air traffic and associated tourism in countries with the highest GHG emissions and historical responsibility for the climate crisis would not only be fair but the most effective and rapid way of tackling aviation global emissions. In this process, we must pay special attention to long-haul flights (Sun *et al.*, 2024). Also, regional disparities in economic dependency of tourism within countries – e.g. the Balearic and Canary Islands in the Spanish state – should be taken into account, in order to secure the wellbeing and livelihoods of historically and systemically marginalised communities.

While reducing the air-traffic-dependent tourism model is crucial to reduce global emissions and tackle overtourism in some of the global tourism hotspots, the overall devastating impacts of the tourism industry driven by capital accumulation should not be overlooked. The consequences of the fast-paced growth of the industry go beyond GHG emissions as outlined in the previous section on touristification in Southeastern Mexico. These multiple negative impacts require an urgent paradigm shift that must accompany the decarbonising of the global economy. Proposals for 'subsidiarity in tourism' (Higgins-Desbiolles, 2023), 'proximity tourism' (Cañada & Izcara, 2021), 'popular tourism' (Cañada *et al.*, 2024) and 'post-capitalist tourism' (Fletcher *et al.*, 2021) are important contributions towards imagining such a shift. Among the elements that could comprise such a new paradigm are: the principle of sufficiency applied to travel (i.e. discouraging excessive consumption); prioritising the needs for rest and leisure of the majority over those of the wealthy minority; and communities deciding the future of their territories and who visits them. The elements for a just transition are readily evident and Stay Grounded is playing its part in advocating and acting to advance this work.

Conclusion

The aviation industry stands at a critical crossroads, grappling with the dual challenges of the climate emergency and the socioeconomic inequalities exacerbated by its growth. The analysis presented here underscores the urgent need for a radical transformation towards degrowth in aviation and tourism, with a focus on climate and social justice. As the impacts of the incipient climate collapse become intolerably intense, it is imperative that we prioritise a just transition that addresses the historical and systemic injustices faced by marginalised communities, particularly in the Global South.

The interconnectedness of aviation and tourism demands a re-evaluation of current practices, emphasising the need to reduce air traffic and shift towards sustainable and equitable alternatives. By recognising the disproportionate contributions of wealthier nations, individuals and multinational corporations to aviation emissions, we can begin to implement reparative measures that support low-emission countries and communities dependent on tourism in transition into more resilient economies.

Ultimately, the path forward points to a post-capitalist future in which we prioritise the wellbeing of people and the planet over profit, fostering a global structural transformation that champions ecological balance, cultural preservation and the rights of Indigenous and marginalised populations. It is only through overcoming the current economic and cultural paradigms that we can degrow aviation and tourism to respond to the climate, ecological and social crises we face. That requires collective action and a strong commitment to justice. This chapter has reported on such action as modelled by the global network that comprises SG, bridging the Global North and the Global South to enact more just futures through dialogue, advocacy and activism.

Notes

(1) Created by Federal Decree on 24 November 1902, as one of the measures to terminate the autonomies and have complete political control of the Yucatán Peninsula.
(2) Although any armed resistance subsisted during this time, memories of the Mexican invasion and subjugation of the Mayan people were alive, fuelling the resentment towards all representations of the Mexican state.

References

Akdağ, G. and Oter, Z. (2011) Assessment of world tourism from a geographical perspective and a comparative view of leading destinations in the market. *Social and Behavioral Sciences* 19, 216–224. https://doi.org/10.1016/j.sbspro.2011.05.126.

Al Jazeera (2023) Flooding death toll soars to 11,300 in Libya's coastal city of Derna. *Al Jazeera*. See https://www.aljazeera.com/news/2023/9/15/flooding-death-toll-soars-to-11300-in-libyas-coastal-city-of-derna (accessed February 2025).

Al Jazeera (2024) Worst drought in century devastates Southern Africa, millions at risk. *Al Jazeera*. See https://www.aljazeera.com/news/2024/10/15/worst-drought-in-century-devastates-southern-africa-with-millions-at-risk (accessed February 2025).

Amouzai, A. and Haddioui, O. (2023) Green Hydrogen in Morocco: Just transition or Greenwashing Neocolonialism? Transnational Institute. See https://www.tni.org/files/2023-10/Green Hydrogen in Morocco_Online_0.pdf (accessed February 2025).

Aragón Falomir, J.A. (2024) Violencia de género en destinos turísticos: Estudio de caso en Cancún, México. *Clepsydra. Revista de Estudios de Género y Teoría Feminista* 27, 33–50. https://doi.org/10.25145/j.clepsydra.2024.27.02.

ATAG - Airport Transport Group Project. (2021) *Waypoint 2050*. See https://aviationbenefits.org/media/167417/w2050_v2021_27sept_full.pdf (accessed February 2025).

Bevacqua, E., Schleussner, C.-F. and Zscheischler, J. (2025) A year above 1.5 °C signals that Earth is most probably within the 20-year period that will reach the Paris Agreement limit. *Nature Climate Change* 15, 262–265. https://doi.org/10.1038/s41558-025-02246-9.

Biofuelwatch (2019) Corsia – A false solution to the very real threat of emissions from aviation. See https://www.biofuelwatch.org.uk/wp-content/uploads/CORSIA-Briefing.pdf (accessed February 2025).

Biofuelwatch (2025) A Global Fat Grab: Examining the push for aviation biofuels – biofuelwatch. Biofuelwatch. See https://www.biofuelwatch.org.uk/2025/fat-grab-report/ (accessed February 2025).

Cañada, E. and Izcara, C. (eds) (2021) *Turismos de Proximidad, un Plural en Disputa*. Icaria.

Cañada, E., Izcara, C., Montovert, B. and Zorzi, M. (2024) Propuestas para el diseño de políticas públicas de turismo popular. Alba Sud. See https://www.albasud.org/publicacion/es/135/propuestas-para-el-diseno-de-politicas-publicas-de-turismo-popularhttps://www.albasud.org/publicacion/es/135/propuestas-para-el-diseno-de-politicas-publicas-de-turismo-popular (accessed February 2025).

Carrillo, E. (2023) Cambia fusiles por hostelería: Sedena busca experiencia única con hoteles del Tren Maya. Forbes México. See from https://www.forbes.com.mx/cambia-fusiles-hosteleria-sedena-experiencia-unica-hoteles-tren-maya/ (accessed January 2025).

Cid, A.S. and Breña, C.M. (2023) México militariza sus playas. Ediciones EL PAÍS S.L. See https://elpais.com/mexico/2023-04-06/mexico-militariza-sus-playas.html (accessed January 2025).

Congreso Nacional Indígena (2022) Declaración del genocidio quintanarroense: La destrucción parcial del pueblo maya rebelde y las memorias de la autonomía. Congreso Nacional Indígena. See https://www.congresonacionalindigena.org/2022/11/15/declaracion-del-genocidio-quintanarroense-la-destruccion-parcial-del-pueblo-maya-rebelde-y-las-memorias-de-la-autonomia/ (accessed January 2025).

Copernicus (2024) Global climate highlights 2024. Copernicus. See https://climate.copernicus.eu/global-climate-highlights-2024 (accessed January 2025).

Delbecq, S., Fontane, J., Gourdain, N., Planès, T. and Simatos, F. (2023) Sustainable aviation in the context of the Paris Agreement: A review of prospective scenarios and their technological mitigation levers. *Progress in Aerospace Sciences* 141, 100920. https://doi.org/10.1016/j.paerosci.2023.100920.

Devine, J. and Ojeda, D. (2017) Violence and dispossession in tourism development: A critical geographical approach. *Journal of Sustainable Tourism* 25 (5), 605–617. https://doi.org/10.1080/09669582.2017.1293401.

Duval, D.T. (2013) Critical issues in air transport and tourism. *Tourism Geographies* 15 (3), 494–510. https://doi.org/10.1080/14616688.2012.675581.

EJAtlas – Global atlas of environmental justice (n.d.) EJAtlas – Global Atlas of Environmental Justice. See https://ejatlas.org/ (accessed February 2025).

EJAtlas (2025) Map of airport-related injustice and resistance. Atlas of Environmental Justice. See https://ejatlas.org/featured/airport-conflict-around-the-world (accessed February 2025).

Fleming, G.G. and Lépinay, I.de. (2019) Environmental trends in aviation to 2050. See https://www.icao.int/environmental-protection/Documents/EnvironmentalReports/2019/ENVReport2019_pg17-23.pdf (accessed January 2025).

Fletcher, R., Blanco-Romero, A., Blázquez-Salom, M., Cañada, E., Murray Mas, I. and Sekulova, F. (2021) Pathways to post-capitalist tourism. *Tourism Geographies* 25 (2–3), 707–728. https://doi.org/10.1080/14616688.2021.1965202.

Gobierno de México (2025) #MañaneraDelPueblo. Gobierno de México. See https://www.youtube.com/watch?v=PWiCy0nz0LE (accessed January 2025).

Gössling, S. and Humpe, A. (2020) The global scale, distribution and growth of aviation: Implications for climate change. *Global Environmental Change* 65, 102194 https://doi.org/10.1016/j.gloenvcha.2020.102194.

Graver, B., Zhang, K. and Rutherford, D. (2019) CO_2 emissions from commercial aviation, 2018. International Council on Clean Transportation. See https://theicct.org/publication/co2-emissions-from-commercial-aviation-2018/ (accessed March 2025).

Graver, B., Rutherford, D. and Zheng, S. (2020) Co_2 emissions from commercial aviation. See https://theicct.org/sites/default/files/publications/CO_2-commercial-aviation-oct2020.pdf (accessed January 2025).

Haßler, A., Dwarkasing, C., Reckmann, E., Sekulova, F., Schneider, F., Iniesta-Arandia, I., Edwards, L., Machler, L., Schmelzer, M., Grebenjak, M., Heuwieser, M., Blázquez Sánchez, N., Bridger, R. and Mingorría, S. (2019) Degrowth of aviation – reducing air travel in a just way Stay Grounded. See https://stay-grounded.org/wp-content/uploads/2020/02/Degrowth-Of-Aviation_2019.pdf (accessed January 2025).

HEÑOI, Stay Grounded, Biofuelwatch and Global Forest Coalition (2022) Producing fuel for other people's planes - A case study on the Omega Green Biofuel refinery in Paraguay. See https://stay-grounded.org/wp-content/uploads/2022/04/EN_agrofuels_case-study_2022.pdf (accessed January 2025).

Hernández, M.G. and de la Calle Vaquero, M. (2023) Touristification. *Oxford Bibliographies*. See https://www.oxfordbibliographies.com/display/document/obo-9780199874002/obo-9780199874002-0278.xml (accessed February 2025).

Hernández Ruiz, R. (2022) Extorsiones en el paraíso: Derecho de piso en cada rincón del Caribe mexicano. *Extorsiones En El Paraíso: Derecho de Piso En Cada Rincón Del Caribe Mexicano*. See https://www.connectas.org/especiales/extorsiones-cancun/ (accessed January 2025).

Herrero, Y. (2019) Forced tourism or: Travelling without Economy class – Stay Grounded. *Stay Grounded*. See https://stay-grounded.org/forced-tourism/ (accessed January 2025).

Higgins-Desbiolles, F. (2023) Subsidiarity in tourism and travel circuits in the face of climate crisis. *Current Issues in Tourism* 26 (19), 3091–3101. https://doi.org/10.1080/13683500.2022.2116306.

IATA (n.d.) Our commitment to fly net zero by 2050. Fly Net Zero. See https://www.iata.org/en/programs/sustainability/flynetzero/ (accessed February 2025).

IATA (2024) Global outlook for air transport. IATA. See https://www.iata.org/en/iata-repository/publications/economic-reports/global-outlook-for-air-transport-december-2024/ (accessed January 2025).

ICAO (n.d.) Low cost carriers (lccs). See https://www.icao.int/sustainability/pages/low-cost-carriers.aspx (accessed February 2025).

ICAO (2016) ICAO Environmental Report 2016. *ICAO*. See https://www.icao.int/environmental-protection/Documents/ICAO%20Environmental%20Report%202016.pdf (accessed January 2025).

INEGI (n.d.) México en cifras. Instituto Nacional de Estadística y Geografía (INEGI). See https://www.inegi.org.mx/app/areasgeograficas/?ag=23#tabMCcollapse-Indicadores (accessed May 2025).

INEGI (2020) Distribución. Quintana Roo. Cuentame INEGI. See https://cuentame.inegi.org.mx/monografias/informacion/qroo/poblacion/distribucion.aspx (accessed January 2025).

Intergovernmental Panel on Climate Change (IPCC) (2022) Annex I: Glossary. In J.B.R. Matthews (ed.) *Global Warming of 1.5°C: IPCC Special Report on Impacts of Global Warming of 1.5°C above Pre-industrial Levels in Context of Strengthening Response to Climate Change, Sustainable Development, and Efforts to Eradicate Poverty* (pp. 541–562). Cambridge University Press.

Jacob, K. (2023) The big melt. *Max Planck Research*. See https://www.mpg.de/21679168/F002_Focus_034-039.pdf (accessed January 2025).

Klöwer, M., Allen, M.R., Lee, D.S., Proud, S.R., Gallagher, L. and Skowron, A. (2021) Quantifying aviation's contribution to global warming. *Environmental Research Letters* 16 (10), 104027. https://doi.org/10.1088/1748-9326/ac286e.

Macronews (2025) Más de 5,300 nuevas habitaciones consolidan liderazgo turístico de q.roo: Gobernadora mara lezama afirma que seguirán promoviendo el destino en la fitur 2025. *Macronews*. See https://macronews.mx/quintana-roo/mas-de-5300-nuevas-habitaciones-consolidan-liderazgo-turistico-de-q-roo-gobernadora-mara-lezama-afirma-que-seguiran-promoviendo-el-destino-en-la-fitur-2025/ (accessed January 2025).

McCausland, R. (2024) Net zero 2050: Sustainable aviation fuels. IATA. See https://www.iata.org/en/iata-repository/pressroom/fact-sheets/fact-sheet-sustainable-aviation-fuels/ (accessed January 2025).

Mexía, F. (2022) Ven y tómate la foto en Cancún ¡con militares! *Azteca Noticias*. See https://www.tvazteca.com/aztecanoticias/foto-cancun-militares-isl (accessed January 2025).

Moreno, L., Ramon, A. and Pedreño, A. (2015) The development of low-cost airlines and tourism as a competitiveness complementor: Effects, evolution and strategies. *Journal of Spatial and Organizational Dynamics* 3 (4), 262–274.

NASA Global Climate Change (n.d.) Global surface temperature. Climate Change: Vital Signs of the Planet. See https://climate.nasa.gov/vital-signs/global-temperature/?intent=121 (accessed February 2025).

Navarro-Drazich, D. and Lorenzo, C. (2021) Sensitivity and vulnerability of international tourism by covid crisis: South America in context. *Research in Globalization* 3, 100042. https://doi.org/10.1016/j.resglo.2021.100042.

Ortega, B. and Ribeiro, M. (2024) An index of the economic dependence on tourism. *Tourism Economics* 31 (3), 426–452. https://doi.org/10.1177_13548166241262836.

Papatheodorou, A. (2021) A review of research into air transport and tourism. *Annals of Tourism Research* 87, 103151. https://doi.org/10.1016/j.annals.2021.103151.

PLACJC and DCJ (2024) ¿Qué son las Falsas Soluciones a la crisis climática? Mapa colaborativo denuncia casos en América Latina y el Caribe. *Mapa de Falsas Soluciones*. See https://www.mapafalsassoluciones.com/ (accessed January 2025).

Ritchie, H. (2019) Who has contributed most to global CO_2 emissions? Our World in Data. See https://ourworldindata.org/contributed-most-global-co2 (accessed January 2025).

Rodríguez, E. (2024) Quintana Roo lidera la tasa de crecimiento poblacional en México. *PorEsto*. See https://www.poresto.net/quintana-roo/cancun/2024/7/12/quintana-roo-lidera-la-tasa-de-crecimiento-poblacional-en-mexico.html (accessed January 2025).

Schleussner, C.-F., Ganti, G., Lejeune, Q., Zhu, B., Pfleiderer, P., Prütz, R., Ciais, P., Frölicher, T.L., Fuss, S., Gasser, T., Gidden, M.J., Kropf, C. M., Lacroix, F., Lamboll, R., Martyr, R., Maussion, F., McCaughey, J.W., Meinshausen, M., Mengel, M. ... Rogelj, J. (2024) Overconfidence in climate overshoot. *Nature* 634 (8033), 366–373. https://doi.org/10.1038/s41586-024-08020-9.

Souza, B. (2023) Atentado a indígenas no PA é novo capítulo de conflito com indústria que quer 'plantar' combustível de avião na Amazônia. Repórter Brasil. See https://reporterbrasil.org.br/2023/08/atentado-a-indigenas-no-pa-e-novo-capitulo-de-conflito-com-industria-que-quer-plantar-combustivel-de-aviao-na-amazonia/ (accessed January 2025).

Stabler, M.J., Papatheodorou, A. and Sinclair, M.T. (2009) *The Economics of Tourism*. Routledge.

Stay Grounded (2018) Map of planned airport projects – stay grounded. Stay Grounded. See https://stay-grounded.org/planned-airport-projects/ (accessed January 2025).

Stay Grounded (2021a) Fact sheet 1 – efficiency improvements. In *Greenwashing Fact Sheet Series*. See https://stay-grounded.org/wp-content/uploads/2021/08/SG_factsheet_8-21_Efficiency_print_02.pdf (accessed January 2025).

Stay Grounded (2021b) Fact sheet 2 – electric flight. In *Greenwashing Fact Sheet Series*. See https://stay-grounded.org/wp-content/uploads/2021/08/SG_factsheet_8-21_Electricity_print_FIN_korr.pdf (accessed January 2025).

Stay Grounded (2021c) Fact sheet 3 – hydrogen flight. In *Greenwashing Fact Sheet Series*. See https://stay-grounded.org/wp-content/uploads/2021/08/SG_factsheet_8-21_Hydrogen_FIN_Korr.pdf (accessed January 2025).

Stay Grounded (2021d) Fact sheet 4 – biofuels. In *Greenwashing Fact Sheet Series*. See https://stay-grounded.org/wp-content/uploads/2021/08/SG_factsheet_8-21_Biofuels_print_Lay02.pdf (accessed January 2025).

Stay Grounded (2021e) Fact sheet 5 – synthetic electro-fuels. In *Greenwashing Fact Sheet Series*. See https://stay-grounded.org/wp-content/uploads/2021/09/SG_factsheet_8-21_Synthetic-E-fuels_print_FIN_A4_Korr.pdf (accessed January 2025).

Stay Grounded (2022a) 6 – Net zero & carbon neutrality. In *Greenwashing Fact Sheet Series*. See https://stay-grounded.org/wp-content/uploads/2021/11/SG_factsheet_9-22_Net-Zero_en_A4.pdf (accessed January 2025).

Stay Grounded (2022b) Common destination: reframing aviation to ensure a safe landing and lay the tracks towards a fair planet. Stay Grounded. See https://reframeaviation.stay-grounded.org/wp-content/uploads/2022/05/Common_Destination_WEB-1-Pager_english.pdf (accessed January 2025).

Stay Grounded (2022c) It's about more than just CO_2. See https://stay-grounded.org/wp-content/uploads/2020/10/SG_Factsheet_Non-CO2_2020.pdf (accessed January 2025).

Stay Grounded (2023a) Aviation: A matter of climate justice. Stay Grounded. See https://stay-grounded.org/wp-content/uploads/2021/11/SG-Climate-Justice-and-Aviation-Factsheet.pdf (accessed January 2025).

Stay Grounded (2023b) Aviation is a health issue. See https://stay-grounded.org/wp-content/uploads/2022/04/SG_Health_EN_Corr.pdf (accessed January 2025).

Stay Grounded (2024) Fact sheet 8 – negative emissions technologies (NETs). In *Greenwashing Fact Sheet Series*. See https://stay-grounded.org/wp-content/uploads/2024/05/SG-factsheet-NETs-May-2024.pdf (accessed January 2025).

Stay Grounded and PCS (2021) A rapid and just transition of aviation: Shifting towards climate-just mobility. Stay Grounded. See https://stay-grounded.org/wp-content/uploads/2021/02/SG_Just-Transition-Paper_2021.pdf (accessed January 2025).

Sun, Y.-Y., Faturay, F., Lenzen, M., Gössling, S. and Higham, J. (2024) Drivers of global tourism carbon emissions. *Nature Communications* 15 (1), 1–10. https://doi.org/10.1038/s41467-024-54582-7

Tandon, A. (2024) Record-breaking Philippines typhoon season was 'supercharged' by climate change. Carbon Brief. See https://www.carbonbrief.org/record-breaking-philippines-typhoon-season-was-supercharged-by-climate-change/ (accessed February 2025).

Teknomers (2022) El idílico resort de playa dirigido por el ejército inflado de Sri Lanka. Teknomers Noticias. See https://teknomers.com/es/el-idilico-resort-de-playa-dirigido-por-el-ejercito-inflado-de-sri-lanka/ (accessed January 2025).

Transport & Environment (2019) Why ICAO and Corsia cannot deliver on climate. See https://www.transportenvironment.org/uploads/files/2019_09_Corsia_assessment_final.pdf (accessed January 2025).

Transport & Environment (2025) Down to Earth: Why European aviation needs to urgently address its growth problem. See https://www.transportenvironment.org/uploads/files/TE_Down_to_Earth_report.pdf (accessed January 2025).

Unicef (2023) Devastating floods in Pakistan. Unicef. See https://www.unicef.org/emergencies/devastating-floods-pakistan-2022 (accessed February 2025).

UNWTO (2020) International tourism highlights, 2020 edition. *UNWTO*. See https://www.e-unwto.org/doi/book/10.18111/9789284422456 (accessed February 2025).

UNWTO (2024) International tourism highlights, 2024 edition. See https://www.e-unwto.org/doi/abs/10.18111/9789284425808 (accessed May 2025).

UNWTO & International Transport Forum (2019) Transport-related CO_2 Emissions of the Tourism Sector – Modelling Results, UNWTO, Madrid. See from https://doi.org/10.18111/9789284416660 (accessed February 2025).

Urban, F., Nurdiawati, A., Harahap, F. and Morozovska, K. (2024) Decarbonizing maritime shipping and aviation: Disruption, regime resistance and breaking through carbon lock-in and path dependency in hard-to-abate transport sectors. *Environmental Innovation and Societal Transitions* 52, 100854. https://doi.org/10.1016/j.eist.2024.100854.

van Westen, R.M., Kliphuis, M. and Dijkstra, H.A. (2024) Physics-based early warning signal shows that AMOC is on tipping course. *Science Advances* 10 (6). https://doi.org/10.1126/sciadv.adk1189.

Vázquez, J. (2025) Aeropuerto de Cancún movilizó 30.4 millones de pasajeros en el 2024. *El Economista*. See https://www.eleconomista.com.mx/estados/aeropuerto-cancun-movilizo-30-4-millones-pasajeros-20250107-740939.html (accessed January 2025).

WP Travel (n.d.) World tourism ranking by country 2024. WP Travel. See https://wptravel.io/world-tourism-ranking-by-country/ (accessed February 2025).

5 Climate Justice in Nepal's Tourism Sector: Confronting Water Scarcity and Systemic Inequities

Nirmal Mani Dahal and Sudhan Subedi

Tourism plays a vital role in Nepal's economy but is increasingly threatened by climate-induced water stress. This chapter critically examines the intersection of water scarcity, tourism and climate justice, highlighting how these dynamics shape the resource access and vulnerabilities of tourism-dependent communities. Nepal's tourism sector is exposed to the dual crises of 'too much' and 'too little' water manifested through floods and droughts amplified by climate change and uneven governance. Using a conceptual research approach, the chapter draws on secondary sources and limited primary input to explore community-based water management models that integrate local knowledge, conservation practices and tourism profit reinvestment. These models demonstrate potential pathways to build water security and resilience within the tourism sector. However, challenges remain in scaling and institutionalising such initiatives. The chapter contributes to the discourse on climate justice by advancing strategies for equitable water governance and sustainable tourism development in Nepal.

Introduction

Climate justice provides a critical framework for understanding how the impacts of climate change and responses to it disproportionately burden vulnerable communities, particularly in the Global South. Among the sectors affected, tourism is especially exposed to climate-related disruptions such as extreme weather events and glacial retreat. As a major economic driver in many regions, the tourism sector demands urgent attention through a climate justice lens (Bigby *et al.*, 2024).

Nepal, located in the Global South, is particularly vulnerable to the impacts of climate change. Given its diverse landscape, rich biodiversity and varied topography, tourism is a significant contributor to the Nepalese economy. Tourism's heavy dependence on water resources, especially for activities such as rafting and boating, as well as for basic services in tourist areas, makes the sector susceptible to climate-related disruptions. Increasingly irregular and extreme weather patterns, including droughts and prolonged dry seasons, accelerate resource depletion and water scarcity. This situation disproportionately affects water-based tourism industries, which face challenges in meeting their operational needs and providing adequate services. Consequently, these pressures can limit their ability to participate in tourism-related economic activities, raising concerns about climate injustice. Addressing water poverty and promoting climate justice in Nepal's tourism sector is, therefore, crucial for sustaining the sector's economic contributions (KC *et al.*, 2021). Khanal (2025) emphasises the importance of community engagement and benefit sharing from tourism, promoting eco-friendly practices and supporting conservation projects for water resource management to foster sustainable tourism. Similarly, Kadayat and Upadhyay (2024) advocate for community-based tourism and ecotourism to attract responsible travellers and enhance sustainability.

However, integrating a climate justice perspective into this sustainable framework is essential. The tourism sector must actively work to address this gap to ensure its continued contribution to the national economy while upholding principles of sustainability (Rastegar & Becken, 2024). Consequently, water resource management strategies should prioritise equitable distribution, acknowledging the disproportionate impact of climate change and water scarcity on the tourism sector. Climate justice necessitates that water management and tourism policies actively mitigate climate change impacts and enhance resilience (Cole *et al.*, 2020). Addressing the interconnectedness of tourism, climate justice and water perspectives is crucial for understanding the challenges and experiences faced by tourism-dependent communities. Therefore, this chapter aims to analyse the intersection of water scarcity and climate justice within Nepal's tourism sector and to advocate for policies that prioritise pathways for climate justice in the industry.

Methods

We used a conceptual research method for this chapter (Xin *et al.*, 2013) drawing on conceptual framework analysis and literature-based theorisation to provide insights and develop a comprehensive understanding into climate justice in the tourism sector of Nepal, focusing specifically on the interface between water and tourism. The conceptual framework analysis approach identified and synthesised basic concepts and ideas from existing literature explaining the dynamics between water and tourism. The insights highlighted systemic inequities and challenges faced by the tourism sector in managing water scarcity within a changing climate. Similarly, the literature-based theorisation approach is used to present innovative approaches, models and strategies for sustainable water management based on the existing tourism research, water management and climate justice in the Nepalese tourism industry. To operationalise this method, we conducted a comprehensive literature review to collect relevant concepts, ideas and knowledge. We analysed case studies and examples from the Nepalese tourism sector and provided theoretical inputs and practical recommendations.

Our analysis is primarily based on secondary data, supplemented by limited primary information to gather information on climate justice, tourism and water management in the Nepalese context. The secondary data sources comprised a wide range of academic and grey literature from multiple disciplines such as climate justice, tourism studies, water management, environmental and tourism policies. These sources were systematically selected to ensure a comprehensive perspective, based on the relevance to the study objective and disciplinary diversity. The literature was categorised thematically (e.g. tourism and resources, policy framework, water challenges and management, climate justice) and we analysed this qualitative content to understand the underlying concepts, relationships and community phenomena.

The primary data were collected through formal discussions with a purposively selected 12 industry professionals in Nepal. Participants were carefully selected based on their roles, experiences and knowledge of tourism management, climate adaptation and justice. The discussions were based on semi-structured interviews that focused on uncovering major themes of policy implementation, climate justice and organisational practices.

In this study, the authors' positionalities (Ateljevic et al., 2005) reflect a dual lens: one shaped by climate change expertise and the other by years of experience in the Nepalese tourism sector. The first author's academic knowledge in climate science informs the analysis of systemic injustices in water access. The first author shaped the chapter's overall structure and methodology, and led the discussion on the tourism sector, water resources in Nepal and climate in(justice) interconnection between water, justice and tourism. Meanwhile, the other author's professional experiences within Nepalese tourism sectors provided valuable first-hand knowledge of tourism policies, case studies, showcases and stakeholder engagement. This background shaped both the research design and analysis, enriching the study with nuanced insights and a strong analytical perspective with the discussion in local operational existences.

The Tourism Sector and Water Resources in Nepal

The tourism sector and its dependence on natural resources in Nepal

Nepal's tourism sector is deeply intertwined with its diverse and abundant natural resources. The towering Himalayas draw trekkers and mountaineers, generating revenue and enhancing Nepal's global recognition as a premier destination (Vaidya, 2023). Beyond the mountain landscapes, Nepal's rich biodiversity and varied ecosystems draw a growing number of nature-based tourists. National parks like Chitwan and Sagarmatha, along with the Annapurna and Kanchenjunga conservation areas, offer significant opportunities for wildlife viewing, flora and fauna observation, trekking and immersive natural experiences (Nepal *et al*., 2022). The Koshi, Gandaki and Karnali River basins and their associated rivers provide platforms for adventure tourism activities such as rafting and kayaking. Similarly, lakes like Phewa, Rara and Shey Phoksundo attract tourists with their serene beauty, offering boating and trekking opportunities (Thakuri *et al*., 2021).

Nepal's diverse landscapes, with altitudes ranging from 60 m to 8848.86 m, further enhance its appeal as a year-round tourist destination. The tourism sector's heavy reliance on these natural resources underscores the critical importance of sustainable resource management practices to ensure the long-term feasibility and resilience of tourism in Nepal. Protecting and preserving these assets from depletion and degradation, particularly in the context of a changing climate, is essential for maintaining Nepal's natural beauty and sustaining the economic benefits it derives from tourism (Bhattarai *et al*., 2021).

Current state of water resources in Nepal: Scarcity and flooding

Nepal's water resources are characterised by two critical extremes: scarcity during the dry season (November to May) and intense flooding during the monsoon season (June to September) (Maskey *et al*., 2023). Despite the presence of numerous perennial rivers, unequal spatial distribution and insufficient infrastructure limit year-round access to water for a significant portion of the population. Furthermore, spring drying and groundwater depletion are increasing due to over-extraction (Adhikari *et al*., 2021). During the monsoon season, the hills experience landslides and debris flows, while the Terai region faces floods and sedimentation, resulting in damage to physical infrastructure and reduced agricultural productivity, with consequent economic and social impacts. Conversely, during the winter and dry seasons, surface water scarcity leads to the over-extraction and consumption of groundwater resources (Nepal *et al*., 2021). This pattern of alternating water extremes (abundance and scarcity) defines Nepal's hydrological landscape, creating a complex challenge that is further exacerbated by the impacts of climate change, particularly through extreme weather events. These events intensify existing challenges, leading to erratic rainfall patterns and prolonged droughts (Dahal et al., 2024). Addressing this dual challenge requires robust, integrated water resource management strategies and effective flood control measures to ensure economic, social and environmental sustainability in Nepal.

Impacts of climate change on tourism and water resources in Nepal

Climate change poses a dual threat to Nepal's tourism industry and water resources. The tourism sector is increasingly vulnerable to significant disruptions from extreme weather events and the degradation of natural attractions, rendering destinations that are heavily reliant on these assets particularly vulnerable. Reduced river flows impacting water-based activities and diminished snowfall in mountain regions exemplify these impacts (Pokharel, 2019).

Changing precipitation patterns, elevated evapotranspiration rates and accelerated glacial melting are contributing to widespread water scarcity, posing serious challenges for water resource management. This shortage has a direct impact on the tourism sector, as hotels, parks, hospitality centres and tourist destinations require substantial water supplies for their operations. Insufficient water availability can negatively affect the overall tourist experience and damage a destination's reputation (KC, 2017).

The intersection of these challenges can disrupt the tourism economy, threaten livelihoods and exacerbate unsustainable resource consumption and conflicts. This underscores the urgent need for both immediate and long-term strategies to address the impacts of climate change, promote responsible practices and ensure the sustainability of both the tourism industry and water resources (Belias et al., 2022).

Strategies for sustainable water management in tourism: Lessons from Nepal

Effective water management is critical for the stability of Nepal's tourism sector, requiring a balance between ecological preservation and economic viability. During peak seasons, high tourist inflows can strain local water supplies, particularly in ecologically sensitive areas where water scarcity is already a concern. Hotels, resorts and recreational facilities consume significant amounts of water for guest services, landscaping and swimming pools, further stressing water resources. Uncontrolled water use can deplete groundwater reserves, reduce river flows and contribute to environmental degradation, negatively impacting both tourism and local communities. Therefore, sustainable water management practices, such as rainwater harvesting and water recycling, can reduce the environmental footprint of tourism and ensure long-term resource availability.

The Government of Nepal has advanced its commitment to integrated water resource management by launching the Kamla River Basin Implementation Action Plan (2025–2045) (Almeida et al., 2024). This strategy seeks to position the Kamla River Basin (Bagmati and Madhesh Provinces) as a model for sustainable water resource management. The basin, encompassing four districts (Sindhuli, Dhanusha, Siraha and Udayapur), is characterised by seasonal extremes of drought and flooding, compounded by severe erosion. With over 600,000 people relying on these water resources, the strategy prioritises sustainable management, improved allocation and ecological stability to enhance local livelihoods. Key activities include the construction of check dams, riverbank stabilisation, controlled riverbed material extraction, groundwater recharge and the implementation of early flood warning systems. These measures aim to ensure water security, enhance agricultural productivity, mitigate disasters and promote regional economic growth (WECS, 2021).

The case of Dhulikhel (Bagmati Province) highlights the water demands of local communities and the complex balance required to support a growing tourism industry. Traditionally reliant on the Bhumidanda water source, Dhulikhel now faces challenges due to upstream community resistance, management conflicts and increasing commercial water demands. Inequalities in water distribution and competing demands from domestic consumption and tourism have further complicated the issue. While the city operates under a socially oriented water board model with support from central and local authorities, ongoing debates regarding tariff adjustments and cost recovery continue to pose stability challenges (Ojha *et al.*, 2020).

Technological advancements, such as rainwater harvesting, greywater recycling and improved irrigation systems, can enhance water use efficiency. Furthermore, strengthening of the legal framework is necessary to balance water use across agriculture, residential needs and tourism. A forward-thinking approach that prioritises infrastructure investments, alternative water sources and conservation efforts can help mitigate future water scarcity (Maniam *et al.*, 2022).

The experiences from the Kamla River Basin and Dhulikhel underscore the need for a holistic and collaborative approach to water management in Nepal. Integrating sustainable water management strategies into Nepal's tourism sector will not only protect vital water resources, but also contribute to long-term economic and environmental resilience.

Tourism operators in Nepal: Innovating with water-efficient practices

Nepal, with its rich natural resources and growing tourism industry, faces a critical challenge in managing its water resources, especially in popular tourist destinations. The combination of rising tourist arrivals and the escalating impacts of climate change has exacerbated water scarcity. Consequently, there is increasing pressure on tourism operators to adopt water-efficient practices. Increasingly, tourists themselves are becoming more conscious of water conservation, prioritising water purity and seeking sustainable travel options, contributing to the growth in responsible tourism (Faulon & Sacareau, 2020).

Rainwater harvesting has emerged as one of the most effective water conservation strategies adopted by tourism operators in Nepal. These systems collect and store rainwater, offering a reliable alternative source for non-potable uses such as landscaping, cleaning and toilet flushing. Several guesthouses and eco-lodges in Nepal have successfully implemented rainwater harvesting in their operations. By reducing reliance on municipal water sources, these operators conserve water and alleviate pressure on often-overburdened public systems (Nepal, 2008). In parallel, the adoption of wastewater treatment technologies is gaining momentum among tourism operators in Nepal. Many facilities are implementing water treatment and recycling systems to process wastewater for non-potable applications. These measures are particularly crucial in regions prone to water scarcity, ensuring that tourism activities do not disproportionately strain local resources.

Furthermore, the adoption of water-conserving technologies is becoming increasingly common in resorts, hotels and lodges. Many tourism businesses are installing low-flow showerheads, faucets and water-efficient toilets. Additionally, hotels are encouraging guests to participate in water-saving initiatives, such as reusing towels and linens. These simple yet effective changes can have a significant impact on long-term water conservation, reducing the tourism industry's overall environmental footprint (Treks, 2025).

Climate (In)Justice in Tourism

Nepal, renowned as a premier travel destination for its rich natural and cultural heritage, is facing increasing challenges from the impacts of climate change, revealing stark inequalities in how various communities experience climate (in)justice (Rastegar & Becken, 2024). The accelerated risk of Glacial Lake Outburst Floods (GLOFs) poses a significant challenge to Nepalese tourism. Likewise, water scarcity represents another critical issue, particularly in the hills and mountains of Nepal. Erratic rainfall patterns and declining snowfall are placing immense strain on water resources, which disproportionately affects local livelihoods and tourist services, thus exacerbating existing inequalities (Ojha *et al.*, 2020).

While these climatic effects have ecological roots, their social, political and economic consequences are equally significant, defining the issue of climate injustice. In the Nepalese tourism sector, the impacts of water shortages and climate-induced disasters are unevenly distributed. The vulnerable groups, such as porters, guides, small lodge operators and residents, are the first to suffer and have limited resources or institutional support to adapt. In contrast, larger and wealthier tourism operators generally have better access to capital, technology and networks to shield themselves from climate risks.

These varied challenges in Nepal illustrate how tourism-related climate impacts differ by location, affecting local livelihoods, infrastructure and heritage, demonstrating clear examples of climate injustice. Addressing climate justice requires nuanced analysis of who is at risk, who is vulnerable and who is protected. These questions are deeply intertwined with Nepal's socioeconomic hierarchies, the marginalisation of certain groups and persistent governance weakness. Additionally, tourism development tends to be top down, with benefits unevenly distributed and local voices, especially those of rural and Indigenous communities, largely excluded from climate adaptation planning and tourism decision making.

Through case studies from across Nepal, this discussion highlights diverse expressions of climate (in)justice in tourism and advocates for fair solutions that support both the environment and local livelihoods. These examples reveal how power imbalances and differing adaptive capacities shape the experiences of tourism workers and communities facing water scarcity, extreme weather and ecological decline. In doing so, the discussion contributes to a broader understanding of how principles of climate justice can and must be integrated into sustainable tourism strategies for Nepal.

Case studies of climate injustices in Nepalese tourism

Climate change has profoundly impacted Nepal's tourism sector, with disproportionate hardship experienced based on geographical, social and economic status (Ritika *et al.*, 2021). A primary manifestation of climate change is the accelerated melting of glaciers, particularly in the Everest and Annapurna regions. Rising temperatures have led to alarming rates of glacial melt, increasing the risk of GLOFs that threaten treks, lodges and tourism-dependent communities (Devkota, 2017).

The increasing frequency of extreme weather events, exacerbated by global warming, has further highlighted climate injustices within the Nepalese tourism industry. The Annapurna snowstorm tragedy of 2014, which resulted in the deaths of numerous trekkers and guides, exemplified the severe impact of climate change and underscored the pervasive climate injustices in the Nepalese tourism sector (Thakuri *et al.*, 2020). During catastrophes, emergency evacuations often prioritise foreign tourists due to diplomatic pressure, insurance considerations and international media attention. As a result, local guides, porters and residents, often equally or even more exposed to danger, are left waiting in uncertainty. Guides and porters, who often come from marginalised and vulnerable communities, are disproportionately exposed to these hazards, with limited or no access to social security or effective early warning systems, placing them at greater risk than wealthier stakeholders in the tourism industry (Devkota, 2017). Local workers, who constitute the core of the tourism value chain, are not insured, equipped with appropriate gear or provided with emergency services even as they remain on the frontlines during disasters. For instance, the case of the 2014 Everest avalanche in Nepal, where 13 Sherpa climbers died, highlights the issue of climate injustice.

The compensation of USD$408 provided to the families of the sherpas is significantly less than the $3000–$5000 these climbers typically earn in a single climbing season (about 2–3 months), revealing a substantial economic discrepancy (Barry & Bowley, 2014).

Water scarcity poses another significant challenge for tourism. Tourist destinations, such as the trans-Himalayan districts of Mustang and Manang, are experiencing declining snowfall and increasingly erratic monsoon patterns, leading to water shortages that impact local livelihoods. Hotels and guesthouses in areas like Kagbeni and Lo Manthang often struggle to provide consistent water services to tourists and are often forced to transport water from remote locations at significant cost. While larger tourism operators can afford to secure sufficient water resources, smaller lodges and local communities face greater constraints, exacerbating existing inequalities (Aase *et al.*, 2010).

Tourism infrastructure development has also contributed to climate injustices through forest degradation. With the expansion of tourism, larger resorts and luxury hotels often encroach upon forested areas, displacing rural and Indigenous communities who depend on these natural resources (Baral *et al.*, 2012). The Tharu community in Chitwan National Park, for instance, has reported reduced access to essential forest products due to the expansion of ecotourism (Kandel *et al.*, 2020).

The livelihoods of rural and Indigenous communities in Nepal are heavily reliant on agro- and cultural tourism. However, uncertain weather patterns, soil erosion and declining agricultural productivity are disrupting agrotourism activities in regions such as Ilam and Palpa. Organic tea and coffee farmers are struggling to maintain crop production levels, leading to financial instability. Consequently, many are compelled to migrate to urban centres or seek employment abroad, further exacerbating socioeconomic disparities (Besky, 2007).

Climate injustices in Nepal's tourism sector highlight the unequal distribution of environmental risks, with the most vulnerable populations bearing the brunt of impacts for which they are least responsible. Addressing these injustices requires inclusive climate policies, greater attention to the needs of vulnerable tourism workers and sustainable tourism strategies that prioritise local resilience. Failure to take decisive action will perpetuate these injustices, undermining the sustainability of Nepal's tourism industry and jeopardising the livelihoods of its most vulnerable communities.

Nepal as a model for inclusive tourism: Promoting water security for all stakeholders

Water security is a fundamental component of sustainable tourism, ensuring equitable access to quality water for all stakeholders, including tourists, local communities and the tourism industry itself (Enriquez *et al.*, 2017). In Nepal, where tourism constitutes a significant source of revenue, increasing water demand, climate change impacts and urbanisation have exacerbated water scarcity challenges (Koirala *et al.*, 2020). Despite these pressures, Nepal is emerging as a leader in inclusive tourism by integrating water security into its tourism policy.

One of the most significant steps toward more inclusive tourism in Nepal is the incorporation of water-conserving practices into tourism development (Kelly & Williams, 2007). Many hotels and eco-lodges in major tourist destinations such as Pokhara, Chitwan and Mustang have installed rainwater harvesting systems to reduce their reliance on external water sources. In regions like the Annapurna and Everest trekking routes, where water scarcity is a persistent issue, tour operators are implementing wastewater recycling technologies to minimise waste and maximise water reuse (Adhikari *et al.*, 2024).

Community-led initiatives have been instrumental in advancing water security in tourism-dependent destinations. In Patan, a community-driven water conservation program has revived traditional water management structures, such as 'hitis' (stone spouts), to ensure fair water distribution among residents, farmers and tourism businesses. This approach not only sustains customary water-sharing arrangements but also ensures that tourism development does not compromise local water access. Similarly, in the Langtang region, local communities have implemented ecotourism guidelines that promote water-saving technologies and contributions to community-level water conservation funds (Joshi, 2015).

Public–private partnerships are also at the forefront of addressing water security challenges in Nepal's tourism sector. Local governments and non-governmental organisations are collaborating with tourism enterprises to implement water-saving policies and infrastructure. For instance, in Pokhara, hotels and restaurants have partnered with local authorities to install water recycling and purification plants, ensuring a stable supply of clean water for both tourists and local residents. Incentives for integrating water-saving practices into businesses have further enhanced the adoption of sustainable water management practices in tourism operations (Gandaki Province Government, 2024).

Technological interventions are being incorporated to mitigate water scarcity exacerbated by climate change and to optimise water efficiency in tourism operations. Several high-altitude resorts have adopted solar water heating systems, reducing their dependence on firewood and minimising environmental impacts. Drip irrigation systems are also being implemented in agro-tourism ventures, such as tea and coffee gardens in Ilam and Palpa, to reduce water loss while maintaining agricultural productivity. These technologies not only conserve water but also enhance the resilience of tourism businesses in water-stressed regions (Dulal *et al.*, 2024; Kala, 2002).

The Government of Nepal is preparing a comprehensive tourism policy through inclusive consultation with relevant stakeholders, prioritising water security, climate justice and environmental protection for long-term sustainability. Climate justice has been recognised as a key priority in tourism policy discussions. This policy defends the water rights of Indigenous people and low-income communities, focusing on preserving natural and cultural heritage, investing in climate-resilient infrastructure, promoting best practice ecotourism and establishing water stewardship standards. The process has involved high-level consultations with industry stakeholders, associations, service providers and workers to ensure broad contributions toward an equitable framework (Tourism mail, 2024). By integrating traditional water management practices, community-centred programs, technological innovations and effective policy mechanisms, Nepal demonstrates the potential for a harmonious coexistence between sustainable tourism and responsible water resource management. The policy aims notably at reducing disproportionate risks, promoting equitable access to water resources and supporting adaptation at the community level. Ensuring that tourism development does not compromise water resources for local communities is paramount to the long-term sustainability of the industry and the wellbeing of those who depend on it. Enhanced coordination among government institutions, the tourism industry and local communities will be critical to scaling up effective water security interventions and bolstering the resilience of Nepal's tourism industry against future challenges. If successfully implemented, this plan can serve as a model for other countries heavily reliant on tourism.

Interconnections between Water, Tourism and Climate Justice in Nepal

The intricate relationships between tourism, water resources and climate justice in Nepal underscore the need for a balanced approach to address both water scarcity and abundance. Decreased snowfall in high-altitude regions and unpredictable rainfall patterns have resulted in water deficits, disproportionately affecting small-scale tourism operators who lack the financial resources to implement necessary adaptations for their clientele. Conversely, Glacial Lake Outburst Floods (GLOFs) and other flood events generate excessive water, causing damage to critical infrastructure, including trekking routes. These extremes highlight the urgent need for equitable and sustainable water management practices to mitigate water-related hazards in the tourism sector and safeguard the interests of all stakeholders (Gurung *et al.*, 2019; Maskey *et al.*, 2023).

Analysing the effectiveness of current policies and practices

Although Nepal has adopted national water policies aimed at enhancing climate resilience, persistent implementation gaps continue to undermine water security. The National Water Plan (2005) and Water

Resources Strategy (2002) framed long-term Integrated Water Resources Management (IWRM) goals, emphasising equitable and sustainable utilisation. These goals are further supported by the National Climate Change Policy (2019) and the Local Adaptation Plan of Action (LAPA). Despite these frameworks, inadequate enforcement and insufficient resource allocation have hampered their effectiveness, especially in rural and high-altitude regions where water availability remains precarious (Regmi & Shrestha, 2018).

The tourism sector has faced significant water-related challenges due to increased tourist activity and changing climatic conditions. The 2018 Sustainable Tourism Policy promotes water conservation, but implementation remains weak. While high-end hotels and lodges are increasingly adopting rainwater harvesting and wastewater recycling systems, small, locally owned lodges often lack the financial resources to invest in such technologies. This creates an uneven playing field; larger businesses benefit from stewardship while small-scale businesses suffer. The absence of robust regulations to ensure equitable water allocation in the tourism sector reflects a lack of inclusive water governance (Baral *et al.*, 2023).

Although progress has been made, water management and climate justice efforts in Nepal remain hindered by enduring structural challenges. While policies such as LAPA and sustainable tourism recognise the importance of equitable water access, implementation and funding constraints limit their broader impact. Persistent disparities in water access between the tourism and agriculture sectors highlight entrenched structural inequalities, disproportionately affecting the most vulnerable communities. However, community-based water conservation initiatives offer a promising pathway towards integrating climate resilience and social justice. Realising this potential requires stronger policy implementation, increased investment in climate adaptation and a more inclusive water governance framework.

Despite these challenges, local adaptation and water conservation efforts have demonstrated considerable success. However, scaling up these initiatives requires additional institutional support and financial resources. While the federal government of Nepal has devolved water management responsibilities to local governments, these entities often lack the technical capacity and funding necessary for effective implementation. International financing mechanisms, such as the Green Climate Fund (GCF), offer potential avenues for supporting water adaptation initiatives but accessing and mobilising these funds is often hindered by complex administrative requirements (Ministry of Finance, 2017). Nepal's access to GCF is facilitated through its National Designated Authority (NDA) under the Ministry of Finance. It coordinates proposals allied with national strategies such as the National Adaptation Plan (NAP). However, due to its limited institutional capacity, Nepal depends on international accredited entities such as the Food and

Agricultural Organisation, the United Nations Development Programme and the Asian Development Bank to implement projects (Nepal, 2023).

Nepal has identified tourism, along with its rich natural and cultural heritage, as a key thematic sector under its National Adaptation Plan (NAP) 2021–2050, allocating US$1.13 billion to support economic and development priorities. The tourism sector currently contributes approximately 8% of GDP in Nepal. However, advancing climate justice within this sector necessitates confronting the disproportionate climate burdens conveyed to tourism workers and communities. By supporting climate justice through the GCF, it can act as a corrective mechanism to address these injustices and these initiatives in the Himalayas would support the development of tourism (Richard, 2016; Schlosberg & Collins, 2014).

Nepal's dependence on international tourism has not resulted in equitable development. Tourism revenue has sidelined the rural mountain communities, while unsustainable practices (e.g. waste mismanagement in trekking regions) are prominent there. Here, the GCF can play a critical role by providing funding to shift to decentralised, climate-resilient tourism that minimises high dependence on international tourists. For instance, GCF funds can be used to safeguard ecosystems essential to ecotourism while diversifying livelihoods through training in digital entrepreneurship or low-impact forms of tourism. By redistributing resources to affected communities and resisting the imposition of globalised extractive tourism, climate justice is better secured (Bainton *et al.*, 2021; Chang *et al.*, 2021).

Showcasing Nepal's innovative solutions in balancing water usage and conservation in tourism

Water is a pivotal resource for Nepalese tourism sector, underpinning everything from hospitality to adventure tourism. However, a changing climate, rising water scarcity along with an increasing number of tourists, are leaving tourist destinations under unprecedented water stresses. Therefore, specific strategies enabling sustainable tourism and climate justice are essential.

A community-based water management model reflects such an approach by focusing local visions of equity, reciprocity and communal stewardship. For instance, in Langtang, this model provides a fair distribution system to lodges, private homes and farmers. The local committee prevents water control by tourism businesses and builds social trust and resource justice to the disadvantaged locals and communities. This bottom-up governance model exemplifies how communities can strengthen water systems, promoting climate justice. In addition, technological interventions and public–private partnerships are also important to manage water resources locally.

Despite these achievements, challenges remain in scaling up water-conserving measures across Nepal's tourism sector. In some areas, tourism development has led to the over-extraction of groundwater, resulting in declining water tables and seasonal water shortages. Small-scale tourism enterprises often lack the financial resources and technical expertise needed to invest in advanced water-conserving technologies. Strengthening policy enforcement, expanding financial incentives and facilitating knowledge transfer among stakeholders in the tourism industry are crucial steps toward overcoming these challenges (Subedi, 2022).

GCF can play a crucial role in supporting these models. However, access to GCF funding for countries like Nepal should prioritise the essentials of promoting climate justice in tourism, especially in the water sector. Its finance mechanisms may not always reach those communities that are highly dependent on tourism and vulnerable to climate change. Therefore, the GCF finance mechanism must be enabled to address these communities to ensure climate just practice.

Unique community-based models that link water resource management with sustainable tourism growth

Nepal's rich cultural heritage and diverse terrain have established it as a prominent international tourist destination, attracting visitors for trekking, adventure tourism and ecotourism. While increased tourist inflows, combined with water shortages, have created significant challenges for resource management, innovative community-based models have emerged effectively to integrate water conservation with sustainable tourism development. These frameworks prioritise local stewardship, equitable resource distribution and long-term environmental sustainability, ensuring that tourism development does not compromise water security.

The Annapurna Conservation Area (ACA) provides a compelling example of localised water management. Recognising the need for sustainable water use in high-altitude trekking areas, local communities have established water user organisations to regulate water supply among tourists, hotels and residents. These communities carefully manage water extraction from glacial streams and springs, maintaining natural replenishment rates. Water rationing is implemented during peak tourist seasons to prevent overuse. Furthermore, rainwater harvesting infrastructure, supported by both international and domestic conservation agencies, enhances water security by reducing dependence on glacial meltwater (Bajracharya *et al.*, 2006).

Chitwan National Park offers another successful model that integrates wetland restoration with ecotourism to sustain vital water ecosystems. Community-based tourism (CBT) programmes actively involve local buffer zone communities in conservation efforts to protect wetlands, which serve as crucial water sources for both wildlife and human populations. Conservation funds, generated from jungle safaris and eco-lodges, are reinvested in wetland restoration projects. These initiatives have facilitated the restoration of natural marshes and ponds, enhanced groundwater recharge and promoted biodiversity. Additionally, environmental education programs teach tourists about wetland conservation and encourage eco-friendly tourism behaviours, such as water conservation in hotels and responsible interaction with water bodies (Ranabhat *et al.*, 2025).

Restoration of Kathmandu Valley's ancient stone spouts ('hitis') is one case of human-directed rehabilitation of water that combines water conservation (Manandhar, 2022) with water stewardship driven by climate justice. Revival of the 'hitis', once abandoned due to urbanisation and groundwater extraction, is now driven by collaborative efforts between tourism bodies, municipal councils and local communities. These spouts minimise deep ground water extraction, provide equitable water access to communities and promote cultural continuity in the guise of traditional water-sharing systems (Khadge & Tiwari, 2014). Used by communities and tourists alike, they minimise environmental degradation as well as contribute to cultural identity, proving how heritage conservation can advance both climate justice and environmental sustainability.

Locally managed irrigation and tourist-driven infrastructure have encouraged water conservation in the dry, water-scarce environment of Upper Mustang. Alternating irrigation ensures equal distribution to homes, fields and inns, and tourist revenues are directed back into the improvement and maintenance of access to water. Locating solar-powered purification points along key tourist stops eliminates dependence on bottled water and localises flexibility. All such collaborative efforts bring about ecological sustainability (Gandaki Province Government, 2024) ensuring climatic justice by providing clean water to marginalised communities affected by the changing climate.

Even with such achievements, lack of adequate funding, poor policy enforcement and the intensified effects of climate change are some of the main drivers that test the sustainability of these interventions. To counteract this, it is essential to move towards climate justice through empowerment of local communities, women, Indigenous peoples and other vulnerable groups to actively participate in water and tourism management. It is vital to improve policy enforcement, support community-led strategies and develop inclusive decision-making processes. In addition to this, facilitating region-specific water conservation interventions, climate-proof investments and knowledge flows between regions that can help upscale such subnational models (Baral *et al.*, 2023). These actions assist in establishing climate resilience by thoroughly addressing inequalities in governance and resource control.

Conclusion

Nepal's tourism sector faces mounting challenges from climate change, particularly in balancing sustainable water management with continued economic growth. Despite its low contribution to global carbon emissions, Nepal experiences significant water scarcity due to climate change and extreme weather, highlighting a clear case of climate injustice. Nevertheless, the country has shown excellent initiative and determination by initiating and implementing community-based water conservation models. These initiatives, which range from localised water management and wetland restoration to heritage preservation, showcase the potential of integrating local knowledge and expertise, community participation and the wise reinvestment of tourism revenues to secure water access and build a resilient tourism sector.

For other water scarce tourist destinations, especially those in the Global South, these models offer valuable insights into how to align tourism development with climate justice principles. By centring community wellbeing, environmental stewardship and equity in climate impact distribution, Nepal's experiences present scalable solutions. Safeguarding fair and sustainable tourism in Nepal requires recognising and addressing the structural climate injustices outlined in this chapter. This includes strengthening water governance frameworks, increasing investment in water conservation and enhancing local capacity through education and awareness. It is equally important to provide incentives for sustainable practices, public awareness and tourism education campaigns. In addition, improved monitoring and assessment systems, cooperation and partnerships are also important. Furthermore, Nepal should determine its position as a leader in sustainable tourism and local water management by integrating climate change and climate justice issues into water management planning and actively advocating for climate justice in global forums.

References

Aase, T.H., Chaudhary, R.P. and Vetaas, O.R. (2010) Farming flexibility and food security under climatic uncertainty: Manang, Nepal Himalaya. *Area* 42 (2), 228–238. https://doi.org/https://doi.org/10.1111/j.1475-4762.2009.00911.x.

Adhikari, S., Dangi, M.B., Cohen, R.R., Dangi, S.J., Rijal, S., Neupane, M. and Ashooh, S. (2024) Solid waste management in rural touristic areas in the Himalaya – A case of Ghandruk, Nepal. *Habitat International* 143, 102994. https://doi.org/https://doi.org/10.1016/j.habitatint.2023.102994.

Adhikari, S., Gurung, A., Chauhan, R., Rijal, D., Dongol, B.S., Aryal, D. and Talchabhadel, R. (2021) Status of springs in mountain watershed of western Nepal. *Water Policy* 23 (1), 142–156. https://doi.org/https://doi.org/10.2166/wp.2020.187.

Almeida, A., Sharma, S., Kansakar, D., Gaudel, P., Foran, T., Penton, D., Cuddy, S. and Gnawali, K. (2024) Kamala River Basin Water Resources Development Strategy: Implementation Action Plan. WECS, CSIRO and JVS. See https://doi.org/10.25919/dxrk-hj66 (accessed October 2025).

Ateljevic, I., Harris, C., Wilson, E. and Collins, F.L. (2005) Getting 'entangled': Reflexivity and the 'critical turn' in tourism studies. *Tourism Recreation Research* 30 (2), 9–21. https://doi.org/10.1080/02508281.2005.11081469.

Bainton, N., Kemp, D., Lèbre, E., Owen, J.R. and Marston, G. (2021) The energy-extractives nexus and the just transition. *Sustainable Development* 29 (4), 624–634. https://doi.org/10.1002/sd.2163.

Bajracharya, S.B., Furley, P.A. and Newton, A.C. (2006) Impacts of community-based conservation on local communities in the Annapurna Conservation Area, Nepal. *Biodiversity & Conservation* 15, 2765–2786. https://doi.org/https://doi.org/10.1007/s10531-005-1343-x.

Baral, N.R., Acharya, D.P. and Rana, C.J. (2012) Study on drivers of deforestation and degradation of forests in high mountain regions of Nepal. *REDD Cell, Ministry of Forest and Soil Conservation, Nepal* Issue 1.

Baral, S., Nepal, S., Pandey, V., Khadka, M. and Gyawali, D. (2023) Position paper on enhancing water security in Nepal. UN 2023 Water Conference, New York.

Barry, E. and Bowley, G. (2014) After Everest disaster, Sherpas contemplate strike. New York Times. See https://www.nytimes.com/2014/04/21/world/asia/after-everest-disaster-sherpas-contemplate–strike.html (accessed April 2025).

Belias, D., Rossidis, I. and Valeri, M. (2022) Tourism in crisis: The impact of climate change on the tourism industry. In M. Valeri (ed.) *Tourism Risk: Crisis and Recovery Management* (pp. 163–179). Emerald Publishing. https://doi.org/https://doi.org/10.1108/978-1-80117-708-520221012.

Besky, S. (2007) Rural vulnerability and tea plantation migration in Eastern Nepal and Darjeeling. Himalayan Research Papers Archive. University of New Mexico.

Bhattarai, K. and Conway, D. (2021) Impacts of economic growth, transportation, and tourism on the contemporary environment. In K. Bhattarai and D. Conway (eds) *Contemporary Environmental Problems in Nepal: Geographic Perspectives* (pp. 563–662). https://doi.org/https://doi.org/10.1007/978-3-030-50168-6.

Bigby, B.C., Smith, J. and Higgins–Desbiolles, F. (2024) *Climate Justice in Tourism: An Introductory Guide*. Travel Foundation.

Chang, H.-S., Su, Q. and Chen, Y. S. (2021) Establish an assessment framework for risk and investment under climate change from the perspective of climate justice. *Environmental Science and Pollution Research* 28 (46), 66435–66447. https://doi.org/10.1007/s11356-021-15708-2.

Cole, S.K., Mullor, E.C., Ma, Y. and Sandang, Y. (2020) 'Tourism, water, and gender' – An international review of an unexplored nexus. *Wiley Interdisciplinary Reviews: Water* 7 (4), e1442. https://doi.org/https://doi.org/10.1002/wat2.1442.

Dahal, N.M., Xiong, D., Neupane, N., Yuan, Y., Zhang, B., Zhang, S., Fang, Y., Zhao, W., Wu, Y. and Deng, W. (2024) Spatiotemporal assessment of drought and its impacts on crop yield in the Koshi River Basin, Nepal. *Theoretical and Applied Climatology* 155 (3), 1679–1698. https://doi.org/https://doi.org/10.1007/s00704-023-04719-3.

Devkota, T. (2017) Climate change and its impact on tourism based livelihood in high mountain of Nepal. *Journal of Development and Administrative Studies* 25 (1–2), 11–23. https://doi.org/https://doi.org/10.3126/jodas.v25i1-2.23435.

Dulal, L.N., Shrestha, R.K., Neupane, P., Malla, R., Bhattarai, S., Neupane, B.R. and KC, A. (2024) Agro–tourism in Nepal: An analysis of Ilam District. *Educational Administration: Theory and Practice* 30 (10) 242–253. https://doi.org/https://doi.org/10.53555/kuey.v30i10.8079.

Enriquez, J., Tipping, D.C., Lee, J.-J., Vijay, A., Kenny, L., Chen, S., Mainas, N., Holst-Warhaft, G. and Steenhuis, T.S. (2017) Sustainable water management in the tourism economy: Linking the Mediterranean's traditional rainwater cisterns to modern needs. *Water* 9 (11), 868. https://doi.org/https://doi.org/10.3390/w9110868.

Faulon, M. and Sacareau, I. (2020) Tourism, social management of water and climate change in an area of high altitude: The Everest massif in Nepal. *Journal of Alpine Research| Revue de Géographie Alpine* 108 (1), 4–12. https://doi.org/https://10.4000/rga.6779.

Gandaki Province Government (2024) *The Study on Status of Water Resources and Future Strategy of Gandaki Province*. Gandaki Province Gandaki, Nepal. http://ppc.gandaki.gov.np/downloadfiles/final-report_Asar-8_nms-1731999481.pdf.

Gurung, A., Adhikari, S., Chauhan, R., Thakuri, S., Nakarmi, S., Ghale, S., Dongol, B.S. and Rijal, D. (2019) Water crises in a water-rich country: Case studies from rural watersheds of Nepal's mid-hills. *Water Policy* 21 (4), 826–847. https://doi.org/https://doi.org/10.2166/wp.2019.245.

Joshi, J. (2015) Preserving the Hiti, ancient waterspout system of Nepal. In *The Fabric: Threads of Conservation* (pp. 1–15). Proceedings of the Australia ICOMOS Conference, Adelaide, Australia.

Kadayat, G.R. and Upadhyay, G.R. (2024) An exploring sustainable tourism development in Nepal. *AMC Journal (Dhangadhi)* 5 (1), 118–134. https://doi.org/https://doi.org/10.3126/amcjd.v5i1.69129.

Kala, R.B. (2002) Off-seasonal vegetable production by drip irrigation system in Palpa District. *Economic Journal of Nepal* 25 (4), 214–219.

Kandel, S., Harada, K., Adhikari, S. and Dahal, N.K. (2020) Ecotourism's impact on ethnic groups and households near Chitwan National Park, Nepal. *Journal of Sustainable Development* 13 (3), 113–127. https://doi.org/https://doi.org/10.5539/jsd.v13n3p113.

KC, A. (2017) Climate change and its impact on tourism in Nepal. *Journal of Tourism and Hospitality Education* 7, 25–43. https://doi.org/https://doi.org/10.3126/jthe.v7i0.17688.

KC, B., Dhungana, A. and Dangi, T.B. (2021) Tourism and the sustainable development goals: Stakeholders' perspectives from Nepal. *Tourism Management Perspectives* 38, 100822. https://doi.org/https://doi.org/10.1016/j.tmp.2021.100822.

Kelly, J. and Williams, P. (2007) Tourism destination water management strategies: An eco-efficiency modelling approach. *Leisure/Loisir* 31 (2), 427–452. https://doi.org/https://doi.org/10.1080/14927713.2007.9651390.

Khadge, S. and Tiwari, S.R. (2014) Conservation of water heritage in Kathmandu Metropolitan City (pp. 452–460). Proceedings of Institute of Engineering Graduate Conference, Kathmandu, Nepal.

Khanal, R. (2025) Sustainable tourism practices in Nepal and its challenges. *SMC Journal of Sociology* 2 (2), 123–147. https://doi.org/https://doi.org/10.3126/sjs.v2i2.74843.

Koirala, S., Fang, Y., Dahal, N.M., Zhang, C., Pandey, B. and Shrestha, S. (2020) Application of water poverty index (WPI) in spatial analysis of water stress in Koshi River Basin, Nepal. *Sustainability* 12 (2), 727. https://doi.org/https://doi.org/10.3390/su12020727.

Manandhar, A. (2022) Valuing water heritage: Reviving water heritage in Kathmandu Valley for sustainable urban water management: A case of Chapagaon Lalitpur. Unpublished Master's thesis, Technical University Berlin.

Maniam, G., Zakaria, N.A., Leo, C.P., Vassilev, V., Blay, K.B., Behzadian, K. and Poh, P.E. (2022) An assessment of technological development and applications of decentralized water reuse: A critical review and conceptual framework. *Wiley Interdisciplinary Reviews: Water* 9 (3), e1588. https://doi.org/https://doi.org/10.1002/wat2.1588.

Maskey, G., Pandey, C.L. and Giri, M. (2023) Water scarcity and excess: water insecurity in cities of Nepal. *Water Supply* 23 (4), 1544–1556. https://doi.org/https://doi.org/10.2166/ws.2023.072.

Ministry of Finance (2017) *Green Climate Fund Handbook for Nepal*. Ministry of Finance, Nepal. https://www.undp.org/sites/g/files/zskgke326/files/migration/np/UNDP_NP-GCF-handbook-for-nepal.pdf.

Nepal, G.O. (2023) *Nepal Country Programme for the Green Climate Fund*. Ministry of Finance, Government of Nepal.

Nepal, S., Tripathi, S. and Adhikari, H. (2021) Geospatial approach to the risk assessment of climate-induced disasters (drought and erosion) and impacts on out-migration in Nepal. *International Journal of Disaster Risk Reduction* 59, 102241. https://doi.org/https://doi.org/10.1016/j.ijdrr.2021.102241.

Nepal, S.K., Lai, P.-H. and Nepal, R. (2022) Do local communities perceive linkages between livelihood improvement, sustainable tourism, and conservation in the Annapurna Conservation Area in Nepal? *Journal of Sustainable Tourism* 30 (1), 279–298. https://doi.org/https://doi.org/10.1080/09669582.2021.1875478.

Nepal, W. (2008) Decentralised wastewater management using constructed wetlands in Nepal. WaterAid at Zambia. See https://coilink.org/20.500.12592/rp4xvk (accessed March 2025).

Ojha, H., Neupane, K.R., Pandey, C.L., Singh, V., Bajracharya, R. and Dahal, N. (2020) Scarcity amidst plenty: Lower Himalayan cities struggling for water security. *Water* 12 (2), 567. https://doi.org/https://doi.org/10.3390/w12020567.

Pokharel, G.S. (2019) Short critical note on water wealth of Nepal. *Journal of Nepal Geological Society* 58, 13–19. https://doi.org/https://doi.org/10.3126/jngs.v58i0.24569.

Ranabhat, S., Khatiwada, S.S. and Rawal, K. (2025) Effect of COVID-19 pandemic on community-based homestay tourism in the buffer zone of Chitwan National Park. *The OCEM Journal of Management, Technology, and Social Sciences* 4 (1), 168–178. https://doi.org/https://doi.org/10.3126/ocemjmtss.v4i1.74759.

Rastegar, R. and Becken, S. (2024) Embedding justice into climate policy and practice relevant to tourism. *Journal of Sustainable Tourism* 33 (10), 2011–2028. https://doi.org/https://doi.org/10.1080/09669582.2024.2377720.

Regmi, B.R. and Shrestha, K. (2018) Policy gaps and institutional arrangements for water resources management in Nepal HI-AWARE Working Paper 16, version 1, ICIMOD, Nepal. https://lib.icimod.org/records/sm0x8-tas08.

Richard, V. (2016) Injecting justice into climate finance: Can the independent redress mechanism of the green climate fund help? Berlin Conference on Transformative Global Climate Governance, 24 May 2016. See https://dx.doi.org/10.2139/ssrn.2796463.

Ritika, K., Giri, I. and Khadka, U.R. (2021) Climate change and possible impacts on travel and tourism Sector. *Journal of Tourism and Himalayan Adventures* 3 (1), 54–62. https://doi.org/https://doi.org/10.3126/jtha.v3i1.39117.

Schlosberg, D. and Collins, L.B. (2014) From environmental to climate justice: Climate change and the discourse of environmental justice. *Wiley Interdisciplinary Reviews: Climate Change* 5 (3), 359–374. https://doi.org/10.1002/wcc.275.

Subedi, A. (2022) Is Pokhara becoming the next Kathmandu in terms of waste management? Onlinekhabar. See https://english.onlinekhabar.com/pokhara-waste-management.html (accessed February 2025).

Thakuri, S., Chauhan, R. and Baskota, P. (2020) Glacial hazards and avalanches in high mountains of Nepal Himalaya. *Journal of Tourism and Himalayan Adventures* 2, 87–104.

Thakuri, S., Neupane, B. and Khadka, N. (2021) Conservation of Gosainkunda and associated lakes: Morphological, hydrochemistry, and cultural perspectives. *Journal of Tourism and Himalayan Adventures* 3 (1), 74–84. https://doi.org/https://doi.org/10.3126/jtha.v3i1.39119.

Tourism mail (2024, December 31) Tourism associations interact on tourism policy 2081. See https://www.tourismmail.com/news/detail/105742/ (accessed May 2025).

Treks, N.S. (2025) Sustainable accomodation. Nepal Sanctuary Treks. See https://www.nepalsanctuarytreks.com/sustainable-accommodation/ (accessed March 2025).

Vaidya, R. (2023) Contribution of mountaineering tourism to Nepalese economy. *Nepalese Journal of Hospitality and Tourism Management* 4 (1), 35–44. https://doi.org/https://doi.org/10.3126/njhtm.v4i1.53313.

Xin, S., Tribe, J. and Chambers, D. (2013) Conceptual research in tourism. *Annals of Tourism Research* 41, 66–88. https://doi.org/https://doi.org/10.1016/j.annals.2012.12.003.

6 Tourism Mobilities, Climate Mobilities and Mobility Injustice in the US Virgin Islands

Mimi Sheller, Leah Trotman, Greg Guannel and Kim Waddell

Based on research as part of the Caribbean Collaborative Action Network (CCAN), a National Oceanic and Atmospheric Administration (NOAA) Climate Adaptation Partnership, this chapter will analyse how complex interacting tourism mobilities and climate mobilities impact the US Virgin Islands (USVI). With three main islands of St. Thomas, St. Croix and St. John catering to slightly different tourism sectors, the USVI is a highly tourism-dependent economy that is also highly vulnerable to climate change impacts including extreme heat, flooding, drought and hurricanes. Within the longer historical trajectories of colonialism, slave plantation economies, US territorial governance and current climate stressors, we consider the relationship between the mobilities of tourism (including air arrivals, cruise ships, yachting, food provision, infrastructural provision, etc.) and the everyday mobilities of Virgin Islanders both on the islands, between islands and to the US mainland. We draw on the theoretical concepts developed by geographer Tim Cresswell of routes, speed, friction and rhythm, along with Sheller's analysis of mobility justice. This enables an elaboration of tourism mobility justice within contexts of climate change, grounded in local experiences of differential mobility hierarchies and power relations of waiting, turbulence, disruption and stalled mobilities.

Introduction

The US Virgin Islands (USVI) are situated within the complex histories, geographies and economies of unevenly experienced (im)mobilities and infrastructure that have shaped the modern Caribbean from the colonial era until today. Long historical trajectories of colonialism, plantation economies and uneven development have shaped climate stressors, vulnerabilities and climate displacement across the Caribbean region, leading many scholars to highlight issues of climate justice (Baptiste & Rhiney, 2016; Rohland *et al.*, 2025; Sheller, 2020b, 2025). Today, the USVI has a highly tourism-dependent economy (USVI Bureau of Economic Research, 2024) which, like much of the region, is experiencing increasing risks due to extreme heat, flooding, drought and hurricanes exacerbated by climate change. However, the impacts of climate change are not evenly distributed but are experienced through class and racial inequalities of vulnerability, exposure and consequences. This chapter will analyse the inequitable impacts of climate change on a highly tourism-dependent economy in the US Virgin Islands through the lens of tourism mobilities (Sheller & Urry, 2004) and mobility justice (Sheller, 2018a), within a climate justice framework.

Around the world, tourism industries are grappling with the impacts of climate change vulnerability and resilience (Becken & Hay, 2012; Dogru *et al.*, 2019; Scott *et al.*, 2012). The USVI, like the rest of the Caribbean region, is vulnerable to many kinds of climate change risks and impacts; we do not propose to review all scenarios here but rather focus on the intersection of climate mobilities and mobility justice in relation to tourism. Broadly understood, 'climate mobilities' refers to the intersection of climate and environmental changes with the diverse forms of human mobility (including migration, displacement, relocation or immobility), as well as the embedding of such (im)mobilities 'in ongoing patterns and histories of movement, and the material and political conditions under which it takes place' (Boas *et al.*, 2022: 3366). For US Virgin Islanders, displacement by climate-related events, in other words, is not a one-way street toward leaving an island home, but rather is often experienced through multiple intervening processes of economic stress, temporary moves, social framing of opportunities and personal hopes and aspirations.

Tourism mobilities are one crucial aspect of expanding the scope of climate mobilities research to observe and examine multiple interrelated mobilities of people, ideas and things in a tourism-dependent economy with many built in injustices. The USVI has a rich and diverse history of forced and voluntary mobility that spans the pre- and post-emancipation periods and extends into the contemporary era in which mobilities into and out of the archipelago are very much shaped by tourism economies. Tourism infrastructure such as cruise ship ports, boat harbours, airports, roads and resort development have shaped the islands, producing uneven geographies of access and inclusion. This infrastructure is built upon the earlier uneven and racialised geographies that emerged from the era of slavery and resistance to it (Olwig, 1985), as well as the mid-20th century industrial economy that included US military installations and government support for the Harvey alumina refinery (1962) and Hess

Oil (1965), one of the largest oil refineries in the hemisphere in St. Croix (Bond, 2021). These patterns of development have produced unequal mobilities and infrastructure, leading to different experiences of climate change vulnerability and impacts.

We refer to such inequitable mobility regimes in terms of mobility justice and injustice. Mobility justice refers to how power and inequality inform the governance and control of movement, shaping everyday patterns of unequal (im)mobility in the circulation of people, resources, energy and information (Sheller, 2018a). We suggest that any analysis of climate justice and tourism in the USVI, and the Caribbean more widely, should begin with an understanding of these uneven histories of differential (im)mobilities and infrastructure, and how they have shaped present patterns of mobility injustice and, hence, climate injustice. This book chapter posits that climate mobilities and mobility justice approaches can be brought together to analyse the relationship between the mobilities of tourism (for example air arrivals, cruise ships, yachting, property development, second home ownership, national park tours, food importation, etc.) and the everyday mobilities of Virgin Islanders both on their islands and between their islands, as well as longer-term movement to the US mainland. Tourists largely benefit from access to transport, hotels, beaches and natural scenery, at the expense of Virgin Islanders who suffer poor transportation and lack of access. In some cases, economic immobility, lack of services and poor infrastructure may cause Virgin Islanders to leave the islands permanently.

Here, we will analyse how complex interacting tourism mobilities and climate mobilities impact the climate resilience of Virgin Islanders and define residential climate vulnerability. We argue that a climate mobilities and mobility justice approach to tourism can help us understand some of the complex interacting factors that are affecting the potential climate resilience and adaptation of USVI communities. We will reflect on climate disruptions of the performances of tourism mobilities and local mobilities in relation to the embodied relations and 'staging' of tourism performances in a highly tourism-dependent economy (Córdoba Azcárate, 2020; Edensor, 2001; Sheller & Urry, 2004). We draw on the theoretical concepts developed by geographer Tim Cresswell of routes, speed, friction and rhythm as crucial elements of 'the politics of mobility' (Cresswell, 2006, 2010), along with Sheller's analysis of mobility justice (Sheller, 2018a). We will analyse complex interacting climate mobilities across multiple scales using theoretical concepts of uneven and differential routes, speed, friction and rhythm, described below. This enables an elaboration of a climate mobility justice analysis of USVI tourism grounded in local experiences of differential mobilities, social hierarchies and power relations that are experienced in terms of phenomena such as waiting, mobility disruption, displacement and stalled mobilities.

Context

Situated between the Atlantic Ocean and the Caribbean Sea, the USVI comprises four main islands (St. Thomas, St. John, St. Croix and Water Island) and over 50 cays. St. Thomas is about 32 square miles, and home to the territory's capital Charlotte Amalie. St. John, which is about 20 square miles, sits 2.5 miles East of St. Thomas and St. Croix, the territory's largest island at about 84 square miles, sits 38 miles south of St. Thomas. Water Island, the territory's smallest residential island at about 0.8 square miles, is less than a mile from St. Thomas. Altogether, the islands are about 137 square miles in total, and transportation between the islands is administered by boat and/or airplane.

The USVI is currently labelled a 'territory' of the US. However, before the purchase of the territory by the Americans in 1917, six other flags flew throughout the islands' storied history: Spanish, Dutch, French, English, the Knights of Malta and Danish (Krigger, 2018). The Danes, however, had the longest period of imperialism (1672–1917), lasting 245 years, and, without a doubt, had the most sweeping social, historical and economic impact on the islands.[1] In highlighting the former colonial power's participation, this paper makes visible Denmark's involvement in the transatlantic slave trade, which has often been made invisible, although we do not fully address key historical events such as emancipation in 1848 and the post-emancipation 'Fire Burn' uprising of 1878. Nevertheless, the legacies of 'consuming the Caribbean' (Sheller, 2003) are evident within the conversion of a Danish plantation colony into a US military asset and platform for extractive industries, including tourism, with profits going elsewhere. The infrastructure of the past sets the scene for the performance of tourism in the USVI today.

Tourism practices in the Caribbean date as far back as the 17th century (Butler, 2006; Weaver, 1998). However, the industry did not achieve its now-coveted regional prominence until the mid-1900s. Before this, early examples of leisure tourism were few and far between due to the activity being limited to the wealthy or those involved in the trade patterns of the region (Toppin-Allahar, 2015). Additionally, with the success of the sugar and other mono-crop industries, from the advent of slavery until the early to mid-1900s, there was no immediate need for local Caribbean governments to diversify their economies and search for revenue elsewhere (Ferguson, 1997). Thus, the tourism industry did not become a large bloc in the economies of Caribbean states until the collapse of the region's sugar, and much later banana, industries in the mid-to-late 1900s (Ferguson, 1997; Monzote, 2013; Pattullo, 1996; Weaver, 1998). Thereafter, tourism became a non-negotiable top priority across the region and remains so today.

The story of modern tourism in the USVI begins in the 1930s. There were concerted efforts by the US federal government to expand the industry; the government intentionally centred the St. Thomas' and St. John's economies on tourism while discussions were had about centring agriculture as the economic driver for St. Croix (Annual Report of the Governor of the Virgin Islands, 1934). Following, hotels and other symbols of tourism were erected. The Virgin Isle Hotel was established in 1950 and had 130 rooms. Before the 1960s, smaller guesthouses with a capacity of 5 to 50 rooms were the norm. On the cusp of the 1960s, now similar to the regional trend, the industry blossomed. Krigger (2018) credits some of this tourism boom with the 1959 Revolution in Cuba. After Castro's ascension to power, Cuba was no longer a popular destination for American tourists. Nearby Puerto Rico and the US Virgin Islands took on much of those tourists and, by 1967, St. Thomas became one of the busiest tourist ports in the world. Statistics from that period better illustrate this drastic increase. In 1962, 131 ships visited St. Thomas holding 57,368 passengers. By 1968, 342 cruise ships had docked in Charlotte Amalie bringing 166,117 passengers onshore. The growth in the number of visitors continues: 2024 visits by cruise ships brought 1.77 million passengers in addition to over 930,000 travellers arriving by air (Global Agents USA, 2025).

Today, the USVI faces a whole host of issues, including a limited economy. The islands are heavily tourism dependent (with additional revenue from sectors such as rum distilling). The territory's economy is also in general decline and reasons for this reduction are numerous. One of the largest oil refineries in the Western hemisphere was based in St Croix, alongside a major alumina refinery. In 2012, the Hovensa oil refinery, one of the largest employers on the island of St. Croix, shut down. At its peak, Hovensa employed nearly 2000 people and, at the time of its closing, contributed US$60M in tax revenue to the USVI (Associated Press, 2012). This closure was not only an immediate economic loss but also a factor in the proliferation of 'brain drain' – the exodus of residents to the nearby continental United States for work elsewhere – as employees joined many others who left the island in search of employment elsewhere, estimated to be as high as 46% of the population (Nugent, 2022). While the oil industry is a crucial driver of global warming, ironically, its departure can also leave behind social devastation. Despite the closure of these polluting industries, their harmful environmental and social impacts remain and are crucial sites of political mobilisation for climate justice in the islands (Bond, 2021).

In 2017, the islands were devastated by two back-to-back Category 5 hurricanes, Irma and Maria, which ceased all tourism on the islands for months and also served as another 'push' factor of emigration. Depopulation following hurricanes has also occurred in other Caribbean islands and is often unevenly distributed in terms of who is driven away, especially affecting non-citizen migrants who provide the lowest-paid labour (Nixon, 2019). Meanwhile, when the USVI territory hoped to rebound in 2020, the coronavirus pandemic took over and, once again, halted the territory's largest revenue maker. There are also other indicators of economic distress: public debt is high – 65% of GDP in 2023 (U.S. Government Accountability Office, 2023), and the island's retirement system has been on the brink of insolvency for many years.

Over the past decade, the territory has suffered a tremendous population decline. The latest 2020 Census details a population decline of 18% from 2010 to 2020. The 2020 US Census Bureau tallies the islands' population at 87,146, which pales in comparison to the 2010 statistic of 106,145 residents. St. Thomas is home to 42,261 residents, St. John has 3881 and St. Croix has 41,004. Reasons for this dramatic population decline vary and have not been systematically and scientifically investigated. However, anecdotally, a lack of economic opportunity, the 2017 hurricanes, high crime rates and a limited higher education system have all been painted as reasons for this decline. These immediate causes, however, should be interpreted in relation to the larger patterns of mobility injustice and climate injustice that we argue have shaped the environmental and economic context as experienced by local inhabitants.

Tourism Mobilities, Climate Mobilities and a Mobility Justice Approach

Tourism remains one of the largest economic sectors in the USVI, yet its benefits are not evenly distributed, nor are its unequal harms. A recent report published by The Travel Foundation (Bigby *et al.*, 2024) explores this inequity and poses the question, 'Whose Needs Come First?' in the tourism industry. As tourists and their desires become more prioritised, locals and their everyday challenges and resource needs are often neglected or completely ignored. Thus, while the sector itself has contributed to the overall growth of the territory's economy, the industry of tourism has many deficiencies, including but not limited to the replication of the colonial relationship between the colony and the metropole and negative environmental impacts. Although specific to Jamaica, Taylor concisely describes this dichotomy best: 'On the one hand, postwar tourism development was meant to relieve the problems of Jamaica's colonial past [specifically, economic problems]. On the other [hand], tourism reinforced many essential features of that original condition and entrapped the island in the clasp of neocolonialism' (1993: 178). Taylor's analysis of tourism's contradictions and later extractive impacts can be applied to the USVI, where government policies have sought to push extractive industries and tourism as the main levers of economic development, while not considering longer-term social consequences on locals and environmental sustainability (Bond, 2021).

The USVI's economic model – its uber-dependence on tourism and tax-free investments (USVI Economic Development Authority, n.d.) – becomes more and more unstable with each passing climate event, threatening the stability of the territory's wider economy and progress. Climate risks in the Caribbean have been historically high. However, for already highly economically vulnerable populations in the territory, these risks are greater. These risks include extreme periods of heat, flooding brought on by excessive rainfall, drought and hurricanes (Hazard Mitigation & Resilience Plan, 2024). Compounding risks exacerbate impacts. Excess heat can lead to a diversity of impacts that can range from heatstroke to road damage. Flooding can lead to business damage and closure, an increase in disease and a limitation of beach access. Hurricanes can lead to similar effects. Guannel *et al*. (2022) found that most urban populations in St. Thomas and St. Croix had high levels of social vulnerability. Climate change compounds USVI social inequality, as it has around the world, especially in (post)colonial contexts (Sultana, 2025).

Mobilities are relevant to climate change research both as a cause of greenhouse gas (GHG) emissions and as an impact when people and communities are displaced. Both mobility types are particularly pronounced in tourism-dependent islands because of the high GHG emissions related to tourism, and because of the high levels of displacement experienced on islands due to tourism development and climate-change impacts such as extreme storms, drought, flooding and infrastructural disruption (Scott *et al*., 2019). Usually, these phenomena are studied separately, with some research focusing on sustainable or regenerative tourism (Cave & Dredge, 2020; Gössling & Hall, 2006), others examining decarbonising transportation (Adey *et al*., 2021a) and others studying climate migration and related displacement phenomenon such as 'green gentrification' (Agyeman, 2013). Sector-specific approaches usually do not pay attention to interactions between transport, tourism and migration. Combined studies of tourism and migration have taken note of complex intersecting (im)mobilities such as residential tourism, retirement migration, second-home tourism, nomadic workers, return migration and labour migration to work in the tourism sector (Bloch & Adams, 2023; Hall & Williams, 2002), but are not necessarily linking these multiple mobilities to climate justice and mobility justice.

The fragility and disruption of transportation infrastructure under current climate conditions and future scenarios suggest that climate risks and resilience will play out in different ways across different social groupings such as tourists and residents. Compounding infrastructural disruptions and emergencies (e.g. power outages accompanied by extreme heat and broken cell phone service; storm-related flooding accompanied by transportation disruptions and lack of access to healthcare, food and water) have uneven and differential impacts. Seldom are all these interacting aspects of climate (im)mobilities brought into one framework that we might call a climate mobility justice approach to tourism. We seek to bring these questions together by paying attention to the interactions between different aspects of tourism mobility, local transportation, displacement and migration, each of which is affected by climate change in a multitude of ways related to the historical and material conditions under which such (im)mobilities take place.

This combined approach enables researchers to examine, for example, how the energy-intensive mobilities of more privileged groups (such as tourists) may be contributing to displacement that exposes less privileged groups (such as local island communities, fishermen and tourism workers) to more extreme environmental impacts, making them more vulnerable to climate change. At the same time, *the loss of tourism*, like the prior loss of fossil fuel industries, can also have negative economic and social impacts that drive displacement, population loss and 'brain drain', weakening the social infrastructure. Tourism affects land use patterns and infrastructural development, as well as local economies and social support systems, in ways that may expose local communities to greater climate risk, increase vulnerability and reduce climate resiliency through impacts on both physical and social infrastructure. In the following sections, we will analyse some of these interactions and impacts, drawing on our Caribbean Climate Adaptation Network (CCAN) research team's observations and preliminary scan of climate adaptation challenges in the USVI.

Operationalising climate mobility justice: Routes, speed, rhythm and friction

Building on our approach to mobility justice, geographer Tim Cresswell has operationalised the politics of mobility through a focus on material and cultural analysis of routes, speed, rhythm and friction. Cresswell observes that:

> When we are faced with the possibility of moving, it is not across a smooth unvariegated plain, but within a topography that includes predesignated paths and roads or simply terrain that is easier to transect. Many of the routes that are open to us have been made possible by established forms of power in earlier times. (forthcoming: 18–19)

The routes for moving across the USVI include ground, sea and air transportation of various kinds, each of which has been established by patterns of power as well as topography. Danish colonial exploitation shaped the ports and sea connections between St Thomas, St Croix, St John and other smaller islands of the archipelago and its neighbours, while geographical features such as hills and mountains, local hydrology, rainfall and runoff also affect patterns of settlement, infrastructures for mobility and how different groups experience or perform the routes, speeds, rhythms and frictions of movement at varying scales.

Routes

In the USVI, the main ports of the plantation slavery era were constructed in Charlotte Amalie, Frederiksted and Cruz Bay, respectively (see Figures 6.1 and 6.2). Until the age of air travel, these were the main entries into the islands and were important stops for 19th-century steamships and 20th-century tourism. Maritime mobilities continue today with people ferries, car ferries, yachting harbours, cruise ship ports (including one that accommodates the largest cruise ship in the world) and some small fishing boats. We note that such routes are not just physical infrastructure but are also symbolic representations of island tourism and how it is performed and staged via embodied movement. The historical colonial port cities, the ferries and yacht harbours, all remain important backdrops for framing the 'picturesque' aspects of Caribbean islands as racialised destinations outside of modernity, which become important stages for tourist consumption, photography and marketing (Sheller, 2003; Thompson, 2006).

Today, tourism travel in the USVI is varied across the islands. St. Thomas has two maritime ports suited for cruise ships in Havensight and Crown Bay. From St. Thomas, locals and tourists regularly travel

Figure 6.1 An overhead shot of the West Indian Company Dock (cruise ship port) in downtown Charlotte Amalie, St. Thomas, USVI. (Credit: Virgin Islands Port Authority)

Figure 6.2 An overhead shot of The Ann E. Abramson Marine Facility (cruise ship port) in Frederiksted, St. Croix. (Credit: Virgin Islands Port Authority)

to St. John, St. Croix, the British Virgin Islands and Puerto Rico by boat or plane; cargo can and will often accompany this journey. The island of St. Thomas also has two airports, the Cyril E. King Airport (CEKA) and the Seaborne Seaplane Terminal. St. Croix has one maritime port suited for cruise ships and another best suited to transporting people and/or cargo to St. Thomas and Puerto Rico. It also has two airports: the Henry E. Rohlsen Airport and the Seaborne Seaplane Terminal. While St. Thomas remains the main island for cruise ship docking and airport arrival, after a 17-year hiatus, St. Croix began welcoming cruise ships in 2018 (see Figures 6.3 and 6.4). St. John does not have an airport or a port to dock cruise ships, but the island attracts private yachts and wealthy tourists from all over the world eager to spend time in exclusive villas. St. Thomas prides itself on more mainstream tourist activities such as water sports, beaches and nightlife. St. Croix has, in more recent years, attempted to rebrand itself as *the* Caribbean destination for food and multi-hyphenate festivals.[2]

Different modes of travel have become part of the chronotope of tourism in the USVI, i.e. symbolic configurations of time and space. Across the islands, the mode of transportation is of two strands: boat or airplane. Boat travel includes cruise ships, people ferries and private,

Figure 6.3 A 1954 orthophotograph of Crown Bay in St. Thomas, USVI. (Credit: Greg Guannel)

chartered boats and yachts. By airplane, tourists can rely on several airlines to provide travel from all over the world to the islands. With an even greater influx of (larger) ships and an increase in commercial airlines and airplanes, the USVI continues to break tourism records. The latest press release (Gilbert, 2025) from the Department of Tourism published in 2025 reveals new records for single-day airplane visitors at 4606 passengers, and a 5.92% increase in weekly arrivals with over 24,000 air arrivals in one week to St. Thomas, and arrivals of 1,770,922 cruise ship passengers in 2024. All commercial jet flights originate from the continental United States or Puerto Rico while smaller (<10 people) passenger planes fly between the USVI and other Caribbean islands. US

Figure 6.4 A 2000 orthophotograph of Crown Bay in St. Thomas, USVI. (Credit: Greg Guannel)

citizens enjoy the luxury of not needing a passport or visa to travel to 'America's Paradise', as the islands are marketed.

When tourists arrive by cruise ship or yacht, the deficiencies in local infrastructure may be perceived as quaint and picturesque: it is part of the experience of visiting somewhere they view as 'exotic' or 'foreign'. Once they are off the ship, tourists are immediately greeted by a gaggle of drivers ready to take them anywhere they wish. Private float planes (seaplanes) or expensive taxis make travel far easier for those who can afford it. This plethora of mobility options for tourism is often in direct conflict with the mobility of locals. For example, when the USVI government prioritises tourism, specifically the cruise ship industry, local fishermen suffer reduced opportunities to access the sea or land their catches. Consequently, the sea from which they fish becomes polluted by large cruise ship waste discharges and marine diesel, further harming the environment and economic livelihoods of traditional and local fishers (Diez *et al.*, 2019). The building of sea walls to protect hotels and beach properties from sea level rise can also lead to increased beach erosion and greater wave action harming unprotected coastlines and the territory's blue economy (Gray, 2023). Here, as tourist mobility increases, fishermen's mobility decreases, affecting the livelihoods of the fishermen and the territory's larger goals of food security, making the entire community more vulnerable to climate disruptions.

The large influx of tourist arrivals also affects local roadways and intra-island travel. It is noteworthy that driving takes place on the left side of the road, but the vehicles are mainly imported from the US market, with steering wheels on the left, which reduces visibility and safety around the steep, narrow curves of hilly parts of the islands. When improvements are made to roadways, especially to access tourist resorts, paving processes lead to an increase in impervious surfaces that, in turn, increase excessive runoff and flash flooding. Indeed, it is important to note that the territory's existing infrastructure is not built to withstand this large increase, even if temporary, of people, whether in terms of water provision and management, energy generation and distribution or waste and sewage management. And, when climate impacts disrupt this infrastructure, it is often locals who feel the effects the most. We explore this further through differing experiences of speed, rhythm and friction.

Speed

Drawing on the work of Paul Virilio, Ivan Illich and others, Cresswell argues that:

> [speed is] at the center of hierarchies of mobility. Being able to get somewhere quickly has traditionally been associated with exclusivity... Speed and slowness are often logically and operationally related... And it is not always high velocities that are the valued ones. Consider the slow food and slow culture movements. (forthcoming: 19)

The Caribbean region has often been valued by tourists as a place for slowing down, marketed via a tropicalised aesthetics of island getaways as an escape from the fast pace of modern life (Sheller, 2014; Thompson, 2006). At the same time, however, islands like those of the USVI were attached to the global logistical networks first of plantation economies, then naval and air ports supporting US military mobility, which later would serve tourists arriving by air and sea. These dominant mobility regimes contribute to the desire for and building of highly automobile-dependent and fossil-fuel infrastructure associated with American modernity (Sheller & Urry, 2006; Sheller, 2018a).

The airports and seaports of the USVI were expanded for US military purposes. In 1941, a Marine Corps Air Facility at Bourne Field in St. Thomas was redesigned as an air station and made part of the Navy Operating Base, St. Thomas, in 1943. This airfield was later handed over to the Department of Interior and eventually became the Cyril E. King Airport used today, under the management of the Virgin Islands Corporation, which also owned sugar plantations and refining facilities, electricity generation and distribution, a saltwater desalination plant and managed the Naval Submarine Base (VI Port Authority). In the mid-1950s, the U.S. Navy also rebuilt the historical dockside quays of Charlotte Amalie, expanding the quayside to serve deepwater cruise ship docks and creating space for a highway.

Air travel brought a new age of tourism routes to the USVI. Airports on St Thomas and St. Croix began as military airstrips before transitioning to civilian airports. Civilian passengers first began visiting St Thomas in 1928 and St Croix in 1950, and both airports have undergone numerous expansions of terminals and runways – most notably, the extension of the runway at the Cyril E. King Airport (CEKA) that was completed in 1992 (Virgin Islands Port Authority, n.d.a, n.d.b). Tourism arrivals increased with the arrival of budget airlines such as JetBlue and Cape Air. The upper end of elite tourism utilises private jet travel and small seaplanes to get to exclusive destinations on St John and other outer islands of the archipelago. Private jet travel is one of the highest GHG emitting modes of transportation, with a small number of elite travellers contributing especially high carbon footprints. Some of the more expensive hotels in the territory (The Westin and Lovango Cay Beach Resort) have chartered private buses and boats to carry passengers from the airport to the hotel and back, bypassing the public ferry system and, once again, privileging the needs of tourists and their desire for convenience and fast service.

Roadbuilding utilised historically constructed routes around the islands, which include extremely steep, narrow, winding switchbacks crossing mountainous terrain in parts of St. Thomas and St. John. The arrival of automobiles and trucks put increasing demands on these small roadways. Island automobilities rely on the importation of vehicles and the ability to maintain them. Cars are expensive and in short supply, and gasoline prices are relatively high. The parts and mechanics are not available to maintain newer car types, so many older vehicles are patched up and kept on the road. The roads themselves are in very poor condition, due in part to flooding impacts and limited local money to maintain them, and this leads to further damage to vehicles and slower driving speed for locals. Public transit is limited to a small network of public buses that cover a small portion of the islands, and a heavily subsidised private system known as Safari buses on St. Thomas and St. John, or Gipsy buses on St. Croix.

Tourists may want to arrive quickly at their destination, but then slow down and 'unwind' once they get there. They may want quick access to a port, and then slow cruising on a yacht. They are also willing to pay a high premium for instantaneous access to taxis, rental cars and other modes of transport. The main point is that tourists expect sovereign control over their mobilities and their speed, which not only places demands on tourism sector workers such as drivers to adjust their own mobilities to be in the right place at the right time, but also can make transportation access slower, more unaffordable and inaccessible for local populations. Unequal infrastructure intersects with and disrupts local expectations about availability, speed and modes of transport. When a cruise ship arrives in port, for example, a proportion of the taxis and Safari buses are diverted to serve the tourists, reducing service for local public transit routes. When climate emergencies occur, it is tourists who have immediate access to the means of evacuation.

Rhythm

One of the ways that the uneven routes and speeds of tourism are experienced locally is through the rhythms that they produce, and how climate disasters impact these rhythms. As Cresswell argues:

> How speed (high or low) is experienced is related to position in social hierarchies and fractured by variables such as class, gender, race, age, ability, and a host of other social markers. Speed (alongside routing, rhythm, and friction) is a key tool in the production of social difference and, as such, is unevenly and asymmetrically distributed. (forthcoming: 116)

In the face of climate extremes, the differential rhythms and frictions in unequal mobilities become particularly acute determinants of climate impacts and their negative outcomes. Weather-related cancellations of air travel, for example, frequently disrupt air travel, and missed flights are a frequent occurrence in the USVI. Thus, there is a perception of unreliability that may shape travel patterns and expectations, and these will be exacerbated by climate change.

Local travellers in the USVI often must bear long waiting times for many mobilities, including waiting for buses, waiting for taxis, waiting for car parts and repairs when needed and waiting in traffic as tourists from cruise ships explore the historic downtowns of the islands. When more than three cruise ships arrive in the territory, which is approximately, on average, once a week, thousands of passengers descend onto the islands, bringing traffic to a crawl. With some 2 million visitors per year visiting three islands of 130 square miles (with probably 75% of these visiting 32 square miles of St. Thomas), traffic and the eventual stopping of normal business activities to accommodate tourist needs is inevitable. The recent Uber Soca Cruise which shut down roads on St. Croix is proof of this (MacAvoy, 2024). The mobility infrastructure is not designed for these numbers and the hilly topography limits the speed and safety of vehicular travel.

A recent post on Facebook by a resident about how tourism impacts their daily commute is a revealing example of how a tourist's mobility can derail the mobility of locals. Like normal, Sarah boarded a Safari bus – similar to a taxi – to begin her day. But, after only a few minutes, she was told to deboard the Safari so that the driver could pick up tourists. On full display is a clear example of the prioritisation of tourists over locals. The tourist's prioritised mobility negatively impacts the mobility of locals in the US Virgin Islands. Locals will often complain about Safaris impeding traffic by stopping in the middle of the road to conduct road tours and show tourists historical landmarks. Of course, it would be unfair to not recognise the economic conditions that push some drivers to prioritise tourists over locals. Drivers often charge tourists more for travel, and drivers see transporting tourists around as more profitable, which can only serve them in this inflation-laden era.

With such conflicting demands for routes and speed, there may be a severe disruption of everyday rhythms. 'Rhythm is an important component of mobility at many different scales', argues Cresswell.

> Rhythms are composed of repeated moments of movement and rest, or, alternatively, simply repeated movements with a particular measure… Rhythm, then, is part of any social order or historical period. Senses of movement include these historical senses of rhythm within them… rhythm is implicated in the production and contestation of social order. (Cresswell, forthcoming: 20)

The local rhythms of everyday life (getting to school, to work, to services and leisure) must work around the dominant rhythms of tourism including the schedules of hotels, the delivery and preparation of food, the leisure activities expected by tourists and the overall seasonal pattern of tourist arrivals and departures.

Rhythm analysis, first developed by Henri Lefebvre, offers an interesting way to understand differential and uneven mobilities in tourism-dependent islands. Food security is a pressing concern of many small islands across the Caribbean region, which have experienced a high risk of disruption to food supply chains due to hurricanes, such as in 2017, dockworker's strikes, like that in 2024, or simply lack of sufficient cold storage. In the USVI, 97% of food is imported and refrigerated cold storage is very expensive (Guo, 2017). Small farmers are trying to meet local needs, for example Ridge to Reef, an organic farm in St. Croix, and We Grow Food Inc., which runs a farmer's market in St. Thomas, but can only meet a small percentage of overall demand. Yet tourists expect fresh food and high variety, so insofar as fresh food is being farmed locally, it may only be accessible at a very high cost, e.g. through hydroponic growing with high energy inputs and expensive temperature and humidity control, as well as the transport of said produce to restaurants. In the case of power outages and a weakened power grid, which are becoming more frequent under climate change scenarios, access to food will be severely constricted for many islanders.

Friction

Ultimately, these combinations of routes, speed and rhythms produce uneven (im)mobilities that exacerbate climate vulnerabilities through various kinds of friction. 'Friction is variably distributed in space and is an important component of a politics of mobility' (Cresswell, forthcoming: 22). We argue that an analysis of these frictions is a crucial part of developing a climate mobility justice approach to tourism because it is through such frictions that people and places become more or less vulnerable or resilient to a changing climate. Sometimes friction can be beneficial – for example, if water flows too quickly downhill over impervious surfaces, it causes localised flooding. We may want culverts and surfaces that slow down water flows and allow them to descend in a slower and more controlled way. We can apply this metaphorically to social infrastructures. Sometimes, a little friction is necessary to capture economic benefits. Virgin Islanders may want tourists to visit, and to slow down and spend some money locally, rather than simply speed through on a quick cruise ship stopover where tourists spend little. In other cases, those labouring in Caribbean tourism economies may find ways of 'resisting paradise' through the withdrawal of their labour (Nixon, 2017) or 'returning the tourist gaze' (Sheller, 2012) in ways that introduce a kind of protective friction against exploitation.

Tourism is also described by Mathilde Córdoba Azcarate (2020) as being 'sticky': localities and governments in tourism-reliant economies may be 'stuck with tourism' because it is such an important economic driver. Yet, when they become too over-reliant on tourism, this may prevent other kinds of economic investments from being made or other longer-term plans from being implemented. Diversified economies are healthier in the face of climate disasters because they are not too dependent on just one sector. When disasters strike, there may be uneven rhythms of recovery in some sectors over others. Relational 'pandemic (im)mobilities' reflect these frictions, seen in unusual patterns of mobility and immobility in relation to tourism (Adey *et al.*, 2021b). While many independent Caribbean island states experienced a collapse of tourism during COVID-19 travel restrictions, as a US territory, the USVI experienced a boost in both air and maritime travel. There was especially a rise in yacht visitations to the territory because other Caribbean nations had closed their borders.[3] Because the USVI remained open, wealthy Americans flocked here to escape lockdowns and isolation. The arrival of tourists, however, produced numerous points of friction locally, i.e. when visitors chose not to wear face masks, putting local service providers and the wider population at risk. When local COVID-19 infection rates rose, tourist behaviour was blamed, and when beaches closed, it deprived residents of one of the only healthy and safe outdoor activities available.

Another key concept that emerged in our discussions of the contemporary condition of the US Virgin Islands was the idea of being 'stalled', i.e. the sense of no progress toward a desirable future. After the 2017 hurricanes, there has been very slow progress in recovery projects, with some homes covered in blue emergency tarps years later. Stalled mobilities may be a pertinent metaphor for climate futures in the US Virgin Islands as a whole. There is a sense of stalled politics and political action, stalled economic prospects, stalled education, stalled voter participation and stalled healthcare, all of which are leading to the phenomena known as 'brain drain', which is when the prime working-age and educated population moves away from the islands to the mainland or elsewhere for better opportunities. This stalled future leaves behind an aging population, who are even more vulnerable to climate change impacts such as extreme heat, sudden flooding and electrical power outages. Stalled mobilities are a way of understanding climate vulnerability in a tourism-dependent economy facing both climate injustices and mobility injustices.

Conclusion

Crucial to all tourism performances is the notion that they can also be contested and redirected. As Tim Edensor first described almost 25 years ago, 'By looking at the contexts in which tourism is regulated, directed and choreographed or, alternatively, is a realm of improvisation and contestation', we can 'consider the constraints and opportunities which shape the ways in which tourist space (here considered as "stages") and performance are reproduced, challenged, transformed and bypassed' (Edensor, 2001: 59). In the face of climate disasters and a growing sense of climate disruption, are there opportunities for tourism-dependent economies to improvise and transform the staging of tourism?

Stalled mobilities suggest the need for workarounds, repair or detours, both physically and conceptually. If roads are wiped out by flooding or landslides, then detours need to be found. If ports are affected by hurricanes, then alternative local provisioning systems need to be mobilised. The impacts of climate disasters as well as the COVID-19 pandemic may have inspired some tourism-dependent economies to open public debate about how they might change toward more sustainable economic models that promote social justice, environmental justice and climate justice (Sheller, 2018b, 2018c, 2020a, 2020b). To be ready for such climate impacts, islands need to begin planning for greater mobility justice, i.e. by ensuring diverse forms of inclusion, accessibility and means of moving that include all classes and sectors of local populations, not just tourists.

On the other hand, in the face of climate disruption, climate migration and the powerful patterns of tourism mobilities, it may be extremely difficult to challenge and bypass the negative impacts of stalled mobilities and ongoing climate mobility injustices. Tourism is not simply an economic investment choice that can be easily replaced with some other development plan, but is deeply ingrained in patterns and infrastructures that have arisen since the colonial era and, as we have argued, are tightly coupled with contemporary routes, speeds, rhythms and frictions.

The era of colonisation and slavery, the building of military infrastructure, the 1960s industrial developments of Harvey Aluminum, Hess Oil and Hovensa and the recent post-hurricane opportunities for tourism: all brought people into the US Virgin Islands. Any economic crash, any hurricane, limited development plans and a general lack of economic opportunities also send (some) people out. New hotels on St Croix, events like the Yacht Boat Show and lines of private jets arriving, stand in contrast to the exodus of youth and talented workforce from the islands. The recent Census shows that the lowest-paid class is migrants from Haiti and the Dominican Republic. The highest paid are Whites. Insofar as inward migration is selective for the wealthy, they have tended to be white; and outward migration has tended to be Black. Different forces shape the routes, speeds, rhythms and frictions that move people in and out, but ultimately, these patterns maintain the same racialised social order and economic hierarchy, which will likely only be intensified by the impacts and vulnerabilities of climate change.

By spelling out more clearly the intersections of climate injustice and mobility injustice within tourism economies in the USVI, we hope to have offered new ways to analyse the current conjuncture. Climate change is going to make life in the USVI more difficult and more costly because it requires new investments by people in their structural safety (strong house, insurance cost), physical safety (cooling, drinking, eating, preparedness) and dealing with shocks (hurricanes, droughts, flash floods, heat waves, diseases, etc.). If US Virgin Islanders are, of necessity, finding ways to withstand the pressures of climate change, the tourist economy's reliance on and deepening of mobility injustices may be the final straw that breaks the ability of local citizens to hold on to a home and remain in these islands.

Acknowledgments

This work was partially supported through the Caribbean Collaborative Action Network (CCAN), a NOAA Climate Adaptation Partnerships Team Grant. Grant Number: NA22OAR4310545. It was also partially supported by NOAA-BIL Project Title: Improving Engagement Methods for Coastal Resilience and Reducing Climate Risk: Bridging Learning Networks From the Urban Northeast (CCRUN) to the US Caribbean (CCAN), (PI) Sheller, Mimi (Worcester Polytechnic Institute), Grant Award Number NA23OAR4310396.

Notes

(1) Of the three main islands, Denmark colonised St. Thomas first in 1672. St. Jan, now known as St. John, was conquered in 1717 and St. Croix was purchased from the French in 1733. An economy based on slavery lasted until the abolition of slavery in 1848. In 1917, the United States of America purchased the DWI for $25 million in gold. According to historical records, locals had mixed feelings about the transfer. The first decade and a half (1917–1931) of US rule was governed by the U.S. Naval Regime (Krigger, 2018).
(2) The island of St. Croix was most recently named 'Caribbean Culinary Destination of the Year' in the 2025 Caribbean Travel Awards. Source: https://viconsortium.com/vi-tourism/virgin-islands-usvi-shines-in-2025-travel-awards--st--croix-named-culinary-destination-of-the-year.
(3) The US Virgin Islands was recently named 'Caribbean Yachting Destination of the Year' in the 2025 Caribbean travel Awards. Perhaps a trickle-down effect of the COVID-19 pandemic. Source: https://viconsortium.com/vi-tourism/virgin-islands-usvi-shines-in-2025-travel-awards--st--croix-named-culinary-destination-of-the-year.

References

Adey, P., Cresswell, T., Lee, J.Y., Nikolaeva, A. and Novoa, A. (2021a) *Moving Towards Transition: Commoning Mobility for a Low-carbon Future*. Zed Books.
Adey, P., Hannam, K., Sheller, M and Tyfield, D. (Eds.) (2021b) Pandemic (Im)mobilities. Special Issue of *Mobilities* 16 (1), 1–19. https://doi.org/10.1080/17450101.2021.1872871.
Agyeman, J. (2013) *Introducing Just Sustainabilities*. Zed Books.
Annual Report of the Governor of the Virgin Islands (1934) University of Florida George A. Smathers Libraries. See https://original-ufdc.uflib.ufl.edu/UF00015459/00009/5j (accessed October 2025).
Associated Press (2012) Major oil refinery to close in US Virgin Islands, *Deseret News*, 18 January 2012. See https://www.deseret.com/2012/1/18/20245330/major-oil-refinery-to-close-in-us-virgin-islands/ (accessed October 2025).
Baptiste, A. and Rhiney, K. (eds) (2016) Special issue on climate justice & the Caribbean. *Geoforum* 73, 17–21. https://doi.org/10.1016/j.geoforum.2016.04.008.

Becken, S. and Hay, J. (2012) *Climate Change and Tourism: From Policy to Practice* (1st edn). Routledge.

Bigby, B.C., Smith, J. and Higgins-Desbiolles, F. (2024) Climate justice in tourism: An introductory guide. Travel Foundation. See https://www.thetravelfoundation.org.uk/wp-content/uploads/2024/07/Climate_Justice_Tourism_v5.pdf (accessed October 2025).

Bloch, N. and Adams, K. (2023) *Intersections of Tourism, Migration, and Exile*. Routledge.

Boas, I., Wiegel, H., Farbotko, C., Warner, J. and Sheller, M. (2022) Climate mobilities: Migration, im/mobilities and mobility regimes in a changing climate. *Journal of Ethnic and Migration Studies* 48 (14), 3365–3379. https://doi.org/10.1080/1369183X.2022.2066264.

Bond, D. (2021) Six part series on St Croix, *St. Croix Source* and the *St. Thomas Source*. See https://www.bennington.edu/news-and-features/climate-justice-begins-places-saint-croix (accessed October 2025).

Butler, R. (2006) *The Tourism Area Life Cycle, Vol. 1: Applications and Modifications*. Multilingual Matters.

Cave, J. and Dredge, D. (2020) Regenerative tourism needs diverse economic practices. *Tourism Geographies* 22 (3), 503–513. https://doi.org/10.1080/14616688.2020.1768434.

Córdoba Azcarate, M. (2020) *Stuck with Tourism: Development, Space and Power in Contemporary Yucatan*. University of California Press.

Cresswell, T. (2006) *On the Move: Mobility in the Modern Western World*. Routledge.

Cresswell, T. (2010) Towards a politics of mobility. *Environment and Planning D: Society and Space* 28 (1), 17–31. https://doi.org/10.1068/d11407.

Cresswell, T. (forthcoming) *The Citizen and the Vagabond: Towards a Politics of Mobility*. University of Minnesota Press.

Diez, S.M., Patil, P.G., Morton, J., Rodriguez, D.J., Vanzella, A., Robin, D.V., Maes, T. and Corbin, C. (2019) Marine pollution in the Caribbean: Not a minute to waste. World Bank Group. https://documents1.worldbank.org/curated/en/482391554225185720/pdf/Marine-Pollution-in-the-Caribbean-Not-a-Minute-to-Waste.pdf.

Dogru, T., Marchio, E.A., Bulut, U. and Suess, C. (2019) Climate change: Vulnerability and resilience of tourism and the entire economy. *Tourism Management* 72, 292–305. https://doi.org/10.1016/j.tourman.2018.12.010.

Edensor, T. (2001) Performing tourism, staging tourism: (Re)producing tourist space and practice. *Tourist Studies* 1 (1), 59–81. https://doi.org/10.1177/146879760100100104.

Ferguson, J. (1997) *Eastern Caribbean in Focus: A Guide to the People, Politics and Culture: Economy: After the Plantation*. LAB. https://ucl.primo.exlibrisgroup.com/permalink/44UCL_INST/18kagqf/cdi_askewsholts_vlebooks_9781909013629.

Gilbert, E. (2025) U.S. Virgin Islands shatters tourism records in 2024. The *Virgin Islands Consortium*. See https://viconsortium.com/vi-tourism/u-s--virgin-islands-shatters-tourism-records-in-2024 (accessed October 2025).

Global Agents USA. (2025) U.S. Virgin Islands breaks tourism records in 2024. *Global Agents USA*. See https://globalagents.us/news/us-virgin-islands-breaks-tourism-records-in-2024 (accessed October 2025)

Gössling, S. and Hall, C.M. (eds) (2006) *Tourism and Global Environmental Change: Ecological, Economic, Social and Political Interrelationships*. Routledge.

Gray, S. (2023) *In the Shadow of the Seawall: Coastal Injustice and the Dilemma of Placekeeping*. University of California Press.

Guannel, G., Lohman, H. and Dwyer, J. (2022) Social and flooding vulnerability in the U.S. Virgin Islands. *Natural Hazards Center*, *Public Health Report Series* 6. https://hazards.colorado.edu/public-health-disaster-research/the-public-health-implications-of-social-vulnerability-in-the-u-s-virgin-islands.

Guo, E. (2017) Feeding the Virgins: The US Virgin Islands are vulnerable to food shortages, but two innovative agricultural projects are getting things growing. *Hakai Magazine*, 17 July 2017. See https://hakaimagazine.com/news/feeding-virgins/ (accessed October 2025).

Hall, C.M. and Williams, A.M. (eds) (2002) *Tourism and Migration. New Relationships Between Production and Consumption*. Kluwer Academic Publishers.

Hazard Mitigation and Resilience Plan (2024) See https://resilientvi.org/hmrp-plan (accessed October 2025).

Krigger, M. (2018) *Race Relations in the US Virgin Islands: A Centennial Retrospective*. Carolina Academic Press.

MacAvoy, K. (2024) Tourism, Police Dept. preview UberSoca J'ouvert and after-party. *The Virgin Islands Daily News*. See https://www.virginislandsdailynews.com/news/tourism-police-dept-preview-ubersoca-j-ouvert-and-after-party/article_fa3f028a-9008-11ef-b97d-f79c32e60094.html (accessed October 2025).

Monzote, R.F. (2013) The Greater Caribbean: From plantations to tourism. *Rachel Carson Center for Environment and Society* 7, 17–24. doi.org/10.5282/rcc/6260.

Nixon, A.V. (2017) *Resisting Paradise: Tourism, Diaspora, and Sexuality in Caribbean Culture*. Reprint edition. University Press of Mississippi.

Nixon, A.V. (2019) When the apocalypse is now: Climate crisis, small island disasters and migration in the aftermath of hurricane Dorian, *Stabroek News*, 9 September 2019.

Nugent, W. (2022) The History of Brain Drain in the U.S. Virgin Islands, *The Virgin Islands Consortium*, 11 September 2022.

Olwig, K.F. (1985) *Cultural Adaptation and Resistance on St. John. Three Centuries of Afro-American Life*. University Presses of Florida.

Pattullo, P. (1996) *Last Resorts: The Cost of Tourism in the Caribbean*. Monthly Review Press.

Rohland, E., García Acosta, V., McDermott, A.G. and Taks, J. (eds) (2025) *The Anthropocene as a Multiple Crisis. Perspectives from Latin America*. Maria Sibylla Merian Center for Advanced Latin American Studies, CALAS, Bielefeld University Press.

Scott, D., Hall, C.M. and Gössling, S. (2012) *Tourism and Climate Change: Impacts, Adaptation and Mitigation*. First edn. Routledge.

Scott, D., Hall, C.M. and Gössling, S. (2019) Global tourism vulnerability to climate change. *Annals of Tourism Research* 77, 49–61. https://doi.org/10.1016/j.annals.2019.05.007.

Sheller, M. (2003) *Consuming the Caribbean: From Arawaks to Zombies*. Routledge.

Sheller, M. (2012) *Citizenship from Below: Erotic Agency and Caribbean Freedom*. Duke University Press.

Sheller, M. (2014) *Aluminum Dreams: The Making of Light Modernity*. MIT Press.

Sheller, M. (2018a) *Mobility Justice: The Politics of Movement in an Age of Extremes*. Verso.

Sheller, M. (2018b) Caribbean futures in the offshore anthropocene: Debt, disaster, and duration. *Environment and Planning D: Society and Space* 36 (6), 971–986. https://doi.org/10.1177/0263775818800849.

Sheller, M. (2018c) Caribbean reconstruction and climate justice: Transnational insurgent intellectual networks and post-hurricane transformation. *Journal of Extreme Events* 5 (4), 1–18. https://doi.org/10.1142/S2345737618400018.

Sheller, M. (2020a) *Island Futures: Caribbean Survival in the Anthropocene*. Duke University Press.

Sheller, M. (2020b) Reconstructing tourism in the Caribbean: Connecting pandemic recovery, climate resilience, and sustainable tourism through mobility justice. *Journal of Sustainable Tourism* 29 (9), 1436–1449. https://doi.org/10.1080/09669582.2020.1791141.

Sheller, M. (2025) Caribbean Islands and the coloniality of climate change: Navigating 'the Anthropocene' through the historical legacies of the Plantation. In P. Noxolo, K. Rhiney and R. Cummings (eds) *The Routledge Handbook of Caribbean Studies* (pp. 35–46). Routledge.

Sheller, M. and Urry, J. (eds) (2004) *Tourism Mobilities: Places to Play, Places in Play*. Routledge.

Sultana, F. (ed.) (2025) *Confronting Climate Coloniality: Decolonizing Pathways for Climate Justice*. Routledge.

Taylor, F.F. (1993) *To Hell With Paradise: A History Of The Jamaican Tourist Industry*. University of Pittsburgh Press.

Thompson, K. (2006) *An Eye for the Tropics: Tourism, Photography, and Framing the Caribbean Picturesque*. Duke University Press.

Toppin-Allahar, C. (2015) 'De beach belong to we!' Socio-economic disparity and Islanders' rights of access to the coast in a tourist paradise. *Oñati International Institute for the Sociology of Law* 5 (1), 298–317. https://opo.iisj.net/index.php/osls/article/view/440.

U.S. Census Bureau (2020) 2020 Island Areas Census: US Virgin Islands. See https://www.census.gov/data/tables/2020/dec/2020-us-virgin-islands.html (accessed October 2025).

U.S. Government Accountability Office (GAO) (2023, June 29) *U.S. Territories: Public Debt Outlook - 2023 Update*. See https://www.gao.gov/products/gao-23-106045 (accessed October 2025).

USVI Bureau of Economic Research (2024) 2023 U.S. Virgin Islands Economy. See https://my.visme.co/view/017w4876-economic-review-2023-usvi-usviber#s1 (accessed October 2025).

USVI Economic Development Authority (n.d.) Services: Tax incentives. See https://usvieda.org/services/ (accessed October 2025).

Virgin Islands Port Authority (n.d.a) Cyril E. King Airport (STT). See https://www.viport.com/cekastt#:~:text=Construction%20of%20the%20current%20airport,dredged%20runways%20in%20the%20Caribbean (accessed October 2025).

Virgin Islands Port Authority (n.d.b) A Brief History of the Virgin Islands Port Authority. See https://www.viport.com/our-history (accessed October 2025).

Weaver, D. (1998) The evolution of a 'plantation tourism' landscape on the Caribbean island of Antigua. *Journal of Economic and Human Geography* 79 (5), 319–331. https://doi.org/10.1111/j.1467-9663.1988.tb01318.x.

7 Carbon Offsetting and Rights of Tourism Hosting Communities in Kenya's Conservancies

Judy Kepher Gona and Lucy Atieno

Wildlife makes significant contributions to Kenya's tourism, which is evidenced in the tourism sector's performance before the COVID-19 pandemic. The development of wildlife tourism is not only concentrated in wildlife protected areas, but also extends to adjacent lands owned by local communities. Our chapter investigates engagement between tourism hosting communities in Kenya's community conserved areas (CCAs) and carbon project developers, to better understand key action situations that play a critical role in community rights. The study combines interviews, field observations, a survey and news media analysis in order to capture interactions between these groups in CCAs where carbon offset projects have been signed, and are either already established or in the process of being initiated. The results are analysed based on Elinor Ostrom's framework for socioecological governance. Analysis centres on action situations emerging from interactions between actor groups, revealing that they structure an engagement approach to foster dominance of external actors, paired with dependency of the local community. We show interaction outcomes where local community dependencies on shared resources are protected, yet their rights to land resources are weakened. In particular, management action that protects landowners' reliance on conservancy grazing zones undermines women's land use rights, herders' access and the younger generation's land ownership rights. With the wrong engagement approach, carbon offsetting, like conservation tourism, can be a tool for community dominance which is characterised by tokenism and leads to more dependence and ceding of rights over resources.

Introduction

Wildlife makes significant contributions to Kenya's tourism, which is evidenced in the performance of the tourism sector before the COVID-19 pandemic. For example, the Kenya Wildlife Service (2012) strategic plan for 2012–2017 reports that wildlife accounts for 90% of safari tourism and 75% of total tourism earnings in the country. Safari tourism continues to be supported by international travel markets, e.g. Europe outbound travel, shown by the Tourism Research Institute (2025) to have holidaying/leisure as prominent purposes of travel, accounting for as much as 71% of arrivals for this segment. A significant proportion of wildlife tourism is organised in conservation areas, which comprise protected areas as well as adjacent lands with community settlements. An important development in conservation areas is the set up of conservancies, where vast areas of community owned lands are also a range (or an extended habitat) for wildlife. A wildlife range is an expanse of area within a habitat, allowing for species movement for their activities, and can be inside a protected area or outside in adjacent lands. In Kenya, community owned lands adjacent to protected areas are wildlife ranges; thus, they are referred to as community conservancies, or community conserved areas or conservancies, when landowners from the community organise as conservancy members, often in an arrangement involving partnerships with external actors.

A study by Bedelian *et al.* (2024) shows that, in Kenya, participation in a conservancy is spatially determined by how close the land owned is to a protected area that has tourism potential. This criterion leaves out a small proportion of landowners from being conservancy members. Land size is also a determining factor for participation and households are not permitted to live on land which they join to a conservancy (Bedelian *et al.*, 2024). Since they have to be outside of the conservancy boundary, some households may need to be resettled temporarily on host parcels of land (Gona & Atieno, 2021).

In the southern part of Kenya, conservancies' combined land area remains smaller compared to the vast adjacent wildlife protected area. Pastoralism is the primary livelihood option for communities in these areas. These pastoralist communities, defined by their nomadic practices of seasonal movement in search of pasture for their livestock, are mainly found in Arid and Semi-Arid Land (ASAL) areas and rangelands of the Rift Valley. Over time, as a result of political and historical dynamics, pastoralist communities in Kenya have settled in ASAL areas around Kajiado, Narok, Laikipia, Samburu, Turkana, Isiolo, Marsabit, Garissa and the Tana River. Some of these ASAL areas form part of Kenya's conserved areas. Pastoralists rely on livestock rearing within the conservancies for sustenance and are therefore highly dependent on the availability of grazing pasture and its sustainable management. Since these communities are involved in grazing practices across multiple generations, their local knowledge has safeguarded grazing ranges.

Besides the use of community lands for grazing pasture, these resources are attracting the interest of multiple other user groups, including tourism investors and carbon offset markets. Carbon offset markets are trading spaces, where buyers pay for carbon credits generated in conservation areas as part of their contribution to solving climate emergencies. This can be attributed to growing recognition of conservation spaces as natural solutions to climate change (Bashir & Wanyonyi, 2024). With increasing emissions from aviation, airlines are among the buyers looking to offset their emissions. Besides these buyers, there are carbon project developers responsible for the setting up of carbon cultivation in viable lands.

Tourism investors had an entry point to community owned land through conservation tourism. Over the years, conservancies have developed to be a centre piece for tourism in Kenya's conservation areas. This has helped local communities living within conservancy areas to forge a strong identity tied to developments in tourism. Currently, there are 205 conservancies registered under the Kenya Wildlife Service (Kenya Wildlife Service, 2024). Their formation is a result of negotiations initiated by actors external to the local landowning community in the conservation area. While these negotiations often aim to establish a common ground framed around environmental protection goals with benefits of sustaining existing livelihoods, they raise justice-related challenges. For instance, the involvement of external actors can lead to concerns about inclusivity, whether all members of the local community have an equal voice in decision making, or whether power imbalances favour external stakeholders over local interests. These factors can impact the fairness of shared resource use and the long-term sustainability of livelihoods addressed in negotiations. Against this background, our chapter explores the rights of local community to these lands, in pursuit of climate mitigation goals through tourism and conservation. Specifically, it addresses the following objectives:

(1) To identify key action situations in conservancy resource system interactions between carbon project developers and landowners hosting tourism developments.
(2) To identify justice concerns that emerge across action situations in conservancy resource system interactions between carbon developers and landowners hosting tourism developments.

The chapter relies on Elinor Ostrom's 'Socio-Ecological Systems' framework to explore interactions among user groups in shared conservancy land resource governance. The framework (see Figure 7.1) provides a structured way to analyse resource governance in the context of resource user interactions within broader socioeconomic and ecological context.

Justice Concerns in Conservation Initiatives

Studies (Cavanagh & Benjaminsen, 2014; Lehmann, 2019; Satyal *et al.*, 2020) show justice-related challenges emerging in conservation initiatives, whose engagement with local communities is on the resources supporting their livelihoods. Cavanagh and Benjaminsen (2014) find that, in projects for carbon sequestration, local communities experience myriad injustices while the project activities are presented as win–win, profitable conservation.

Injustices are also evident where impacts of conservation activities on local community are not uniform (Bedelian *et al.*, 2024), an issue that raises concerns on equity. For carbon offset projects, injustices may go unnoticed when developers on location maintain conceptual and geographic separation between offset consumers and actual sites for carbon sequestration as a way to shield their activity from scrutiny (Cavanagh & Benjaminsen, 2014). There is thus a need to address such challenges through environmental justice approaches, as conservation and carbon offset issues relate to the environment.

Environmental justice means that everyone should be equally protected from environmental harm and have access to a clean and healthy environment (Thom & Mah, 2020). Environmental justice is traditionally defined along dimensions of distributive justice, procedural justice, recognition justice and transformational justice (Thom & Mah, 2020). Here, we reference the definitions provided by Thom and Mah (2020) for these different forms of justice. Distributive justice in conservancies relates to reach of environmental benefits to stakeholders and reducing environmental burdens for all. Procedural justice is concerned with equal opportunities for participation in conservancy decision making. Recognition justice describes respect for local culture. Transformational justice entails the effect of change on conservancy ecosystems in a way that fosters more just power relations. In this chapter, we focus on justice through the lens of upholding local community rights to shared environmental resources. In particular, their existing rights to own, to use, to access and to draw income from resources of interest, prior to external actor engagement with these resources. The upholding of these rights, when exposed to situations of different actor interactions that entail power asymmetries and diverging interests, is important for every dimension of justice.

The entry of carbon offset programs in Kenya's conservancies builds on existing engagement with the local landowning community and other external actors. Carbon project developers propose 'triple-win solutions for climate change mitigation, biodiversity preservation, and local socioeconomic development', which Cavanagh and Benjaminsen

(2014: 55) note to be a usual strategy employed by conservationists to validate the need for their activity. These triple win points portray carbon development projects as morally and socially acceptable to green investors (Cavanagh & Benjaminsen, 2014), a view that is also projected to the communities they engage with through conservancies.

Conservancies in Kenya are either community, private or group organised (Damania *et al.*, 2019). Tourism partners, conservation nongovernmental organisations (NGOs), land-owning communities and private individuals are key players jointly conceiving the formation of conservancies in Kenya. Irrespective of the parties who negotiate or conceive the formation of a conservancy, tourism earnings are of high strategic importance to each. A 2018 conservancy survey confirms that tourism, through leasehold, bed night and conservation fees, accounts for more than 60% of conservancy income (Damania *et al.*, 2019). The bed night fee is a proportion paid per occupied bed in the conservancy; and a conservation fee is paid per visitor as an additional payment for conservation purposes. Besides earnings, conservation measures in tourism have wielded authority to control use of grazing land by the pastoral community, to enable pasture growth to an optimum level that supports wildlife to thrive and also to maintain the availability of pasture to local herds of livestock. These contributions to conservancy income and control wielded over conservancy resources suggest the dominance of tourism interests, an engagement point that current carbon offset projects in conservancies are building on.

Today, developers interested in tapping into soil carbon within tropical savanna ecosystems in conservancies are leveraging existing engagements with landowning communities, negotiated earlier with other actors. Soil has high carbon storage capacities, and this is optimal when not decimated by management practices (Georgiou *et al.*, 2022). Vast areas of land are needed for soil carbon projects. Trees, grass lands and soil, important for carbon storage, are resources which landowning communities have varying dependencies on, as well as rights to, including ownership, usage and access. Degradation of these environmental resources, due to unsustainable use patterns, as well as climate change impacts, makes conservation measures increasingly necessary. Thus, negotiators put forth the urgency of engagement with land-owning communities based on the need for environmental protection.

Carbon offsets do not decrease carbon emissions but instead may pay for the capture and storage of carbon dioxide from the atmosphere to prevent it from contributing to climate change (Chandler, 2024). The payments are an additional cost incurred by offset buyers to enable their activities to have a lower carbon footprint. Part of these payments benefits local communities as a compensatory measure for their resource use. Indeed, a study by Shinbrot *et al.* (2022) confirms that economic security was the greatest benefit for local people participating in a carbon offset project. This matches up with tourism in community conservancies which promises a share of its earnings to local communities. As such, both tourism and offset programs in conservancies are commoditised nature solutions. Even so, Holmes and Cavanagh (2016) show that market-based ideas in conservation are tightly coupled with authority and control.

Socio-Ecological Systems (SES) Framework

Ostrom's SES framework shows that communities can self-manage common pool resources, like land, in a sustainable way. This was the observed reality in the community of landowners in Kenya's pastoralists areas, where grazing resources, despite shared use, were not depleted and there was sustained livestock rearing across generations. Conservation initiatives in pastoralists areas extend the boundary of community beyond landowners and their households, to introduce private actors in tourism and carbon offset projects, each with varying influences on the sustainable management of shared resources in conservancies. Research shows that power asymmetries are a key factor in the strings of challenges and failures marring successes in conservation projects that feature a community engagement aspect. For instance, a study by Dawson *et al.* (2024) indicates that ecological outcomes are less favourable in situations where local communities have a comparably lower control in governance systems for conserving biodiversity. In SES, power asymmetries are evident in interactions between actors. Ostrom positions these interactions as 'Action Situations' (AS), explaining the situations as 'the social spaces where individuals interact, exchange goods and services, solve problems, dominate one another, or fight' (2011: 11).

In ecosystem level conservation initiatives, focus on soil carbon sequestration promises outcomes for climate change mitigation, with co-benefits to sustain key resources that communities are dependent on. Given the lack of a legal framework for soil protection and governance in Africa (Ruppel & Ginzky, 2021), we arrive at the important question: *How are dependencies on and rights to these resources protected?* Already studies have pointed out exploitations in carbon offsetting. For example, Cavanagh and Benjaminsen (2014) write about a failed carbon offset program at Mt. Elgon National Park in Uganda, whose development involved violent dispossession of the local community. The offset program maintained a conceptual and geographic disconnection between offset markets and actual carbon project sites, allowing exploitations in carbon offsetting initiatives (Cavanagh & Benjaminsen, 2014). These burdens for community relate to costs for climate action, which Rastegar and Becken (2024) categorise as a distributive justice concern in tourism. Whether these outcomes were unintended or not, Cavanagh and Benjaminsen (2014) view gaps in the approach to community engagement as an enabler of injustices.

Carbon project developers are the latest entrants to conservancies, proposing vegetation cover management to allow carbon sequestration.

This way, conservancies are gaining traction as resource systems with multiple users. Carbon developments in conservancies are presented as climate change mitigation efforts, with outcomes benefitting local communities. Just like tourism projects in conservation areas, financing from carbon projects is pitched as enhancing conservation and additionally contributing to the mitigation of global warming. Carbon projects promise to sustain pasture at optimal levels while also mitigating climate change, in a way that has benefits for the sustainability of local people's livelihoods in livestock rearing. Moreover, mitigation efforts in carbon projects have a long-term effect in cushioning such communities from serious risks that climate change poses to them. Considering the connection between environmental risks and justice concerns (Dias *et al*., 2021), mitigation successes of carbon projects can contribute to climate justice. In Brazil's urban Amazon, a study by Dias *et al*. (2021) has shown that flood-related disasters aggravate the violation of social and environmental rights. Thus, the interactions of carbon developers with landowners within a conservancy context with established community-tourism interactions, converges to outcomes with rights implications. However, climate change discussions in tourism research rarely address matters of justice directly (Rastegar & Becken, 2024). This chapter contributes to overturning this omission.

Methods

While the study covered all registered conservancies in Kenya, discussions here are narrowed to those where tourism is a major driver of conservation activities and carbon offset projects have been initiated. These include conservancies under the Northern Rangelands Trust in Northern Kenya, which is recognised as a model for regional associations (Bashir & Wanyonyi, 2024), and those affiliated with the Maasai Mara Wildlife Conservancies Association (MMWCA) in South West Kenya.

We used a mixed-methods approach to investigate community engagement in carbon offset projects, combining qualitative data from interviews, surveys, field observations and analysis of mainstream news media articles.

Interviews

Fieldwork was conducted in one of the conservancies, allowing for direct observations and interviews. To protect the confidentiality of participants and organisations, specific details about the locations of field data collection are not disclosed. Some interview participants felt that the information shared was sensitive and declined audio recordings. In November and December 2024, we conducted interviews with conservancy managers who were targeted as beneficiaries of carbon development projects in conservancies within Kenya. The interviews focused on perceived priorities, successes and failures of these projects and implications for these on the local community.

Semi-structured interviews were conducted with eight conservancy managers, who had firsthand knowledge of carbon projects implemented in their areas of work. While some participants were comfortable with audio recording, others declined, citing concerns over privacy given the sensitive nature of the topic. In these cases, the researcher conducting the field work took written notes during the interview to ensure that the information shared was accurately captured. Recorded interviews were transcribed and analysed thematically to highlight emerging issues.

Field observations were made by the first researcher while conducting interviews and administering surveys. During these observations, informal conversations with local people were also conducted to gain a broader understanding of the community dynamics and context surrounding the influx of initiatives to support conservation in community areas. Field notes were taken to document these interactions and provide insights that supplemented the survey and interview data.

Survey

A survey was given to local community members to gauge information exchange between them as target beneficiaries and implementing organisations for these projects. An initial survey was administered to 41 landowners from three conservancies, who were identified as beneficiaries of ongoing carbon projects in conservancies within Kenya where soil carbon projects were developed. The purpose of the survey was to gather general insights into the participants' engagement as beneficiaries in carbon projects. The survey consisted of both closed- and open-ended questions. Data from this survey were analysed to identify common insights into specific topics, as well as evaluating sentiments behind responses.

Media articles

To further contextualise the perspectives obtained from the interviews and survey, news articles were incorporated into the study as secondary data. News articles that were publicly available online from mainstream media houses were sourced, with the aim of acquiring reports on carbon project developments in conservation areas. This resulted in 26 articles being selected for their relevance to carbon projects in community conservation areas. The articles were analysed for media framing of community engagement, revealing key actors and their interactions within carbon projects, as well as emerging issues in carbon project developments over the years. Publicly available news media reports were primarily drawn from conservancies in the northern Kenya and Maasai Mara areas. These sources provided insights into conflicts and legal disputes surrounding the development of carbon projects in conservancies.

Given the sensitive nature of the topic, ethical considerations were paramount. The anonymity and confidentiality of all participants were upheld, as well as their geographic placement, and informed consent was obtained prior to participation in interviews and surveys. Participants were made aware of their right to decline audio recording, with alternative methods of data collection (e.g. written notes) provided where necessary.

Results

The findings are organised by the key concepts of the Socio-Ecological Systems governance framework developed by Elinor Ostrom (2011), with a focus on action situations to highlight justice concerns. The frameworks key components include key actors, resource system, governance system, ecosystem services and action situations (Figure 7.1).

Key actors

Within the conservancy set-up, key actors with direct resource use are landowners, tourism investors and auxiliary investors such as carbon project developers. There are several other distant actors whose links to the resource is mediated through these primary actors. News media reports revealed the important role of land courts and human rights groups in addressing and intervening in land-related conflicts in conservancies.

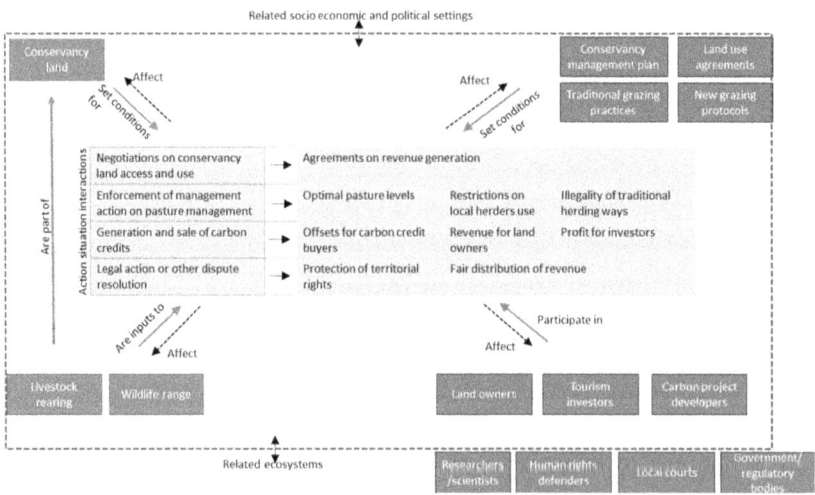

Figure 7.1 Conservancy land resource use system, centring interactions and outcomes of key action situations, based on components of Ostrom's SES framework

Conservancy land resource system

Conservancies in Kenya have been established to safeguard pastoral lifestyle and livelihoods, protect wildlife and provide reliable household income. Community conservancies are in areas where the local community often holds land-owning rights, allowing them various rights to land resources.

Prior to the setting up of conservancies, pastoralist communities, as the sole primary resource users, viewed grazing areas as ancestral lands. Traditionally, they have ownership rights to these lands. The communities living in these areas are keen on pasture management, as it provides feed for their livestock. Their grazing practices are incorporated in nomadic lifestyles, allowing parcels of grazing land to regenerate after some period of time. They also have cattle *bomas* (enclosures) (Figure 7.2) within household spaces to protect their herds. Some landowners use thorn and electric fences for these enclosures in order to keep wild animals away from their herds.

The entry of tourism and carbon development actors transforms what communities consider as ancestral lands to conservancy lands. The local community's increasingly structured engagement with tourism and conservation partners for conservation purposes has brought in an additional dimension of the commercial value of conservancy land: 'We give out Sh6.8 million [Kenyan shilling] annually to our members as leases to their land' (Kahongeh, 2022: n.p.). Over and above these transformations is the land adjudication role of government agencies that involves allocating titles to families and restraining free or communal access to pasture. The result is a complex system of resource access and use, interactions and benefits. These put into question what governance approaches need to be in place for system components to work together for positive outcomes for ecosystem health for grazing and range lands.

Figure 7.2 Mobile *boma* in one of the conservancies visited for field research. File photo taken by Judy Kepher Gona

Governance systems

The set-up of conservancies in Kenya was initiated through the following approaches:

(1) **Tourism investors negotiated conservancies.** These are conservancies initiated by investors fronting revenue opportunities through wildlife protection providing a guaranteed monthly income for every landowner who contributes land. Land is leased from landowners by a community landholding company and investors enter into a lease agreement with the landholding company to run tourism lodges or camps. Additionally, a bed night fee is paid to landowners on whose land these lodges and camps are located. Investors pay for registration of leases to cushion against sales and other transactions on the said land (Damania *et al.*, 2019). This system is preferred where land is adjudicated into private title, which is the case with most Maasai Mara conservancies in the south west part of Kenya. These conservancies are more independent, although they may receive grants from time to time for large capital expenditure projects.

(2) **NGO negotiated conservancies.** Northern Kenya community conservancies are mainly NGO negotiated. These conservancies are frequently registered as Trusts and are extensively dependent on fundraising, making them vulnerable to the risk that is associated with aid. This system is often implemented where land is communally owned, under single title. The accruing benefits, mainly license fees from tourism, are paid annually into a community kitty. The community leaders use these funds as approved by a committee. Household direct benefits are limited, and the system is vulnerable to abuse by leaders.

(3) **Community engineered conservancies.** These are conservancies set up by the community without NGO or private sector engagement. The conservancy is registered as a non-profit organisation (NPO). The landowners have diverse operations/programs including tourism, capacity building programs in tourism and training for income generating activities. They are supported by diverse philanthropic organisations and philanthropists are the investors. Lease fees are paid regularly to households, who receive other benefits, as mentioned above. The conservancies are extensively dependent on fundraising, making them vulnerable to the risk that is associated with aid.

(4) **Privately conceived conservancies.** These conservancies are initiated by individuals with large tracts of land who register a conservancy for land protection and wildlife protection and may use tourism as a vehicle for financing the operations of the conservancy. These are found in former white settlement areas including the Laikipia, Naivasha and Nakuru areas, adjacent to the Maasai Mara ecosystem. These conservancies receive grants, often from international donors.

Conservancies in Kenya are organised into governable units using diverse legal instruments like companies limited by guarantees, trusts or associations. Key to governance are issues of equity in benefit sharing, transparency, disclosure, inclusion and the accountability of all actors. The governance systems are both contractual and relational, ensuring that traditional systems of governance are acknowledged and used.

Formalised governance systems come into use with conservancy set up, to guide operations and the management of land resources among a growing number of user groups. Conservancies have land lease agreements with landowners from local communities. Conservancies also operate under the guidelines of a conservancy management plan, which is approved by the Kenya Wildlife Service. Resource users also propose other guidelines, for example grazing protocols. Grazing plans are developed by the conservancy in collaboration with the grazing committee, as noted by Ngotho: 'It is local people who draw up grazing plans using traditional knowledge, with 20 percent of the funds going to grazing committees' (2023: n.p.).

Land leased from the local community is then zoned for areas that are grazeable and also dual use areas with neighbouring communities. There are periodic checks on vegetation for invasive species and grass and soil health, especially where the soil is extremely loose or the ecosystem is fragile, to determine rotation grazing patterns in these zones. There are also periodic counts of livestock to determine if numbers are within the carrying capacity of the land. Each household, based on its acreage in the conservancy, gets an allocation for livestock to graze in shared spaces. Ideally, systems of governance of pasture, livestock, land, water sources, wildlife and carbon are designed to deliver equitable benefits to all actors and sustainable livelihoods for the landowners.

Ecosystem services

Livestock rearing and wildlife protection are the main factors that benefit diverse actors sharing resources from conservancy land. Livestock serves as a bank reserve for pastoral communities and therefore numbers are significant factors for wealth and cultural pride. This makes pasture management for livestock rearing an important pillar in conservancy management. From conservancy visits, we noted that approaches to pasture management included: reducing numbers of livestock; improving breeds at the household level; and managing herds for landowners for crossbreeding to accelerate the numbers of new breeds, thereby reducing pressure on land and pasture. Livestock rearing has immediate benefits for household income and food security as a primary livelihood option for local pastoral communities. With the introduction of multi-decision-making processes, this resource is no longer private. Instead, livestock rearing decisions, for example where to graze or the size of a herd to have on an area, are now subject to collective decision making involving conservancy stakeholders external to households.

Wildlife protection supports biodiversity and has recreational and aesthetic value benefitting the attractiveness of tourism to the conservancy. In the Maasai Mara region, many households are supported by tourism livelihoods: 'Eighty per cent of households in this area depend solely on proceeds from tourism as a source of their livelihoods. They work as drivers, tour guides and hotel staff. Others sell curios to tourists' (Kahongeh, 2022: n.p.). As wildlife pays for the use of land and resources found on it through tourism, it indirectly influences all other resource, land, water, wildlife, livestock and pasture. This creates the potential for resource conflict when tourism exerts unnecessary pressure on sources of water and forests shared between tourism, wildlife, livestock and households. Environmental degradation has risks for livestock loss for pastoralist communities. As this is the lifeline for pastoral communities, they face greater economic insecurity because degradation likewise affects tourism, which is an alternative revenue stream for them. Degradation of environmental resources in conservancies has been linked to diminished value of the tourism product. The irony of what drives this problem is highlighted in a policy guide by the East African Wild Life Society (2017), mentioning that wildlife in the Maasai Mara National Reserve is threatened by its own attractiveness, i.e. to tourism. This follows from disturbances from increasing numbers of tourist facilities within and around the reserve, as environmental stresses from these lead to a decline of wildlife populations.

Abundance of wildlife populations in conservation areas is an integral component of safari tourism. Besides threats from tourism development, the risk for the wildlife population worsens with other stressors, for example drought driven by climate change, which is likely to destabilise earnings from safari tourism. As noted during the COVID pandemic, 'tourism went a bit down, and most of the people were not having a steady income stream, so having the carbon project was an alternative source of income, especially to landowners, and to the people around the community' (personal communication, interview three on 19 November 2024). The limitations for tourism earnings due to the effects of climate change on wildlife populations was one of the reasons for conservancies to welcome alternative financing options to support local communities who are dependent on tourism income. This paves the way for private entities to develop carbon projects on conservancy lands.

Action situations

From our data, we identify four key interactions between resource users, entailing decision processes and having outcomes that raise concerns for landowner's justice.

Negotiations on conservancy land use

Key engagement mechanisms for landowners in carbon projects have been through community meetings and workshops. Carbon developers have had regular consultations with landowners in the conservancies where we conducted our survey, to gather input and feedback on project design, planning, implementation and for information exchange. Our survey included statements where respondents were asked to choose from the options: strongly agree, agree, disagree or strongly disagree. Most statements received widespread agreement, with respondents predominantly selecting 'agree' or 'strongly agree'. Some statements elicited varied responses, particularly with a minority of respondents selecting 'disagree'. These variations indicate divergence in opinions or perceptions among the respondents. Notable disagreements are on statements relating to information flow from the local community to carbon project developers. In order to illustrate these findings, statements with notable disagreement are shown in Table 7.1.

Community input into carbon projects has been secured through meetings including annual general meetings, board of trustees' meetings and other stakeholder's meetings, which provide an engagement platform with landowners. A majority of survey respondents felt comfortable raising questions, noting that they had opportunities to ask questions on projects too. Also, many respondents felt that their insights were taken

Table 7.1 Survey statements with noted disagreement from respondents, $N = 41$

Statement	Strongly agree	Agree	Disagree
I received regular updates from implementing organisations about the project I was involved in as a beneficiary.	15	23	3
The project information was shared in a language and format that was easy for me to understand.	14	26	1
The information provided by the implementing organisation give me a full balanced understanding of both potential benefits and negative outcomes for our community associated with the project.	15	25	1
The information provided by the implementing organisation was sufficient for me to understand the changes expected at its completion.	10	30	1
There were enough opportunities to ask questions about the project to the implementing organisation.	9	31	1
I was comfortable raising my concerns or questions with the implementing organisation.	11	29	1
The questions I asked were adequately addressed.	6	30	5
The project allowed me to share my insights and expertise, and I felt that these were taken seriously.	2	32	7
Other project beneficiaries had the opportunity to share their input or feedback on the project and its implementation.	7	32	2

seriously. Some responses from the survey showed that landowners' inputs or opinions about carbon projects given out of their own volition, or without invitation, were not taken seriously.

Negotiations are formalised with non-disclosure agreements, as a precondition for conservancies to join carbon projects. Outputs for these processes include formalised documents to guide interactions for the common use of conservancy land resources. Legal agreements between landowners and carbon project teams confirm approvals for projects in conservancies. Signing the non-disclosure agreement was put ahead as a condition for participating in a carbon project within the conservancy:

> Before you become a team of the carbon project, you must be leasing land to a conservancy. And once you are part of the team, you are able to consent to be part of the team, by having the proper documentation, and legal agreements of the land. (personal communication, interview one on 18 November 2024)

There was follow up to convince conservancies to sign up to the non-disclosure agreement in order to be part of the carbon offset initiative being set up. Project partners introduced the project, informing landowners about project objectives, potential benefits and risks. This was a priority need considering that the local community had no prior information about carbon sequestration projects: 'Our people did not know what carbon was, we don't have a word for it in our local language but now we have created awareness' (Marete, 2022: n.p.). Following the efforts to raise awareness, landowners can then voluntarily opt to participate in carbon projects.

Formal approvals, all documented, are obtained from community assemblies or other relevant decision-making bodies. For example, Annual General Meetings for the landowner's trust intentionally have ongoing carbon projects as an agenda item for discussion, evidenced through signed minutes of the meetings. In conservancies, Landowner Committee meetings, meetings with community members and Board of Trustees meetings are other formal avenues in which discussions of developments from ongoing carbon projects indicate landowner's continuing approval of these arrangements. Most important is that approval is a continuous process with regular and inclusive community consultations to ensure that landowner concerns and feedback are addressed.

Community meetings were held in order to ensure that the community is: aware of the project; knows how it will benefit them; knows how the project will be implemented in the community and conservancies; and knows how the community will be involved in making the project successful. Additionally, there has been capacity building training to ensure the community understands how the project will benefit them: 'If community is able to know what is happening. Are they skilled enough to talk about carbon. Do they really know what is happening, that can easily help to measure the success?' (personal communication, interview five on 4 December 2024).

At the most basic level, negotiations address the various interests in conservancy land use, noting that local communities hold the ownership rights. Interview data show overlapping interests among primary actors in sustainable use of land resources. Landowners are found to be keen on maintaining healthy ecosystems, conserving biodiversity and ensuring long-term land productivity. They see actions for environmental conservation as helpful in restoring biodegraded land, improving biodiversity and also ensuring the sustainability of the rangeland ecosystem. As one interviewee explained the benefits: 'Well conserved lands, healthy grasslands/forests, recovered eroded lands, good grazing natural patterns, healthy soil, improved forest lines, improved livelihoods of the community' (personal communication, interview one on 18 November 2024). This interest in maintaining a healthy grassland ecosystem is likewise shared by carbon project developers and tourism investors. For carbon project developers this enables optimal soil carbon storage, whereas for tourism investors it supports wildlife abundance. Interests in maintaining healthy grass cover is thus a common ground that unites landowners, carbon project developers and tourism investors in shared goals.

Enforcement of management action on pasture management

Pasture management, e.g. through rotational grazing, allows grass to grow fast and consequently enables soils in conservancies to continually sequester carbon. Rotational grazing addresses concerns of overgrazing by local herders, in order to conserve grasslands. Both tourism and carbon developers are backing the idea to control grazing on conservancy lands, emphasising its benefits for optimal pasture regeneration. Part of management action to restore degraded lands is through grazing plans, a welcome idea consistent with the landowner's priorities of 'fattening herds through the improved grazing plans' (personal communication, interview four on 19 November 2024). As the interviews indicated, benefits for conservation goals in carbon projects are equally benefits for livestock rearing activities:

> Sustainable grazing will help to balance land use for livestock and for conservation, because through the carbon project, one of the activities is rotational grazing, and to be able to do this, was to help balance the land use, and for other conservation uses. (personal communication, interview six on 5 December 2024)

Through casual conversations in the field sites, it was observed that conservancies support efforts aimed at improving cattle breeds, to enhance their ability to withstand drought, build resilience to diseases and increase meat and milk production. Thus, improving breeds within conservancy herds is responsive to climate change.

As part of grazing committees, local communities decide on the patterns of grazing to be followed by herders at conservancies. For example: 'The community divided the area into six grazing blocks of 2000 acres each where they practise rotational grazing. There are community committees that assist in the implementation of the grazing plans' (Marete, 2022: n.p.). Negative environmental impacts may be driven by herders' non-compliance with grazing protocols, resulting in overgrazing on some sites: 'Overgrazing has damaged soils, resulting in a significant reduction in grass that grows back each year' (Ondieki, 2024a: n.p.). Grazing zones are developed from agreements to control landowners' cattle herds from overgrazing, as a precautionary measure to protect the grasslands. Controlled grazing has some sense of restriction to pastoralists' access to sections of grazing blocks: '…all herders must stay away from the entire area under grazing management to allow grass to grow to maturity' (Marete, 2022: n.p.). This restriction stays due to the introduction of exclusive tourism zones and dual use areas for conservation and grazing, with external oversight to access grazing fields.

The sense of restriction to access through zoning of spaces in conservancies indicates that local rules on nomadic grazing have been replaced by rotational grazing. Whereas under ancestral lands practices local pastoralists could move freely with their herds in search of grazing pastures, they now have 'no go zones' within their territorial borders. Our field observations confirm that conservancies promote a form of settled pastoralism, as a change from nomadic pastoralism. Cattle *bomas* used initially are being replaced. Herds still move from place to place but are settled within grazing zones.

Besides helping to sequester soil carbon, pasture management helps to address problems faced by the local community. One example is securing resources that become scarce during drought: '…there is a need for the practice of controlled grazing in a bid to save livelihoods during droughts… We have a buffer zone being managed by grazing committees where grazing is controlled to allow rangeland regeneration' (Marete, 2022: n.p.). Drought driven resource conflict is a recurring challenge facing nomadic pastoralist communities in Kenya's conservation areas. As Ondieki notes: 'Conflicts with other pastoralists usually occur when they invade areas in search of pasture during long dry spells' (2024a: n.p.). Unfortunately, interventions to address these conflicts have not been successful:

> For a generation, the government has applied force to enforce order in the northern counties, only for violence to return a few months later in a vicious cycle that grows worse by the year. This is now changing. Attitudes to livelihoods and conservation are shifting. Guns are going silent and wildlife thriving. (Kahongeh, 2022: n.p.)

Carbon projects are seen to be instrumental in addressing resource conflicts in these areas: In northern Kenya's arid and semi-arid lands where cattle rustling and resource-based conflicts have been endemic for decades, carbon trading is bringing hope and stability to the region (Kahongeh, 2022: n.p.; Ondieki, 2024b: n.p.). These projects open up opportunities to diversify livelihoods, and cushion communities from severe experiences of scarcities during climate extremes, in addition to being an alternative income source.

Generation and sale of carbon credits

Economic successes of carbon projects are indicated by successful carbon credit generation and sales. As media reported, 'We are now aware that practicing planned grazing is not only good for our livestock but also earns money' (Marete, 2022: n.p.). An expected outcome for this is a fair or equitable distribution of carbon project revenues. Furthermore, there is the possibility to use earnings for developing local businesses and enterprises, such as sale of motorcycle spare parts and livestock trading by young men (Kahongeh, 2022: n.p.; Ondieki, 2024b: n.p.), or women creating artworks from beads (Ondieki, 2024b: n.p.). These options support alternative sources of income besides livestock rearing, which are vital because livestock husbandry is sensitive to climate change impacts. Landowners view economic benefits from carbon projects as additional income, building on the revenue streams they already have, as opposed to replacing such streams from their livelihoods. Whereas landowners could initially get a guaranteed income from tourism, they now additionally earn from sale of carbon credits. As Ngotho reported: 'Each member is required to set aside 21 acres for conservation purposes, which recently earned them sh.70 million from carbon credit buyers' (Ngotho, 2023: n.p.). Benefits are shown for diverse members of the community: '…carbon is then sold as offsets on the global carbon market to earn the women and other community members' sustainable income, support the conservation, protect endangered wildlife and address climate changes' (Ondieki, 2024a: n.p.).

Earnings from the sale of carbon credits fund several development initiatives within conservancies. Benefits for the local community are presented as aid, for example to fund healthcare, schooling and the water supply (Ondieki, 2024a: n.p.; Ondieki, 2025: n.p.). This support in the form of aid is relevant because such communities are marginalised and lack the financial capacity to access services in times of need. As Ondieki reported: '…many cancer patients in the region often sell livestock- their sole source of livelihood, to cater for medical expenses - plunging families into deeper financial despair' (Ondieki, 2025: n.p.). International partners and carbon credit buyers seeking to offset their emissions come to the aid of a local community, for example on health matters. This way, news media reporting portrays these international organisations as having the power to support, or power to come to the rescue of, a needy, marginalised community; for example, when their aid helps to ease the burden of inadequate healthcare.

It is worth highlighting the risk of over dependency on project revenues by local communities in conservancies. As they had lived through the recent downturn period of project revenues in conservation tourism, they are cautious about their economic vulnerability if carbon project revenues decline or cease. One conservancy manager raised concern about what happens in the case of project failure and investors not fully recovering their investments. The response he received was that this was a non-issue, since the financing provided was in the form of grants (personal communication, interview five on 4 December 2024).

Legal action or other dispute resolution methods

One internal challenge plaguing conservation tourism success is the on-and-off resistance of local communities to project steps. These resistances not only slow down or cause delays in project processes, but also suggest rights violations to, or justice concerns by, local landowning communities. Protests from landowning communities were evident in news media articles, with reporting of court cases. Conflicts involve land disputes, disproportionate distribution of revenue and lack of transparency by carbon developers.

Land tenure insecurity features prominently in the interviews as a potential risk to landowners involved in carbon projects. There are emerging conflicts over ownership of land, with some community members, transformed to squatters on their own land, now finding their way to courts and land tribunals (Ondieki, 2025: n.p.). This happens where community members transfer their land rights to brokers, or new parties, interested in carbon trading, e.g. title deeds, or a land lease signed for some period of years. These offset projects cover vast areas of land with leases spanning several years. In a system where ancestral land is passed down through generations, some younger members of the families are waking up to a new reality and have to watch as contracts were signed dictating the use of their livelihood resources. It is important to ensure clear, legally recognised ownership for participating landowners. Land lease agreements should be renewed at regular intervals to maintain ownership by the community.

Carbon projects developments have been taken to court by local communities over unfair payments. Project developers take advantage of local communities when the engagement approach is not transparent. One report stated:

> What we are being paid is too little. There was a time when we had a court case and we were asked how the amounts were calculated; we were clueless. We were then asked how many tonnes of carbon we sell annually; we had no idea. (Ondieki, 2025: n.p.)

The indirect link between the communities and foreign carbon offset markets creates a lack of transparency in proceeds, paving the way for disproportionate revenue distribution. For this reason, indirect engagement leads to opaque agreements in terms of community engagement. In one case where a contract between community and private entities was revoked by the local government authority, it is reported that carbon offset deals were signed by communities who did not understand the market: 'private entities are signing opaque agreements with community group ranches and conservancies to buy carbon credits from them' (Ngotho, 2023: n.p.).

The legal disputes indicate injustices emerging from the engagement model with tourism hosting communities in conservancies. The action situations point to rights violations for local communities, which align with the views of Survival International, an international Indigenous rights organisation. Survival International (cited in Muchui, 2023: n.p.) faults a soil carbon sequestration project for: 'interference with community land rights, traditional grazing systems, lack of adequate public participation, and a flawed carbon verification process'. Other disputes with landowners are over boundaries, across which herders trespass in search of grazing pasture for their cattle. The conflicts over boundaries not only suggest scarce grazing pasture, but also a likelihood that herders maintain an increasing number of livestock.

Discussion

In community conserved areas, conservation initiatives bring both benefits and challenges for those involved, often raising justice concerns. Our study finds that the dominance of external conservation actors, paired with a dependency of the local community, can lead to violation of landowners' rights when priorities differ. Conservancies relying on tourism provide financial compensation to locals, yet intense forms of community revolts indicate economic benefits alone are insufficient. Most precisely, this spotlights the need for political empowerment in order for local communities to have an influence on project developments. Compensation is usually based on participation in conservancies and carbon sequestration projects, excluding some community members whose lands are distant from tourism areas (Bedelian *et al*., 2024). When local communities have equal control with other actors in environmental governance, ecological outcomes are much better (Dawson *et al*., 2024). We discuss here how this needed equality is shaped by conservation group dominance and local dependencies.

Dominance of conservation actors

Local communities in conservancies traditionally managed livestock through social trusts, sharing and nomadic grazing, ensuring sustainable pasture use. These systems fostered pastoral livelihoods across generations due to social accountability. However, engagement with private entities introduces non-disclosure agreements, diminishing social trust. Top-down conservation initiatives lack social accountability and participation in conservancies can separate communities from resources, as noted by Bedelian *et al.* (2024) regarding access to grazing land and water. Gona and Atieno (2021) highlighted the increased workload for women due to resettlement away from water sources.

Conservation driven by tourism and carbon sequestration maintains boundaries that can infringe on communities' rights to resources. While compensation is provided, restrictions on land use and possible dispossession arise, weakening local authority. Rotational grazing, managed traditionally, is sometimes overlooked by governance strategies, leading to challenges and protests.

To justify restricted access, local communities are portrayed as harmful to the environment, marginalising their input. Land resources are problematised, with activities framed as illegal, as seen in Bedelian *et al.*'s (2024) work mentioning illegal grazing. Resource conflicts reinforce negative stereotypes of pastoral communities, diverting focus from systemic issues. This subtly removes locals from conservation roles, allowing external control. Opaque negotiations and unfair benefit distribution consolidate power among private actors, a trend observed by Bedelian *et al.* (2024). Sene (2023) and Jujonas and Seekamp (2020) also note that economic benefits often favour a privileged few, neglecting marginalised communities. These issues resonate with findings that conservation projects undermine community rights, exacerbating inequities. Moreover, carbon offset projects marginalise women's needs and limit access to essential resources.

Inter-generational equity is also a concern, as conservancy set-ups often overlook the land inheritance systems of pastoral communities. Land inheritance and ownership transition systems among pastoral communities make ownership fluid as every male who goes through initiation and successfully navigates the *moran* stage is entitled to inheritance. The *moran* stage is an age set system for young men under the pastoralist Maasai system to undergo initiation to adulthood. Under ancestral land systems, land as a form of wealth was transferred to younger generations in a family lineage inheritance. Resulting conflicts and litigation from this flaw highlight the need for youth involvement in sustainable and equitable projects. Governance gaps further enable exploitation of community rights, with unclear structures and a lack of transparency regarding beneficiaries' legal responsibilities. Cavanagh and Benjaminsen (2014) note that brokers in the carbon offset market may conceal project effects by disconnecting consumers from sequestration sites.

Dependency of the tourism hosting community

Community compensation from tourism provided steady income, but when tourism collapsed during the pandemic, carbon development became a key alternative for conservancy earnings. However, existing conservancies tied to tourism suffered ongoing exploitations, shown in conflicts over grazing boundaries. If carbon offset projects use tourism community structures without addressing these issues, injustices may worsen. Satyal *et al*. (2020) argue that conservation initiatives can only be partially just due to deep rooted inequalities. In patriarchal communities like those in Maasai Mara, women's exclusion from land ownership limits their participation in conservancies (Bedelian *et al*., 2024; Gona & Atieno, 2021). When carbon offset projects emphasise benefits and take an aid approach, the opportunity to remedy patronage systems that exclude women can be diminished and the system of oppression perpetuated.

Carbon project developers driven by market demand for offsets enter conservancies emphasising pasture management for carbon sequestration, aligning with local conservation goals. Upon starting project implementation with conservancies, they identify the developmental needs of the local communities, most of which are marginalised and unmet. Recognising this gap, they appear to redirect their attention to becoming more vocal and prominent in charitable deeds for the local community. This effectively flips their position from being needy of carbon cultivation supported by local community resources, to the local community being needy of their aid. This 'aid' dimension legitimises projects, even as they shift emissions reduction responsibility from the Global North to the Global South (Cavanagh & Banhaminsem, 2014; Lehmann, 2019). Such dependency can overshadow environmental goals and perpetuate unequal partnerships, similar to the 'saviour syndrome' witnessed in volunteer tourism (Anderson *et al*., 2021; Bandyopadhyay, 2019; Helmick, 2022).

Local pastoralists are shown to lack agency in resolving resource conflicts, relying on external help. While tourism and carbon projects offer alternative incomes and control grazing for profit, they can create unhealthy dependency. This raises concerns about the long-term reliance on compensatory measures and potential permanent loss of land rights. Strong safeguards are needed to protect traditional livelihoods.

Overemphasising community dependency on tourism suggests a weak position, especially if donor aid is withdrawn. Focusing mainly on economic rewards overlooks the need for political and psychological empowerment (Battilani & Strangio, 2021). When focus on the rewards from tourism is the predominant basis for community engagement, conservation outcomes are likely to incline towards injustices. We therefore refer to the call to 'socialise tourism' (Higgins-Desbiolles *et al*., 2022) as a baseline necessity for identifying tourism communities' rights and benefits in Kenya's wildlife conservancies. This approach puts the focus on tourism being 'answerable to the society in which it occurs' (Higgins-Desbiolles, 2020: 617).

Conclusion

Project developers have evidence of participatory processes to engage communities. Nonetheless, the dominance of conservation actors and the dependencies of tourism hosting communities lead to power asymmetries that would be problematic in addressing differences between groups without stepping on the rights of the weaker party. In particular, differences in what successes of carbon offset projects mean for different groups come into question. Gains on carbon market earnings go hand in hand with the expansion of carbon offset projects, and in an engagement where community is patronised, this means insecure land tenure for them. On the other hand, compensatory measures can result in landowners furthering their interests in grazing, as they can afford to buy more herds. Therefore, successes of the project are likely to enable more pressure on grazing pastures from landowners increasing herd numbers. Failures to understand the cultural lens that community applies to foresee successful project outcomes is a blind spot that will challenge the current approach to pasture management.

Carbon offset projects, like conservation tourism, can be a tool for domination of the community defined by tokenism, leading to more dependence and ceding of rights over resources. It is important that such conservation initiatives drawing on land use address the future demand for land for human settlement, settled pastoralism and other needs from generation to generation. Disproportionate sharing of revenue, displacement from resource access and exclusion from decision making, accumulate to form environmental injustices that communities face in conservancy partnerships involving tourism and carbon offsetting. Community interactions with powerful tourism and carbon project developers will only be effective if negotiations offer protection to other dimensions of resource rights, beyond rewards and benefits of resource use.

References

Anderson, K.R., Knee, E. and Mowatt, R. (2021) Leisure and the 'white-savior industrial complex'. *Journal of Leisure Research* 52 (5), 531–550. https://doi.org/10.1080/00222216.2020.1853490.

Bandyopadhyay, R. (2019) Volunteer tourism and 'the white man's burden': globalization of suffering, white saviour complex, religion and modernity. *Journal of Sustainable Tourism* 27 (3), 327–343. https://doi.org/10.1080/09669582.2019.1578361.

Bashir, M.A. and Wanyonyi, E. (2024) Winning space for conservation: The growth of wildlife conservancies in Kenya. *Frontiers in Conservation Science* 5. https://doi.org/10.3389/fcosc.2024.1385959.

Battilani, P. and Strangio, D. (2021) Mass tourism and social sustainability: Insights from the Italian and French coasts. In P. Battilani and C. Larrinaga Rodríguez (eds) *Coastal Tourism in Southern Europe in the XXth Century* (pp. 37–56). Peter Lang.

Bedelian, C., Ogutu, J.O., Homewood, K. and Keane, A. (2024) Evaluating the determinants of participation in conservancy land leases and its impacts on household wealth in the Maasai Mara, Kenya: Equity and gender implications. *World Development* 174, 1–13.

Cavanagh, C. and Benjaminsen, T.A. (2014) Virtual nature, violent accumulation: The 'spectacular failure' of carbon offsetting at a Ugandan National Park. *Geoforum* 56, 55–65. https://doi.org/10.1016/j.geoforum.2014.06.013.

Chandler, D.L. (2024) Explained: Carbon credits. MIT Climate Portal. See https://climate.mit.edu/posts/explained-carbon-credits (accessed March 2025).

Damania, R., Desbureaux, S., Gael, S., Pasquale, L., Mikou, M., Gohil, D., Bharat, S. and Mohammed, F.P. (2019) *When Good Conservation Becomes Good Economics: Kenya's Vanishing Herds*. World Bank.

Dawson, N.M., Coolsaet, B., Bhardwaj, A., Booker, F., Brown, D., Lliso, B., Loos, J., Martin, A., Oliva, M., Pascua, U., Sherpa, P. and Worsdell, T. (2024) Is it just conservation? A typology of Indigenous peoples' and local communities roles in conserving biodiversity. *One Earth* 7 (6), 1007–1021. https://doi.org/10.1016/j.oneear.2024.05.001.

Dias, V. M., Soares, P P.d.M.A., Brondizio, E.S. and Cruz, S.H.R. (2021) Grassroots mobilization in Brazil's urban Amazon: Global investments, persistent floods, and local resistance across political and legal arenas. *World Development* 146, 105572. https://doi.org/10.1016/j.worlddev.2021.105572.).

East African Wild Life Society (2017) *Policy Pathways Towards Enhancing Sustainable Management of Maasai Mara National Reserve*. East African Wild Life Society.

Georgiou, K., Jackson, R.B., Vindušková, O., Abramoff, R.Z., Ahlström, A., Feng, W., Harden, J.W., Pellegrini, A.F.A, Polley, H.W., Soong, J.L., Riley, W.J. and Torn, M.S. (2022) Global stocks and capacity of mineral-associated soil organic carbon. *Nature Communications* 13, 3797. https://doi.org/10.1038/s41467-022-31540-9.

Gona, J.K. and Atieno, L. (2021) Pastoral women participation in community conservancies in Maasai Mara. In M. Novelli, E.A. Adu-Ampong and M.A. Ribeiro (eds) *Routledge Handbook of Tourism in Africa* (pp. 429–441). Routledge.

Helmick, L. (2022) White saviorism: An insider perspective. *Art Education* 75 (3), 9–13.

Higgins-Desbiolles, F. (2020) Socialising tourism for social and ecological justice after COVID-19. *Tourism Geographies* 22 (3), 610–623. https://doi.org/10.1080/14616688.2020.1757748.

Higgins-Desbiolles, F., Bigby, B.C. and Doering, A. (2022) Socialising tourism after COVID-19: Reclaiming tourism as a social force. *Journal of Tourism Futures* 8 (2), 208–219. https://doi.org/10.1108/JTF-03-2021-0058.

Holmes, G. and Cavanagh, C.J. (2016) A review of the social impacts of neoliberal conservation: Formations, inequalities, contestations. *Geoforum* 75, 199–209. https://doi.org/10.1016/j.geoforum.2016.07.014.

Jujonas, M. and Seekamp, E. (2020) 'A commons before the sea': Climate justice considerations for coastal zone management. *Climate and Development* 12 (3), 199–203.

Kahongeh, J. (2022) How community conservation is silencing gunfire in northern Kenya. See https://nation.africa/kenya/health/how-community-conservation-is-silencing-gunfire-in-northern-kenya-3868824 (accessed November 2024).

Kenya Wildlife Service (2012) *Strategy 2012-2017*. Kenya Wildlife Service.

Kenya Wildlife Service (2024) *Kenya Wildlife Service Strategic Plan 2024-2028*. Kenya Wildlife Service.

Lehmann, I. (2019) When cultural political economy meets 'charismatic carbon' marketing. A gender-sensitive view on the limitations of Gold Standard cookstove offset projects. *Energy Research & Social Science* 55, 146–154.

Marete, G. (2022) How carbon credits cash is changing pastoralists lives. See https://nation.africa/kenya/news/how-carbon-credits-cash-is-changing-pastoralists-lives-4021592 (accessed December 2024).

Muchui, D. (2023) Community defends carbon offset projects in northern Kenya. See https://nation.africa/kenya/counties/meru/community-defends-carbon-offset-projects-in-northern-kenya-4196204 (accessed November 2024).

Ngotho, S. (2023) Lenku quashes 'opaque' deals on carbon credits. See https://nation.africa/kenya/counties/kajiado/lenku-quashes-opaque-deals-on-carbon-credits-4260672 (accessed November 2024).

Ondieki, G. (2024a) Samburu women take lead in restoring degraded pastures. See https://nation.africa/kenya/health/samburu-women-take-lead-in-restoring-degraded-pastures-4538356 (accessed December 2024).

Ondieki, G. (2024b) Guns go silent in bandit-ravaged regions as locals embrace carbon trading. See https://nation.africa/kenya/health/guns-go-silent-in-bandit-ravaged-regions-as-locals-embrace-carbon-trading-4693366 (accessed November 2024).

Ondieki, G. (2025) Conservancy eases cancer patients financial burdens. See https://nation.africa/kenya/health/conservancy-eases-cancer-patients-financial-burden-4904588 (accessed February 2025).

Ostrom, E. (2011) Background on the institutional analysis and development framework. *Policy Studies Journal* 39, 7–27. https://doi.org/10.1111/j.1541-0072.2010.00394.x.

Rastegar, R. and Becken, S. (2024) Embedding justice into climate policy and practice relevant to tourism. *Journal of Sustainable Tourism* 33 (10), 2011–2028. https://doi.org/10.1080/09669582.2024.2377720.

Ruppel, O.C. and Ginzky, H. (2021) *African Soil Protection Law: Mapping Out Options for a Model Legislation for Improved Sustainable Soil Management in Africa - A Comparative Legal Analysis from Kenya, Cameroon and Zambia*. 1st edn. Nomos Verlagsgesellschaft mbH & Co. KG.

Satyal, P., Corbera, E., Dawson, N., Dhungana, H. and Maskey, G. (2020) Justice-related impacts and social differentiation dynamics in Nepal's REDD+ projects. *Forest Policy and Economics* 117, 102203. https://doi.org/10.1016/j.forpol.2020.102203.

Sene, A.L. (2023) Justice in nature conservation: Limits and possibilities under global capitalism. *Climate and Development* 16 (9), 838–847. https://doi.org/10.1080/17565529.2023.2274901.

Shinbrot, X.A., Holmes, I., Gauthier, M., Tschakert, P., Wilkins, Z., Baragón, L., Opúa, B. and Potvin, C. (2022) Natural and financial impacts of payments for forest carbon offset: A 14 year-long case study in an indigenous community in Panama. *Land Use Policy* 115, 106047. https://doi.org/10.1016/j.landusepol.2022.106047.

Thom, D. and Mah, A. (2020) Introduction: Tackling environmental injustice in a post-truth age. In D. Thom and A. Mah (eds) *Toxic Truths: Environmental Justice and Citizen Science in a Post-Truth Age* (pp. 1–28). Manchester University Press.

Tourism Research Institute (2025) *Annual Tourism Sector Performance Report 2024*. Tourism Research Institute.

8 The Nexus of Environmental Justice and Potential for Sustainable Tourism under Colonial Occupation in Palestine

Mazin B. Qumsiyeh and Andrea Bibee

In Palestine, environmental justice and tourism are intertwined and both are impacted by prolonged colonial occupation which asserts control over most elements of daily life. An apartheid system was created to further the colonisers' goals and interests, which in the case of Palestine pertains to reshaping the land while removing the indigenous people. On top of that, Palestine is experiencing the shared worldwide strain and impacts on natural resources including from climate change, habitat destruction, overexploitation of natural resources, invasive species and pollution. These five challenges are multiplied and exacerbated because of the Israeli occupation and colonisation of Palestine. Through structured interviews with five tourism experts and 10 tourism operators plus a literature review, we identified key challenges to and opportunities for sustainable tourism in Palestine. We show how responsible international tourism and expanded local tourism (especially alternative tourism) can aid in achieving environmental justice, human rights and even sustainability.

Introduction

In Palestine, environmental justice and tourism are intertwined and both are impacted by prolonged colonial occupation which asserts control over most elements of daily life. An apartheid system was created to further the colonisers' goals and interests, which in the case of Palestine pertains to reshaping the land while removing the indigenous people. On top of that, Palestine is experiencing the shared worldwide strain and impacts on natural resources including from climate change, habitat destruction, over exploitation of natural resources, invasive species and pollution. These five challenges are multiplied and exacerbated because of the Israeli occupation and colonisation of Palestine. Through structured interviews with five tourism experts and 10 tourism operators plus a literature review, we identified key challenges to and opportunities for sustainable tourism in Palestine. We show how responsible international tourism and expanded local tourism (especially alternative tourism) can aid in achieving environmental justice, human rights and even sustainability.

While other countries have seen people travel to perform religious duties, they were often travelling within the same country (e.g. India). By contrast, Palestine has drawn religious pilgrims from many countries for nearly two millennia. By the 4th century AD, Christianity had spread in a global fashion and across many jurisdictions, which created what can properly be referred to not only as religious pilgrimage but religious tourism to Palestine (Isaac *et al.*, 2016; Mohamed & Suleiman, 2011). As expected, the Israeli occupation of the West Bank (including Jerusalem) and Gaza in 1967 resulted in drastic shifts in the management of tourism here. Israeli authorities have quickly ensured that tourists are met by Israeli tour companies and guides to the exclusion of the local Palestinians (Kelly, 2016). Partly in response, and partly to diversify and grow income, Palestinians developed new forms of tourism ranging from ecotourism to geotourism to cultural tourism to dark tourism (Abahre & Al-Rimmawi, 2023; Isaac, 2010a, 2010b; Kassis, 2013; Kassis *et al.*, 2015). Indeed, there is a trend of tourists coming to Palestine to be in solidarity with Palestinians and to bear witness to the injustices that Israel imposes in the daily lives of those it impacts. This phenomenon deserves special attention. Dark tourism and justice tourism are intertwined and become important to the mission of advancing justice and human rights in Palestine.

Background

Palestine is located between Europe, Asia and Africa with an area of 27,000 km². The 1967 occupied Palestinian Territories constitute 22% of historic Palestine (West Bank: 5879 km², Gaza: 378 km²) (UNCTAD, n.d.). Geologically, the area belongs to the African tectonic plate which, as a result of plate collision, resulted in having the northern part of the Great Rift Valley located here (including the lowest point on Earth at the Dead Sea). Palestine is the western part of the Fertile Crescent, where humans first developed agriculture some 12 millennia ago (Eshed *et al.*, 2004; Zeder, 2008). This also allowed for rich human development, denoted by the fact that the Eastern Mediterranean region is thought of as the 'Cradle of Civilization' (Lopes & Almeida, 2017). The deep connection of these indigenous populations to the land of Palestine underscores a continuous presence that predates modern political boundaries by millennia (Qumsiyeh, 2004).

The unique geography and geology have given Palestine more biological diversity than some countries ten times its size. The diverse habitats cover five ecozones: the central highlands, the semi-coastal region, the eastern slope, the Jordan valley and the coastal region. Palestine also spans five phyto/bio-geographical regions (Mediterranean, Irano-Turanian, Saharo-Arabia, Coastal and Sudanese). The indigenous Palestinians lived in all parts of the country in some 1300 villages and towns in relative harmony with each other and with nature. The population in the first decade of the 20th century was 850,000, with various religious persuasions (3% Jewish, 13% Christian, 80% Muslim and 4% other). The land was owned or, as is typical of many indigenous societies, known to be used by local people for millennia. Since 1948, hundreds of Palestinian villages and towns have been destroyed and a huge refugee population was created by the nascent state of Israel (Pappe, 2006; Qumsiyeh, 2004).

As in other countries, environmental decline in Palestine impacting people and nature stems from climate change, habitat destruction, overexploitation, pollution and invasive species. Also, as in many other countries, colonialism played a significant role in this decline. The first to warn of the potential threat to the Palestinian's environment was Ives (1950). More recent studies show ominous evidence of decline (see Braverman, 2021, 2023; EQA, 2022; Qumsiyeh & Abusarhan, 2020, 2021; Qumsiyeh *et al.*, 2014; Tal, 2002). Wars and weapons also devastate the landscape (Qumsiyeh, 2024; Yin *et al.*, 2025). Case examples include:

(1) refugees and displaced Palestinians in the West Bank and Gaza create significant environmental and economic pressures (more studies are needed here);
(2) relentless actions of Israel to prevent Palestinians from accessing their water sources resulted in sealing of springs which had devastating environmental impacts for people and wildlife;
(3) destruction of Hula Lake and wetlands (included removing the local villagers and the disappearance of 219 species); and
(4) diversion of water from Lake Tiberias devasted the Jordan River basin (see Messerschmid, 2008, 2012, 2014).

These issues impact sustainability and tourism. For example, the destruction of the Jordan River ecosystem impacted potential ecotourism in that area and resulted in the shrinking of the Dead Sea – the lowest point on Earth and an inland salt lake recreational area – which diminished its long-term potential as a tourist destination (see Salameh & El-Naser, 2008).

Methods

We conducted a meta-analysis and synthesis of the literature associated with environmental justice and tourism in Palestine using our own Palestine Institute for Biodiversity and Sustainability (PIBS) archives and Prof. Qumsiyeh's ongoing work as an educator in Tourism and Hotel Management at Bethlehem University. We also did a number of interviews with five key officials in the Ministry of Tourism and with ten tour operators over the period of June to December 2024. The series of questions we asked focused on their understanding of the nexus of tourism with environmental justice under conditions of occupation/colonisation and what challenges and opportunities they see pertaining to tourism in Palestine (again with a focus on environmental justice). Data from the literature and from the semi-structured interviews were analysed and culminated in a summary focusing on the challenges and opportunities for sustainable tourism in Palestine. We also offer a brief case study of the work of the Palestine Institute for Biodiversity and Sustainability (https://www.palestinenature.org/) focusing on its impact *vis a vis* environmental and climate justice. These diverse sources of data have been analysed to spotlight issues of climate justice and environmental justice in a Palestine under prolonged occupation and oppression.

Findings

Challenges

We asked 15 individuals to list in order of significance the main challenges to sustainable tourism in Palestine. We then analysed the data to develop a list of recommendations. The five challenges (that were included by eight or more of the respondents) in order were:

(1) Wars and conflict which produce things like the drying up and deterrence of tourism and the destruction brought on by war (see Qumsiyeh, 2024).
(2) The ongoing practices of the Israeli occupation: this included things like denying mobility of tour guides; an inability to organise tours effectively (unless you are a Jewish Israeli tour company); diverting tourists away from indigenous Palestinian communities; and significant fragmentation of key tourist sites, denying uniform access to them (for example, Nazareth in Galilee and Bethlehem in West Bank). The Segregation Wall in particular even separates Jerusalem from Bethlehem which is its suburb a mere 10 kilometres away; this impacts not only people but even wildlife and biodiversity (Husein & Qumsiyeh, 2022). The Israeli occupation has led to restrictions on movement and access to sites (including, for example, an everchanging and unpredictable imposition of hundreds of checkpoints), which can deter potential

visitors and limit the economic benefits that tourism could provide to Palestinian communities. The latter especially impacts local tourism.
(3) Inadequate infrastructure (roads, transportation networks, parking spaces, etc.).
(4) Excessive commercialisation and commodification of tourism and its negative impact (including waste).
(5) Disconnect between the local population and the tourists. Much of the tourism, especially to religious sites, is essentially visiting a location while avoiding meeting local people. Israeli tour guides even actively discourage interactions with local Palestinians. Thus, Christian tourists often visit places like the Church of the Holy Sepulchre and the Church of the Nativity in Bethlehem while not interacting with or knowing much about local Palestinian Christians in those areas. Tourists engage with sites primarily as spectators of relics rather than as participants in a living cultural and spiritual heritage (see ATG, 2010).

Remarkably, not one of the 15 people interviewed mentioned tourism's environmental damage (e.g. pollution, use of water, dependence on jet fuel, etc.). This is an area generally ignored in the rush to increase tourism for economic benefit (see Milano *et al.*, 2024, for a review). At PIBS we are focused on the environment and are located near the Church of the Nativity, which gets hundreds of thousands of tourists each year. As such, we engage in studies like this one to promote potential sustainability in various aspects of human activities in the region. The threats must be studied and dealt with realistically (see below under opportunities).

Human-induced climate change will drastically affect the Arab world (Verner, 2012). A World Bank study shows that the impacts. including water resource decline. will be drastic by 2030 (World Bank, n.d.). In the West Bank and Gaza, demand will double, while supply will shrink dramatically. This is coupled with population growth and habitat destruction leading to 'unliveable' conditions (Kuttab, 2024; Verner, 2012). The damage from climate change is compounded by environmental injustice in Palestine (Al-Haq, 2021; EQA, 2020; Salem, 2011; Weizman & Sheikh, 2015). As a result of climate change, the environmental issues discussed above could get exponentially worse in the occupied Palestinian territories. One study suggests that even more scarcity of water, caused by the intertwining factors of climate change and military occupation, could see a rise in incidences of diarrhoea, cholera and dehydration (Mimi *et al.*, 2009). Clearly, unregulated and unfettered tourism driven strictly by a desire for an increased number of tourists is 'toxic' (Pezzullo, 2009). Rational planning needs to be implemented to ensure tourism can face the coming challenges. The next set of findings explore what opportunities exist for sustainable tourism that brings benefits to the people with minimal or no impact on the environment.

Opportunities

The same representative tourism stakeholders were asked about opportunities to address the challenges noted above. The six most frequent answers given were:

(1) International support towards liberation and self-determination to allow freedom of movement would allow tourists to come to Palestine and be guided by locals.
(2) Improved public spaces (e.g. roads, parking lots, green spaces, etc.).
(3) Further development of alternative tourism (e.g. niches such as ecological, cultural, hiking, political/solidarity). Given the unique ecology and diverse landscapes present in Palestine, ecotourism has become an increasingly popular motive for visiting the region. One can easily explore five distinct eco-zones in a span of 70 km. In 2007, Bethlehem University hosted a summit on ecotourism in the West Bank and identified three main benefits of marketing this style of tourism for the international visitor: economic stimulation, increasing or preserving biodiversity and movement towards national development goals.
(4) Better local and national (i.e. the Palestinian Authority) support for local communities (relating to tourism). For example, some respondents asked for fewer taxes, fewer regulations and more incentives for small family businesses.
(5) Many souvenir makers informed us of exploitation by sellers/marketers, where more than 90% of a souvenir's price may go to sellers' commissions and to kickbacks to tour operators and bus drivers, who bring tourists to the shops. This necessitates controls and administrative support from those developing regulations and standards for this sector of the economy.
(6) Promote domestic and Palestinian diaspora tourism which can become a more reliable source than international tourism, especially during times of conflict or instability (see Novelli *et al.*, 2012).

Only one of the people we interviewed spoke obliquely of sustainability in tourism but it was only in the context of stating 'how we increase the amount of tourists while we have limited space (including in hotels)… it is impossible?'. When specifically asked, hardly anyone knew about the impact of tourism on the environment, including: use of water, pollution, travel's contribution to greenhouse gases, etc. (see also Becken & Scott, 2024; Milano *et al.*, 2024). Several tour operators told us that we need programs to teach about such things as environmental justice and climate justice after we explained to them briefly what these terms mean. One stated: 'At the diploma program we talked briefly about ecotourism but only that there are nice natural areas and we can bring tourists to see them…. no one [in the diploma program] explained to us about why or how we should do sustainability projects'.

These insights asserted to us the significant need for education but also for government regulation pertaining to sustainability in tourism. One cannot expect those who are financially benefitting from tourists to worry much about the actual impact of tourism on climate change or other sustainability issues (see Peeters & Papp, 2024). One alternative in Palestine is to expand ecotourism and other sustainable forms of tourism, including domestic and diaspora tourism, while decreasing more damaging forms of tourism (see also Qumsiyeh & Handal, 2018; Tabash, 2017).

Another area to highlight is the growing role of many institutions that work for environmental justice such as the Land Research Center (https://www.lrcj.org/en), the Applied Research Institute (ttp://ARIJ.org), Al-Haq (https://www.alhaq.org/) and our Palestine Institute for Biodiversity and Sustainability (PIBS http://palestinenature.org). PIBS was founded in 2014 and has operated mostly through volunteer efforts and by individual donations. Its achievements in research, education, conservation and community engagement towards sustainable human and natural communities are exemplary (Qumsiyeh, 2023a; Qumsiyeh & Amr, 2020; Qumsiyeh et al., 2017). The institute works to empower local communities by fostering food sovereignty, benefitting from nature while protecting it and resistance strategies to achieve environmental and human rights goals (Qumsiyeh, 2023b; Qumsiyeh & Abusarhan, 2020, 2021; Qumsiyeh et al., 2017, 2022, 2024). Here are some of PIBS' accomplishments relating to environmental justice and conservation:

- Published more than 150 research papers in peer reviewed journals in many areas, including fauna, flora, conservation measures, environmental injustice, sustainable agriculture, climate change, protected areas and human rights. Some of those papers can be seen at palestinenature.org.
- Published more than 300 other articles of a more lay person nature in magazines, journals and key information websites.
- Launched a Mobile Educational Unit (MEU), which is the first mobile exhibit of its kind in Palestine (see Figure 8.1). In two years before the genocidal war commenced in late 2023, we visited 46 schools, benefitting 6011 students in marginalised communities who were unable to visit us.
- Enhanced awareness of tens of thousands of visitors through visits to our Natural History and Ethnography Exhibits and the Botanical Garden (see Figure 8.2).
- Produced various educational modules, such as interactive games (see Figure 8.3), tailored to the Palestinian community, brochures and posters on environmental topics including adaptation and mitigation of climate change to distribute to schools together with animated videos.
- Established 20 environmental clubs in West Bank schools and helped empower university students to work for environmental issues.

174　Part 3: Case Studies in Climate (In)Justice in Tourism

Figure 8.1 Students at a marginalised school holding our climate change brochures that articulate challenges and solutions. The students are in front of our mobile museum/educational centre. (Credit: PIBS)

Figure 8.2 Children visiting one of the exhibit areas of the natural history museum and learning about the geography of Palestine with the issue of desertification (see Qumsiyeh et al., 2014). (Credit: PIBS)

Figure 8.3 Two children competing in a game we created that teaches them about conservation and sustainability. (Credit: PIBS)

- Held over 1800 workshops, half of them local and half to over 40 countries, benefitting more than 15,000 participants.
- Led the efforts for the new Protected Area Network and National Biodiversity Strategy and Action Plan (NBSAP) (PIBS, 2022).
- Provided leadership and acted as a model for national, regional and global actors in areas like environmental justice and challenging settler colonialism *vis a vis* the environment.
- Set up a successful animal rehabilitation unit.
- Engaged in community service activities that benefited especially marginalised communities (including women, farmers and Bedouins).

An example of a successful integrative pilot project that our institute implemented was to work with four local communities to the west of Bethlehem and south of Jerusalem to promote eco-friendly agriculture, local tourism and ecotourism to the tourists already coming to Bethlehem for religious tourism. Some of the data can be found in Qumsiyeh *et al.* (2023, 2024). We are now expanding this work, with support from a National Geographic Society grant and other sources, to eight communities and creating the first Palestinian biosphere reserve (on biosphere reserves, see Bouamrane *et al.*, 2019). This reserve will draw both domestic tourists and some of the international tourists already coming to Palestine (for pilgrimage, for example) to help protect the environment while also promoting sustainable practices to the local people.

Discussion

From our intensive interviews and via the literature review, we can state that religious tourism remains the most significant component of Palestine's tourism landscape, drawing visitors to its historical and religious sites, such as Nazareth, Bethlehem and Jerusalem. The impact of religious tourism on the environment is of concern. Increased foot traffic to sensitive historical sites can lead to environmental degradation, including erosion and pollution. The potential of tourism development in Palestine is very high if one deals with environmental justice and other decolonising issues.

It is natural that local interviewees prioritised ending Israel's monopoly on tourism, which would solve other challenges to normalising tourism away from this power disparity. As noted above, there was significant agreement on the challenges faced by locals in ensuring more sustainability in tourism. Likewise, there was agreement on the opportunities to improve the tourism sector in the direction of sustainability (which is also related to climate and other environmental justice issues). What was not expected is the finding that tourism operators had little knowledge of the challenges and opportunities they face in relation to environmental issues such as climate change.

The Tourism Panel on Climate Change (TPCC) noted that 'climate justice has not been explored sufficiently in the tourism context' (TPCC, 2023: 18). Important pronouncements like the Davos Declaration on Climate Change and Tourism (UNWTO, 2007) overlooked issues of climate justice and only exhort tourism stakeholders to 'secure financial resources to help poor regions and countries'. Similarly, the Glasgow Declaration on Climate Action in Tourism (2021) focuses on pathways to reducing greenhouse gas emissions in the tourism industry and supporting commitments to reach 'net zero'; it is silent on climate justice concerns. As our research has shown, it is possible to engage in people centred climate change adaptation and mitigation in Palestine despite the context of injustice (Qumsiyeh *et al.*, 2022; UNDP, 2010).

Sustainable tourism in Palestine, like other activities, is not possible unless the local people's interests are taken into account and in a meaningful way. Nature-based solutions are thus critical to sustain both human and natural communities (Turner *et al.*, 2022). If done properly, the development of alternative tourism in Palestine, including eco-tourism, can help both people and nature. Alongside improved adjustments to current tourism, focusing on more local and eco-friendly approaches would help significantly in issues of sustainability. If we take Bethlehem as an example, over 3 million tourists visited the birthplace of Jesus during periods of calm. But most of those tourists did not stay in Bethlehem due to having been brought by Israeli tour companies and led by Israeli tour guides. They spent very little money locally. Less than 8% stayed in Bethlehem overnight. It would have a significant positive economic impact for the local Palestinians if these tourists actually stayed and shopped in the hotels, restaurants and souvenir shops in Bethlehem; however, this boon in business would raise other issues, for example, increased water shortages. Israel takes most of the Palestinian water (Messerschmid, 2012, 2014) and with ongoing impacts from climate change and an already significant decline in rainfall, we expect the allocation of water to local Palestinians to decrease substantially while the local population continues to grow.

Environmental justice and climate justice are related concepts that address aspects of environmental and social issues. Both concepts share a concern for equity and fairness, and ensuring protection of both people and nature. These concepts emphasise the need to address systemic inequalities and empower affected communities. The climate justice concepts grew out of the environmental justice movement and took off as a discipline of its own, especially after the first Climate Justice Summit at The Hague during the Convention of the Parties 6 (COP6) meeting of the United Nations Framework Convention on Climate Change (UNFCCC) (Schlosberg & Collins, 2014). Attention to climate justice has grown in the past 25 years to become central to the environmental justice movement and to encompass global issues including mitigation, adaptation and international equity (Evans-Agnew & Aguilera, 2023; Schlosberg & Collins, 2014).

Since October 2023, two years have passed with hardly any tourists in Bethlehem as Israel engaged in a genocide, ethnic cleansing and ecocide initially in Gaza, and then in Lebanon and the West Bank. Bethlehem itself was isolated and tourism dried up as the Israeli military expanded its repressive measures and invasion of Palestinian villages, towns and refugee camps. Without international laws being enforced to bring justice during a genocide, then it is difficult to imagine those same laws and actors helping to bring about environmental justice in Palestine. For now, this is mostly being done on the ground, day by day, in grassroots and indigenous ways.

Alternative tourism models have emerged in Palestine in response to the challenges posed by traditional religious tourism, alongside the occupation and environmental injustice. These models focus on community engagement, cultural exchange and environmental sustainability, and emphasise local participation, and hence the creation of a more equitable distribution of tourism's benefits. Initiatives such as eco-tourism and cultural tourism encourage visitors to engage with local communities, fostering a deeper understanding of Palestinian culture and the sociopolitical context. Alternative tourism seeks to address the environmental impacts associated with mass tourism by promoting responsible travel practices. For instance, initiatives that encourage visitors to participate in local agricultural practices or community projects can help to enhance the sustainability of tourism while providing economic benefits to local residents (Aswita *et al.*, 2023). By integrating environmental considerations into tourism development, alternative tourism can contribute to climate justice by ensuring that local communities are not only beneficiaries of tourism but also active participants in its management and sustainability.

Similarly, justice tourism has gained traction as a means of raising awareness about the sociopolitical issues facing Palestinians (see Kassis, 2013; Kassis *et al.*, 2015). This form of tourism encourages visitors to engage with the realities of life under occupation, providing a platform for advocacy and solidarity with local communities. Currently, this style of tourism has a very small climate change footprint in Palestine. While over three million tourists came to the occupied Palestinian territories in 2022 on religious tourism (PCBS, 2024), our institute estimates that, at best, less than 5000 justice tourists and solidarity activists came in that year (unpublished data of PIBS). Such 'alternative tours' that highlight the impact of the occupation on daily life also study the impact of the occupation on the environment and raise awareness of climate change and other environmental challenges (Qumsiyeh & Abusarhan, 2021).

We also have a significant opportunity to draw some of the religious tourists to simply see other things *while* they are here to increase their understanding of and participation in the stewardship of the environment. For example, our institute PIBS at Bethlehem University has a museum of Natural History and Ethnography which is focused on highlighting the rich local biodiversity and the environmental challenges and solutions (Qumsiyeh, 2017, 2023a; Qumsiyeh *et al.*, 2017). This becomes a form of non-violent, popular resistance to the oppression of the occupation (Qumsiyeh, 2021). In this way, the interplay of tourism and environmental justice is tied to the potential for sustainability (Lee & Jamal, 2008; Whyte, 2010). In Palestine, under this prolonged occupation and colonisation, the interplay is most acutely felt because of the imbalance in power dynamics (Alatout, 2006; Qumsiyeh & Albaradeiya, 2022). International law can play a role in addressing situations of prolonged occupation and colonisation, such as in the case of Palestine (Koutroulis, 2012). There is also a body of international environmental law that is relevant to our cause (Sands & Peel, 2012) especially in light of the ongoing ecocide (Qumsiyeh, 2024; Qumsiyeh & Abusarhan, 2020, 2021). Most recently, in July 2024, the International Court of Justice (ICJ) ruled on several issues raised by the UN General Assembly (UNGA) regarding the impacts caused by Israeli policies and practices through its occupation, including on natural resources and environmental harm (ICJ, 2024). The World Court determined that, not only was Israel's use and disposal of hazardous waste in the Occupied

Territories a violation of multiple international humanitarian laws and treaties, the entire occupation itself was deemed illegal.

Our findings suggest that the main challenges faced can likely be minimised through a multi-faceted approach to tourism in Palestine that increases both economic stimulation and global solidarity to build pressures to dismantle the largest impact on environmental and other justice issues in the region: the colonial displacement and settler occupation of Palestine. Knowing that religious tourism is the biggest draw of international visitors to the region, there needs to be a shift in their experience while in the West Bank to include more interaction with and contribution to the local people and economy. There must be a change in marketing and regulations that allows for more opportunities for tourists to directly interact with those who are caretaking these sites and stewarding the land – the Palestinians. Through low climate impact activities, such as visiting the Palestine Museum of Natural History or hiking in local agricultural areas, such as Battir, the religious tourists could start to become more connected to the land and the people. Rather than simply passing through on their way to one holy site, they could build appreciation for the sovereignty of the indigenous population by witnessing the traditional methods of tending to the land and sites that they are here to visit. As the positive impact coming from tourism increases, the local authorities would theoretically be motivated to improve current regulations to maximise profits and protections for the local industries. Ultimately, however, final decision making comes from the occupation powers which do not act in ways to ease or improve conditions for the local population (see, for example, OHCHR, 2023). This is why every decision, including regarding tourism, should be made with the end goal of dismantling the occupation.

Increasing awareness of and opportunity for alternative tourism in Palestine is crucial to this end. Each tourist that makes a solidarity visit to Palestine, regardless of why they are drawn to the area – the nature, the religious sites, ancient history, research, etc. – can create an exponential impact on global solidarity upon their return home. Indeed, most recently, bestselling US author Ta-Nehisi Coates documented just 10 days that he spent experiencing alternative tourism in the West Bank in his newly released book *The Message* (Coates, 2024). This firsthand account, reflecting on what he witnessed, has caused ripple effects across the United States and much of the West; indeed, his book was a top seller during the week of Christmas 2024, and has a waitlist in libraries across the country. Every tourist brings a story back with them and in Palestine much of that story is shaped by the occupation and the palpable injustice and awe felt by international visitors when they are allowed to get off the Israeli tour buses and really spend time with the communities of Palestine. If we start to approach tourism as a key to increasing the global movement for the liberation of Palestine, then this will inevitably include a more sustainable living environment, as the occupation has the number one impact on the health of every living organism in the region.

Practically, this means getting creative within the current challenges and not wasting any time. The opportunities identified in the respondent interviews include changes that would not only improve tourism but improve local experiences as well. Increased accessibility and improved commercial regulations will benefit all. While tourism is being drastically limited during the current genocide, making local strides for local people, which will inevitably benefit tourism once it is returned to its pre-war numbers, is an option. Working on local sustainability practices with or without tourists involved will only make it easier when the tourists return. Similarly, petitioning local administrators with tangible data for improved market regulations will serve the local economy regardless of the consumers. Ultimately, the challenges and opportunities exist with or without tourists and, while this pause is occurring, it might be an ideal time to reassess and make digestible steps towards improving conditions while solidarity grows globally, even from afar, even without setting foot in Palestine. The world has never been able to witness these conditions of steadfast resistance in the face of naked colonial violence in such a clear way before. Millions of people worldwide have expressed solidarity with the people of Palestine. This presents a platform for further growth in the justice and solidarity tourism movement that supports Palestine as these witnesses maybe drawn to come and visit a liberated and sustainable Palestine.

Conclusion

In Palestine, we are still experiencing active colonisation and oppression. At least two of the people we interviewed argued that addressing tourism's environmental and climate justice issues would have to wait until after liberation is achieved. Yet, we argue that work can be done in preparation for post-conflict peace building and better natural resource management (Bruch *et al.*, 2016; Stahn *et al.*, 2017) and that a good model of this is found in the work undertaken by PIBS (see above).

The challenges articulated in the findings can be faced by taking advantage of the opportunities we outlined from our research study, advancing both environmental and climate justice through these processes. Others have proposed pathways to advance toward these forms of justice by combining the power of law (local and international law) with the power of the people (local action); analyses suggest this combination can be effective (e.g. Maru, 2023; Qumsiyeh, 2023b).

The tourism sector in Palestine, including ecological, religious and alternative niches of tourism, faces significant challenges due to the ongoing and prolonged Israeli occupation and apartheid. This complicates efforts to promote sustainable practices and equitable benefits for local communities. One tool for challenging this apartheid system is through creative forms of tourism that contribute to self-emancipation. This includes areas like expanding connectivity and local tourism (as a form of resistance and resilience). It includes recruiting internationals to visit or enhancing existing visitor itineraries by enticing them off Israeli controlled circuits and into visiting other areas (like Palestine's natural areas and museums).

Those visitors who witness the occupation and the impact it has on all living creatures in the region help grow solidarity, get involved and bring closer the goal of peace and sustainability for both human and natural communities. Increased awareness of the situation on the ground for plants, animals, bodies of water and most importantly the indigenous people, is one of the prongs that is needed to bring about a global campaign to end Israeli impunity and create accountability for the ongoing violations of myriad laws of man and nature. Bringing tourism into an occupied territory that is already experiencing a shortage of resources due to the Israeli restrictions and misuse of nature on top of climate change is a challenge and the cost versus benefit must be regularly reassessed. However, building global awareness of and resistance to the occupation is vital for any sustainable next steps. Environmental justice *in* Palestine is part and parcel of overall justice *for* Palestine.

Acknowledgements

We are grateful to the National Geographic Society for the grant funding for our work in the Bethlehem area with communities to develop a biosphere reserve plan that includes addressing issues of environmental justice. Thanks also to the editorial team for editing this manuscript.

References

Abahre, J. and Al-Rimmawi, H. (2023) Geotourism in Palestine. In M. Allan and R. Dowling (eds) *Geotourism in the Middle East* (pp. 249–261). Springer. https://doi.org/10.1007/978-3-031-24170-3_15.

Al-Haq (2021) Climate Oppression: A Major Tool to Establish and Maintain Israel's Apartheid Regime over the Palestinian People and Their Lands: Submission to the Office of the High Commissioner for Human Rights (OHCHR) Pursuant to Human Rights Council Resolution 47/24 'Human Rights and Climate Change'. Al-Haq.

Alatout, S. (2006) Towards a bio-territorial conception of power: Territory, population, and environmental narratives in Palestine and Israel. *Political Geography* 25 (6), 601–621. https://doi.org/10.1016/j.polgeo.2006.03.008.

Alternative Tourism Group of Palestine (ATG) (2010) Come and see: A call from Palestinian Christians. See https://www.atg.ps/resources/file/pages/Guidelines.pdf (accessed March 2025).

Aswita, D., Apriana, E., Herlina, Samuda, S. and Abubakar (2023) Ethno eco-tourism: Utilizing nature and culture for more sustainable tourism development. *Sociology and Anthropology* 111, 12–20. https://doi.org/10.13189/sa.2023.110102.

Becken, S. and Scott, D. (2024) Tourism and climate change stocktake: A call to action. *Journal of Sustainable Tourism* 32 (9), 2018–2038. https://doi.org/10.1080/09669582.2024.2390577.

Bouamrane M., Dogsé, P. and Price, M.F. (2019) Biosphere reserves from Seville, 1995, to building a new world for 2030: A global network of sites of excellence to address regional and global imperatives. In M. Reed and M. Price (eds) *UNESCO Biosphere Reserves* (pp. 29–44). Routledge.

Braverman, I. (2021) Environmental justice, settler colonialism, and more-than-humans in the occupied West Bank: An introduction. *Environment and Planning E: Nature and Space* 4 (1), 3–27. https://doi.org/10.1177/2514848621995397.

Braverman, I. (2023) *Settling Nature: The Conservation Regime in Palestine Israel.* University of Minnesota Press.

Bruch, C., Slobodian, L., Nichols, S.S. and Muffett, C. (2016) Facilitating peace or fueling conflict? Lessons in post-conflict governance and natural resource management. In C. Bruch, C. Muffett and S. Nichols (eds) *Governance, Natural Resources and Post-Conflict Peacebuilding* (pp. 953–1040). Routledge.

Coates, T. (2024) *The Message*. Random House.

EQA (2020) Nationally Determined Contributions (Palestine) submitted to United Nations Framework Convention on Climate Change (UNFCCC). See https://bit.ly/3cCVWSq (accessed March 2025).

EQA (2022) Sixth National Report to the Convention on Biological Diversity. See https://chm.cbd.int/database/record?documentID=257520 (accessed March 2025).

Eshed, V., Gopher, A., Gage, T.B. and Hershkovitz, I. (2004) Has the transition to agriculture reshaped the demographic structure of prehistoric populations? New evidence from the Levant. *American Journal of Physical Anthropology* 124 (4), 315–329. https://doi.org/10.1002/ajpa.10332.

Evans-Agnew, R.A. and Aguilera, J. (2023) Climate justice is environmental justice: System change for promoting planetary health and a just transition from extractive to regenerative action. *Health Promotion Practice* 24 (4), 597–602. https://doi.org/10.1177/15248399231171950.

Glasgow Declaration on Climate Action in Tourism (2021) See https://www.unwto.org/the-glasgow-declaration-on-climate-action-in-tourism (accessed July 2024).

Husein, D. and Qumsiyeh, M.B (2022) Impact of Israeli segregation and annexation wall on Palestinian Biodiversity. *Africana Studia* 37, 19–26.

International Court of Justice (ICJ) (2024) Summary of the advisory opinion of 19 July 2024. See https://www.icj-cij.org/node/204176 (accessed March 2025).

Isaac, R.K. (2010a) Moving from pilgrimage to responsible tourism: the case of Palestine. *Current Issues in Tourism* 13 (6), 579–590. https://doi.org/10.1080/13683500903464218.

Isaac, R.K. (2010b) Alternative tourism: New forms of tourism in Bethlehem for the Palestinian tourism industry. *Current Issues in Tourism* 13 (1), 21–36. https://doi.org/10.1080/13683500802495677.

Isaac, R., Hall, C.M. and Higgins-Desbiolles, F. (eds) (2016) *The Politics and Power of Tourism in Palestine*. Routledge.

Ives, R.L. (1950) The Palestinian environment. *American Scientist* 38, 85–104.

Kassis, R. (2013) The struggle for justice through tourism in Palestine. In L.A. Blanchard and F. Higgins-Desbiolles (eds) *Peace Through Tourism: Promoting Human Security Through International Citizenship* (pp. 225–240). Routledge.

Kassis, R., Solomon, R. and Higgins-Desbiolles F. (2015) Solidarity tourism in Palestine: The Alternative Tourism Group of Palestine as a catalyzing instrument of resistance. In R. Isaac, C.M. Hall and F. Higgins-Desbiolles (eds) *The Politics and Power of Tourism in Palestine* (pp. 37–52). Routledge.

Kelly, J.L. (2016) Asymmetrical itineraries: Militarism, tourism, and solidarity in occupied Palestine. *American Quarterly* 68 (3), 723–745. https://doi.org/10.1353/aq.2016.0060.

Koutroulis, V. (2012) The application of international humanitarian law and international human rights law in situation of prolonged occupation: Only a matter of time? *International Review of the Red Cross* 94 (885), 165–205.

Kuttab, E. (2024) Reframing war: Women, sanctions, and impoverishment in Gaza. *Journal of Middle East Women's Studies* 20 (2), 252–260. https://doi.org/10.1016/j.wsif.2025.103081.

Lee, S. and Jamal, T. (2008) Environmental justice and environmental equity in tourism: Missing links to sustainability. *Journal of Ecotourism* 7 (1), 44–67. https://doi.org/10.2167/joe191.0.

Lopes, H.T. and Almeida, I. (2017) The Mediterranean: The Asian and African roots of the cradle of civilization. In B. Fuerst-Bjelis (ed.) *Mediterranean identities—Environment, Society, Culture* (pp. 3–25). InTech.

Maru, V. (2023) A pathway to climate and environmental justice. *American Journal of Law and Equality* 3, 103–149. https://doi.org/10.1162/ajle_a_00060.

Messerschmid, C. (2008) What price cooperation? Hydro-hegemony in shared Israeli/Palestinian groundwater resources. In A. Aliewi, K. Assaf and A. Jayoussi (eds) Proceedings of the 1st International Conference on Sustainable Development and Management of Water in Palestine. Amman, Jordan, August 2007. El-Bireh: HWE- UNESCO-PWA.

Messerschmid, C. (2012) Nothing new in the Middle East–reality and discourses of climate change in the Israeli-Palestinian conflict. In J. Scheffran, M. Brzoska, H. Brauch, P., Link and J. Schilling (eds) *Climate Change, Human Security and Violent Conflict: Challenges for Societal Stability* (pp. 423–459). Springer.

Messerschmid, C. (2014) Hydro-apartheid and water access in Israel-Palestine: Challenging the myths of cooperation and scarcity. In M. Turner and O. Shweiki (eds) *Decolonizing Palestinian Political Economy. Rethinking Peace and Conflict Studies* (pp. 53–76). Palgrave Macmillan.

Milano, C., Novelli, M. and Russo, A.P. (2024) Anti-tourism activism and the inconvenient truths about mass tourism, touristification and overtourism. *Tourism Geographies* 26 (8), 1313–1337. https://doi.org/10.1080/14616688.2024.2391388.

Mimi, Z., Mason, M. and Zeitoun, M. (2009) Climate change: Impacts, adaptations and policy-making process: Palestine as a case study. Issam Fares Institute for Public Policy & International Affairs, American University of Beirut. See http://hdl.handle.net/20.500.11889/4258 (accessed March 2025).

Mohamed, B. and Suleiman, J. (2011) Challenges of religious tourism in Palestine. *Journal of Tourism, Hospitality & Culinary Arts* 3 (3), 25–38.

Novelli, M., Morgan, N. and Nibigira, C, (2012) Tourism in a post-conflict situation of fragility, *Annals of Tourism Research* 39 (3), 1446–1469. https://doi.org/10.1016/j.annals.2012.03.003.

Office of the High Commissioner for Human Rights (OHCHR) (2023) The human rights situation in the Occupied West Bank and East Jerusalem. See https://www.ohchr.org/sites/default/files/documents/countries/palestine/2023-12-27-Flash-Report.pdf (accessed March 2025).

Palestine Central Bureau of Statistics (PCBS) (2024) Palestinian Central Bureau of Statistics and the Ministry of Tourism and Antiquities issue a joint press release on the occasion of the World Tourism Day, 27 September 2024. See https://www.pcbs.gov.ps/site/512/default.aspx?lang=enandItemID=5833 (accessed March 2025).

Palestine Institute of Biodiversity and Sustainability (PIBS) (2022) National Biodiversity Strategy and Action Plan. See https://tinyurl.com/ntnkjarc (accessed March 2025).

Pappe, I. (2006) *The Ethnic Cleansing of Palestine*. One World Publications.

Peeters, P. and Papp, B. (2024) Pathway to zero emissions in global tourism: Opportunities, challenges, and implications. *Journal of Sustainable Tourism* 32 (9), 1–27. https://doi.org/10.1080/09669582.2024.2367513.

Pezzullo, P.C. (2009) *Toxic Tourism: Rhetorics of Pollution, Travel, and Environmental Justice*. University of Alabama Press.

Qumsiyeh, M.B. (2004) *Sharing the Land of Canaan*. Pluto Press

Qumsiyeh, M.B. (2017) Nature museums and botanical gardens for environmental conservation in developing countries. *Bioscience* 67 (7), 589–590. https://doi.org/10.1093/biosci/bix011.

Qumsiyeh, M.B. (2021) Challenging colonization: Building sustainable human and natural communities in Palestine. Radical Ecological Democracy. See https://www.radicalecologicaldemocracy.org/challenging-colonization-building-sustainable-human-and-natural-communities-in-palestine/ (accessed March 2025).

Qumsiyeh, M.B. (2023a) Developing institutions that serve national goals: Case study of the Palestine Institute for Biodiversity and Sustainability. *Al-Quds Journal for Natural Sciences* 1 (3), 6–10.

Qumsiyeh, M.B. (2023b) Environmental justice in Palestine: Rights of natives to their environment versus colonial onslaught. Security in Context. See https://www.securityincontext.org/posts/environmental-justice-in-palestine-rights-of-natives-to-their-environment-versus-colonial-onslaught (accessed March 2025).

Qumsiyeh M.B. (2024) Impact of Israeli military activities on the environment. *International Journal of Environmental Studies* 81 (2), 977–992. https://doi.org/10.1080/00207233.2024.2323365.

Qumsiyeh, M.B. and Abusarhan, M.A. (2020) An environmental Nakba: The Palestinian environment under Israeli colonization, *Science For the People* 23 (1). https://magazine.scienceforthepeople.org/vol23-1/an-environmental-nakba-the-palestinian-environment-under-israeli-colonization/ (accessed March 2025).

Qumsiyeh, M.B. and Abusarhan, M. (2021) Biodiversity and environmental conservation in Palestine. In M. Öztürk, V. Altay and R. Efe (eds) *Biodiversity, Conservation and Sustainability in Asia* (pp. 1–21). Springer.

Qumsiyeh, M.B. and Albaradeiya, I.M. (2022) Politics, power, and the environment in Palestine. *Africana Studia* 37, 9–18. https://ojs.letras.up.pt/index.php/AfricanaStudia/article/view/11963.

Qumsiyeh, M.B. and Amr, Z.S. (2020) Protection of endangered ecosystems via establishing museum research and education facilities: Experience from Palestine and proposal for the Arabian Gulf. *Museums in the Middle East Journal* 1, 29–32.

Qumsiyeh, M.B., Bassous-Ghattas, R., Handal, E.N., Abusarhan, M, Najajreh, M.H. and Albaradeyiyah, I. (2023) Biodiversity conservation of a new protected area 'Al-Arqoub', South Jerusalem Hills, Palestine. *Parks Journal* 29, 33–42, https://portals.iucn.org/library/efiles/documents/IUCN-Parks-NS-vol29-001.pdf#page=33.

Qumsiyeh, M.B. and Handal, E.N. (2018) Ecotourism: Opening a natural window to Palestine. *This Week in Palestine* 244, 38–42.

Qumsiyeh, M.B., Handal, E., Chang, J., Abualia, K., Najajreh, M. and Abusarhan, M. (2017) Role of museums and botanical gardens in ecosystem services in developing countries: Case study and outlook. *International Journal of Environmental Studies* 74 (2), 340–350. https://doi.org/10.1080/00207233.2017.1284383.

Qumsiyeh, M. B., McHugh, C., Shaheen, S. and Najajrah, M.H. (2024) Bio-cultural landscape and eco-friendly agriculture in Al-Arqoub, South Jerusalem, Palestine. *Agroecology and Sustainable Food Systems* 48 (10), 1489–1513. https://www.palestinenature.org/research/Bio-cultural-landscape-and-eco-friendly-agriculture.pdf.

Qumsiyeh, M.B., Saeed, R., Najajreh, M.H., Katbeh-Badr, N., Ikhmais, H., Simonett, O, Mackey, A. and Libert, M.E. (2022) Environmental education and climate change in a colonial context. *Africana Studia* 37, 109–121, https://ojs.letras.up.pt/index.php/AfricanaStudia/article/view/11970.

Qumsiyeh, M.B., Zavala, S.S. and Amr. Z.S. (2014) Decline in vertebrate biodiversity in Bethlehem, Palestine, *Jordan Journal of Biological Sciences* 7, 01–107.

Salameh, E. and El-Naser, H. (2008) Restoring the shrinking Dead Sea – The environmental imperative. In F. Zereini and H. Hötzl (eds) *Climatic Changes and Water Resources in the Middle East and North Africa. Environmental Science and Engineering* (pp. 453–468). Springer.

Salem, H.S. (2011) Social, environmental and security impacts of climate change on the Eastern Mediterranean. In H. Brauch et al. (eds) *Coping with Global Environmental Change, Disasters and Security* (pp. 421–445). Springer. https://doi.org/10.1007/978-3-642-17776-7_23.

Sands, P. and Peel, J. (2012) *Principles of International Environmental Law*. Cambridge University Press.

Schlosberg, D. and Collins, L.B. (2014) From environmental to climate justice: Climate change and the discourse of environmental justice. *Wiley Interdisciplinary Reviews: Climate Change* 5 (3), 359–374. https://doi.org/10.1002/wcc.275.

Stahn, C., Iverson, J. and Easterday, J. (2017) *Environmental Protection and Transitions from Conflict to Peace: Clarifying Norms, Principles, and Practices*. Oxford University Press.

Tabash, M.I. (2017) The role of tourism sector in economic growth: An empirical evidence from Palestine. *International Journal of Economics and Financial Issues* 7 (2), 103–108. https://dergipark.org.tr/en/pub/ijefi/issue/32035/354450.

Tal, A. (2002) *Pollution in a Promised Land: An Environmental History of Israel*. University of California Press.

Tourism Panel on Climate Change (TPCC) (2023) Tourism and climate change stocktake (S. Becken and D. Scott, ds.). See https://tpcc.info/stocktake-report/ (accessed March 2025).

Turner, B., Devisscher, T., Chabaneix, N., Woroniecki, S., Messier, C. and Seddon, N. (2022) The role of nature-based solutions in supporting social-ecological resilience for climate change adaptation. *Annual Review of Environment and Resources* 47, 123–148. https://doi.org/10.1146/annurev-environ-012220-010017.

UNCTAD (n.d.) Background: The question of Palestine. See https://unctad.org/topic/palestinian-people/The-question-of-Palestine (accessed March 2025).

United Nations Development Program (UNDP) (2010) Climate Change Adaptation Strategy and Programme of Action for the Palestinian Authority, https://bit.ly/2UhA5tq (accessed March 2025).

United Nations World Tourism Organization (UNWTO) (2007) The Davos Declaration on Climate Change and Tourism. See https://cetesb.sp.gov.br/proclima/wp-content/uploads/sites/36/2014/05/declaracion_davos.pdf (accessed Match 2025).

Verner, D. (ed.) (2012) *Adaptation to a Changing Climate in the Arab Countries: A Case for Adaptation Governance and Leadership in Building Climate Resilience*. The World Bank.

Weizman E and Sheikh, F. (2015) *The Conflict Shoreline: Colonialism as Climate Change in the Negev Desert*. Steidl

Whyte, K.P. (2010) An environmental justice framework for indigenous tourism. *Environmental Philosophy* 7 (2), 75–92. https://www.jstor.org/stable/26168043.

World Bank (n.d.) Water Resources Management. See https://www.worldbank.org/en/topic/waterresourcesmanagement (accessed March 2025).

Yin, H., Eklund, L., Habash, D., Qumsiyeh, M.B. and Van Den Hoek, J. (2025) Evaluating war-induced damage to agricultural land in the Gaza Strip since October 2023 using PlanetScope and SkySat imagery. *Science of Remote Sensing* 11, 100199. https://doi.org/10.1016/j.srs.2025.100199.

Zeder, M.A. (2008) Domestication and early agriculture in the Mediterranean Basin: Origins, diffusion, and impact. *Proceedings of the National Academy of Sciences* 105 (33), 11597–11604. https://doi.org/10.1073/pnas.0801317105.

9 Affective Solidarity in Melting Destinations: Stepping Forward, Standing With and Staying Connected to Climate Justice

Monica Nadegger and Carina Ren

Using two melting destinations in Greenland and Tyrol as empirical sites of exploration, we exemplify how climate solidarity interferes with tourism development and strategies through citizen interventions and social movements. By situating the different approaches of affective solidarity and climate justice in the two destinations, we unravel the intersecting and diverging colonial, tourism and development legacies still prevalent and emerging in current business models set on expansion, conquering and domination in melting destinations. We discuss who and what is included in decision-making processes in tourism development, what tensions these activities spur locally and what alliances form. We link the empirical examples to discussions on how affective solidarity can nourish collective action in the fight for climate justice. Based on these examples, we build on three moves of stepping forward, standing with and staying connected as a methodological approach to guide and build affective solidarity across melting destinations.

Introduction

Climate and related socioenvironmental changes are increasingly becoming visible across the globe. The global surface temperature has already increased more than +1 °C, soon surpassing the boundary condition of 1.5°C of the Paris Climate Agreement (Intergovernmental Panel on

Climate Change (IPCC), 2023). The concentration of greenhouse gases such as CO_2 or methane linked to human activity is steadily increasing, with climate change affecting those who are most vulnerable and contributing the least, through hot weather extremes, droughts, heavy precipitation and subsequent loss of land, agriculture and, eventually, safe places to live and thrive (IPCC, 2023). In a tourism context, cold-weather destinations are impacted and challenged by warming temperatures, snow scarcity, shorter stretches of cold temperatures, thawing permafrost and more dramatic weather events. This has led to what we term 'melting destinations'.

Melting destinations are places of tourism, in which snow, ice or cold weather, otherwise serving as a prerequisite for its tourism offer and, more importantly, for the sustaining of everyday life and the wellbeing of its residents, are withering. In this chapter, we draw insights from Greenland and Tyrol, using illustrative examples to compare and bridge two seemingly very distant and different destinations. The Indigenous island nation of Greenland, an Arctic region, is subject to warming of the climate up to four times the global pace (Rantanen et al., 2022). Across the country, warming temperatures, less and more unstable ice, permafrost thaw and more unpredictable and extreme weather have been seen to impact tourism in various and profound ways (Ren et al., 2024). Similarly, while not warming quite as fast as Arctic destinations, mountainous destinations, such as Tyrol located in the Austrian Alps region, are also experiencing climate change, with shrinking glaciers and unreliable snowfall adversely effecting the winter tourism sector. Therefore, in this chapter, we will introduce and explore two such melting destinations – Greenland and the Tyrolean Alps – and show how struggles for climate justice and the formation of climate solidarity are distinct yet interconnected. Most often, climate justice is explored either conceptually or through the exploration of singular cases or destinations. Thus, this chapter offers unique insights using two illustrative examples to compare and bridge different and also similar experiences as destinations.

By situating the different approaches of affective solidarity and climate justice in the two destinations, we unravel the intersecting and diverging colonial, tourism and development legacies still prevalent and emerging in current business models set on expansion, conquering and domination in melting destinations. By analysing the adversity that both destinations face and new conversations and relations that emerge from such struggles, we can better understand the dynamics of climate justice in relation to tourism across melting destinations and speculate about how climate solidarity may be built between them. As argued by Ren and Nadegger (2026) new ways of connection and comparison in tourism may foster a common repertoire of 'staying with the trouble' (Haraway, 2016) of climate changes, both at a local and global level. We are thus curious about what we can learn from struggles for climate justice and solidarity that take place in and across these destinations.

To this aim, we proceed in the following way. We start by exploring the key pillars of climate justice in tourism, focusing on the potential of climate justice as a movement of shared civil protest. We then introduce the concept of affective solidarity (Baxter, 2021; Hemmings, 2012; Vachhani & Pullen, 2019) as a key pillar for the emergence of collective civil protest and situate it within the context of climate justice in the two melting destinations. We use examples of two citizen protest movements, one *for* tourism and one *against*, as illustrations to unfold and discuss the three moves of affective solidarity: *stepping forward, standing with and staying connected* (Baxter, 2021; Walters & Butterwick, 2017). Stepping forward entails, for instance, speaking out to bring dissonance and its roots in oppressive systems to the fore. Standing with connects to practices that showcase and provide support, resources and alliance to those stepping forward, while acts of staying connected engage in a continuous commitment and connections around a shared cause (Walters & Butterwick, 2017). These methodological approaches help to study social movements around climate justice in tourism and build affective solidarity across melting destinations.

Climate Justice in Tourism

> A working definition of social justice is work toward a world of equity, dignity, and basic rights through democratically organized social spaces. Environmental justice is work toward a world of equity, dignity, and rights with respect to ecological conditions and decision making processes. Socioenvironmental justice, then, is work toward a world that accounts for equity, dignity, and rights of all members of the interlocking human and environmental material and discursive conditions. (Rose, 2014: 267)

As exemplified in Rose's (2014) statement, climate justice in tourism combines several elements: (1) social justice (addressing and dismantling oppressive systems like colonialism, Eurocentrism and racism to resolve inequalities between communities) and (2) ecological justice (confronting fossil-fuel dependency through environmental protection and restoration that includes all life on Earth, see also Rastegar *et al.*, 2023). The aim of climate justice approaches in tourism is then to 'expose the *root causes* of climate change to address and *dismantle these systemic issues* and structures in different ways' (Sultana, 2022: 119).

The *root causes* of climate injustice in tourism are entangled with growth-driven extractivism, carbon-intensive elitism and structural oppression. The logic of growth evident in capitalist neo-liberal approaches entails a relentless focus on capital accumulation and resource extraction underpinning current tourism practices, prioritising profit over ecological limits (Blanco-Romero *et al.*, 2023). Evidently, the related high-carbon, elitist lifestyles concentrated in the Global North and West countries perpetuates unequal ecological footprint (Gössling *et al.*, 2010; Higgins-Desbiolles, 2024). Similarly, deep-rooted issues of colonialism, racism and sexism (Juskus, 2023; Sealey-Huggins, 2018; Sheller, 2021) enable systemic othering and devaluation of human and non-human others (Qian, 2022; Rastegar, 2022) which, in turn, perpetuates exploitative tourism paradigms. This includes continued appropriation of land that belongs to Indigenous or marginalised communities (Harbor & Hunt, 2021; Lazic & Della Lucia, 2024).

These *systems of oppression* exacerbate existing inequalities now and in future tourism activities amid the climate crisis, reaching beyond and, at the same time, connecting destinations and regions through shared vulnerability and responsibility. As Zhou *et al.* (2024) show in their meta-analysis on the influence of climate factors on tourism demand, countries with very little responsibility, carbon contribution and decision-making power (such as Small Island Destinations, low-income countries, etc.) will be adversely impacted. In contrast, wealthier high-income countries (typically in the Global North and West) might see increased travel demand, accompanied by higher carbon emissions. As they conclude, the future development of both tourism demand and climate justice fuels a 'vicious cycle of climate injustice, highlighting the disparities in climate-related social costs globally' (Zhou *et al.*, 2024: 1762). Importantly, to study climate (in)justices means not only to focus on 'unequal impacts of climate change on vulnerable destinations/peoples' (Scott & Gössling, 2022: 206), it also requires considering and overcoming such injustices for a just transition to more equitable and sustainable forms of tourism (Rastegar & Ruhanen, 2023b).

The types of climate justice to tackle the aforementioned vicious cycle can be categorised along several main pathways of action, such as distributive justice (equal distribution of harm and amenities), procedural justice (laws, regulation and policies and their democratic formation and constitution) and recognition justice (the recognition of places, identities and their dignity) (Chen, 2023; Rastegar, 2022; Rastegar & Ruhanen, 2023a, 2023b). However, in this chapter, we want to zoom in on the collective action and shared civil protest needed to dismantle these systemic issues. Hence, we look at climate justice as a social movement.

Climate justice has always built on grassroots initiatives, which nourish local and global alliances and emerge in shared civil protest (Chen, 2023; Juskus, 2023; Sultana, 2022). Sharing a history with environmental justice, the terminology and a plea for a climate-just world emerged in the racialised class struggle in the United States when marginalised communities resisted the disproportionate exposure to pollution and toxic waste (Chen, 2023), resisting the labelling of their spaces, homes and neighbourhoods as 'sacrifice zones' (Juskus, 2023). Further building on these activist roots, climate justice has since become a 'galvanizing force', driven by Black, Indigenous and People of Colour (BIPOC), women, youth and LGTBQIA+ activists (Sultana, 2022) who fight for (and have always fought for) justice for all human and more-than-human flourishing on Earth.

To further nourish climate justice as a resistance movement and collective action for change, local and global multispecies alliances are required. But how can such alliances emerge, persist and unfold? In the next section, we turn to solidarity as one way to understand the momentum behind climate justice as a lively and collective praxis (Sultana, 2022).

Building affective solidarity for climate justice in tourism

Solidarity can be defined as 'the intricate bonds that bond a group or community, often rooted in a shared sense of identity and belonging among its members' (Bazzani, 2024: 1633). Similarly, Guia refers to social solidarity as the 'social cohesion of a particular community, a type of natural solidarity between members of the same kinship or culture or other characteristics that members share. The shared characteristics produce social bonds among solidary members, together with positive collective moral obligations' (2021: 379). Solidarity is often seen as rooted in a pre-defined identity, a geographically bound context with shared governance or rules (for example, in a nation state), or in small, tightly knit communities with direct forms of belonging and connectedness (Bazzani, 2024). However, such conventional framing falters in addressing the complex, interrelated and unequal root causes and consequences of the climate crises and the recognition of such in striving for climate justice (Sealey-Huggins, 2018). As Bazzani (2024) emphasises, collective welfare outside of nation states or multi-state agreements is difficult to enforce from a governmental perspective. In addition, direct forms of solidarity become more challenging in a global society (see also Rastegar, 2022) as the climate crisis and its consequences are temporally and spatially dispersed and responsibility and limitations are not 'equal'. This is the case with historical and current forms of tourism that feature neo-colonial relations and Western exploitation (Higgins-Desbiolles, 2022). Future paths look very different for different communities that are positioned so unequally.

With tourism as a globally stretched and locally embedded phenomenon, social solidarity that builds on kinship and, ultimately, reconciliation requires connections 'between communities of people' (Higgins-Desbiolles, 2024: 480). It also involves asking with whom and what we can make peace by recognising 'mutual vulnerabilities and mortality' (Higgins-Desbiolles, 2024: 480). We propose the concept of 'affective solidarity' (Baxter, 2021; Hemmings, 2012; Vachhani & Pullen, 2019) as a promising way forward to 'think-with' climate justice in tourism. This concept emphasises an active form of care that includes more-than-human wellbeing and the right to flourish (Rastegar, 2022; Rastegar & Becken, 2024) in and across melting destinations.

Building on feminist reflexivity and politics, Hemmings describes affective solidarity as seeking 'solidarity with others, not based in a shared identity or on a presumption about how the other feels, but on also feeling the desire for transformation out of the experience of discomfort, and against the odds' (2012: 158). The affect that sparks such a desire for transformation can range from rage, anger, frustration and grief to connection and hope. Such affect can not only travel through social relations but can also be anchored in the more-than-human materialities of everyday life: domestic homes, landscapes, craft, other-than-human companions, materialities and technologies.

These affects guide us to a realisation that the way one ought to be treated (like a human being with equal rights) and the hope and possibilities one ought to have (like a career, autonomy over one's own body, flourishing and vital land and nature) do not match lived experiences of sexism, exploitation and discrimination (Baxter, 2021; Hemmings, 2012). This dissonance then might turn into a feeling of injustice, the desire to voice this injustice and to reach out to others who might feel the same injustices and want to build an alternative pathway collectively. Such '[f]lows of affect help align and intensify relations between people in solidarity' (Baxter, 2021: 901), which then build the basis for collective action towards a 'sustainable feminist politics of transformation' (Hemmings, 2012: 148).

While these studies mainly focus on gendered instances of oppression (Baxter, 2021; Hemmings, 2012; Vachhani & Pullen, 2019), we see affective solidarity as a potentially powerful concept for climate justice in tourism, focusing on the dissonances in the possibilities and autonomy of livelihoods and their erosion through systems of oppression. In relation to climate justice in tourism, and as we illustrate further down, affective solidarity emerges in melting destinations where melting snow or ice threatens local livelihoods. However, it may also be suppressed in places where existing infrastructure and histories of domination obscure the need for transformation.

Affective solidarity as a feminist commitment strives for connection across differences by honouring a multiplicity of worldviews and approaches that are necessarily based in local attachment and value different experiences while reaching out to form global alliances. Such a connection through affective solidarity should not oppose or even erase the political and oppressive conditions of current climate injustices (such as colonialism) and the vulnerability of marginalised communities that come with them (Sheller, 2021) or try to create a unified and dominant voice of 'what is right or wrong'. Instead, it is a plea to enact climate justice as a critical praxis by *'stepping forward'*, *'standing with'* and *'staying connected '* (Baxter, 2021; Walters & Butterwick, 2017). These three feminist moves bring affective solidarity into practice by highlighting dissonance, providing support and finding allies, as well as maintaining a commitment to causes of climate justice (Walters & Butterwick, 2017).

We now move to two illustrative examples of such affective solidarity from Greenland and the Tyrolean Alps. We look at ways in which 'stepping forward', 'standing with' and 'staying connected' (Baxter, 2021; Walters & Butterwick, 2017) could emerge in melting destinations with accounts from citizen protest in both places.

A Brief Tourism History and Context of Melting Destinations: Greenland and the Tyrolean Alps

Greenland

The first place we introduce is Greenland, an emerging destination covering stretches across the High Arctic, Arctic and sub-Arctic. While the island nation is a massive mass of land and ice, it is only inhabited by some 57,000 people, distributed across 17 towns and 54 smaller coastal communities. Greenland encompasses a huge territory with small, dispersed settlements with limited and costly accessibility (Ren *et al*., 2024). The development of tourism goes hand in hand with the gradual process of Greenlandic decolonisation from its early beginnings in 1953, when Greenland transitioned from a Danish colony into becoming a Danish county, thus allowing for leisure-based travels that had hitherto been forbidden (Hegelund, 2009). In that sense, tourism – starting from the ability to travel to Greenland in 1953 until the opening of Greenland's first civilian transatlantic airport in 2024 – is closely connected to the emancipation of an Indigenous nation.

Today, Greenland is a niche and off-the-beaten-track adventure destination. As an Arctic island, it is remotely situated from main tourism markets, difficult and costly to access by plane or cruise ship and offers only a limited range of experiences and services within a very poorly developed and limited distribution system (Rambøll Consulting, 2014). But things are changing very quickly. Greenland has received massive global attention due to recent geopolitical events, one of which being Trump's 2019 and 2025 statements of wanting to 'buy Greenland'. With an international airport opened in 2024 in Nuuk, two new airports planned in the tourist 'capital' of Ilulissat in 2025 and the cruise hub of Qaqortoq in 2026, tourism numbers are expected to grow in years to come. While the development of Greenlandic tourism is discursively framed by its authorities as motivated by a wish to create value 'for all of Greenland', Markussen and Ren (2023) have pointed to severe regional inequalities and injustices reproduced within current systems and rooted in a colonial past (Leoni, 2019). This shows that 'without equality, a sustainable development is unlikely to take place, and negative impacts will certainly threaten the future' (Leoni, 2019: 75).

Tyrol

In the context of the European Alps, Gobiet *et al.* show that 'the number of snow-abundant winters by the end of the century would represent a 1-in-30 year's event, compared to 8 in the reference climate (1960 and 1991)' (2014: 1147). Tyrol, as an example of an alpine winter destination, has been profoundly transformed and shaped by a massive ski tourism industry for over a century. However, as snow increasingly turns to rain during the winter high season, smaller ski areas at lower elevations are beginning to experience the toll of climate warming (Steiger & Stötter, 2013; Steiger & Scott, 2020) and, like Greenland, are becoming melting destinations. At the same time, snow-secure terrains with glacier ski resorts and high-altitude skiing facilities equipped with state-of-the-art snowmaking technology are seeing an increased concentration of the market and potential future gains in winter tourism (Steiger *et al.*, 2019).

At the same time, the alpine tourism model here has parallels to the international quest of dominating and conquering places beyond Europe, but instead of crossing oceans, it extends to remote Alpine peaks (Mathieu, 2022). With the building blocks of the industrial revolution, increased consumption and production and the hope for technological innovation, the economic growth through tourism in the Alps has produced marginalisation of other sectors and practices such as agriculture or local heritage (Mazza, 2023) and contributed to the steep rise in carbon emissions worldwide (Steiger *et al.*, 2024). The 'success' story of Alpine ski tourism is hence also nested in both local and global injustices.

We have briefly sketched situations in which a warming climate can offer new possibilities for tourism development, is nested in (colonial) heritage or creates different 'winners' and 'losers' in these two melting destinations. What the reports of situated climate changes show from these melting destinations is that all these changes are entangled with issues of (in)justice, however, not always in clear and straightforward ways. At times, just decision making is even blurred or made impossible. We now draw on examples from both destinations, where these underlying injustices and the hope for a climate-just future have nurtured affective solidarity and come to the fore.

Affective Solidarity in Melting Destinations

Daring to slow and scale down development: Choosing tourism over mining in Greenland

One of the most prominent debates in the past decades in Greenland revolves around the extraction of enormous deposits of rare earth minerals (including uranium, fluoride and thorium) discovered in South Greenland. The findings coincided with the lifting of a decade-long moratorium on mining radioactive elements in 2013 by the Greenlandic

Government, which had gained institutional control over mining (from Denmark) in 2009. If mined, these extremely lucrative deposits would provide a substantial contribution to economically secure Greenland's further independence from Denmark and help create a new and prosperous future for its residents. But the question is: Should Greenland allow uranium mining, which would threaten the livelihood of local inhabitants as well as the region's fragile ecosystems? Discussions have raged locally and nationally on what to decide.

Studies on local views on uranium extraction spoke of concerns but also a positive view of the possible opportunities from mining in an area marked by depopulation and decline (Bjørst, 2016). Ren and Bjørst (2016) show how mining companies, with the help of prospectuses, videos and other visual and written material, seek to establish a positive vision of mining, which 'yields a promise of a docile, productive and risk-free landscape' (2016: 296). While seeking to communicate the landscape as hospitable to mining exploration, what was omitted was how contamination and restricted access would also render the landscape unwelcoming, unproductive and unsafe for local farmers, hunters and other humans and non-humans dwelling on and living off the land.

While many spoke in favour of mining (for its contributions to national economic growth to spur self-determination, possible regional development), many arguments also existed against it (including pollution, contamination and regional degeneration). To create a counter-narrative to mining, adversaries of mining extraction introduced tourism as a more sustainable alternative, for instance, pointing to hiking and angling tourism in local streams. What was made clear, however, was that mining and tourism futures were mutually exclusive in South Greenland; there was a clear choice between developing one or the other (Ren & Bjørst, 2016). Because, as asked in a newspaper article describing the mining debate, 'how many tourists would like to fish with a uranium mine in their back garden?' (Andersen, 2015, our translation).

At the time of writing, Ren and Bjørst (2016) did not have high hopes for tourism in Narsaq, faced with big mining companies, stating that: 'Unfortunately, the tourism future cannot compete with the scale offered by the mining future: a large scale. Tourism only offers modest and uncertain revenue and does not lend itself well to decade-long promises of steady income' (2016: 298). But it turned out that their predictions were wrong.

After an election ended in victory for the party Inuit Ataqatigiit (IA), which went to the polls on the promise of banning uranium extraction, the ban was reintroduced in 2021. While a mining project could have created jobs and billions in revenue for Greenland, an opinion poll revealed that a majority considered the environmental consequences and risk of an environmental disaster to be too great (Voller *et al*., 2021). In this regard, Mute B. Egede, the leader of IA, stated: 'People's living conditions are one thing. And then there is our health and the environment ... Things are

Figure 9.1 Demonstration against uranium mining. © Lill Rastad Bjørst Used with permission

going on, and now the population has stood up and demanded that it stop' (Voller *et al.*, 2021: n.p.). Through different acts of opposition – voting, ruling and physically protesting (see Figure 9.1) – Greenland citizens, politicians and locals formed an alliance to reintroduce the national ban on uranium extraction.

Today, tourism is gradually developing in South Greenland. For example, small-scale sheep farmers have opened their homes as bed and breakfasts (B&Bs) to hikers. These alternative accommodations provide the sheep farmers with additional financial opportunities under climate changes. Wennecke *et al.* shared experiences of a local tourism entrepreneur:

> Being a sheep farmer probably isn't as good as it used to be. Perhaps just ten years ago, things were much better. Now we have the drought, where rain just doesn't come. But then we thought that we wanted to have something to do with tourism to make more money so that we didn't have to move from our farm. Particularly because it's a family farm, and we want to stay here, even though it doesn't give that much. But we want to be here. (2019: 56)

By standing with the land, these farmers and Greenlandic voters pushed back a major corporation whose activities would have rendered liveable and partly tourism-based futures impossible.

The shutting down of larger prospects for mining also meant that tourism development soon overturned mining as a major economic priority for South Greenland and Greenland as a whole. In November 2024, the nation's very first Tourism Act was passed to accommodate growing tourism. The two main and most contested points in the law were Greenlandic ownership, enforcing a two-thirds majority of Greenlandic ownership in tourism companies operating in Greenland and the implementation of cruise and accommodation taxes. In his 2025 New Year address, Prime Minister Mute B. Egede pointed out that the law outlines:

> [...] the framework for a development that clears a path for and considers the country's population. Sustainability is not only about the environment but also about ensuring that the Greenlandic people control their future. What is central to the law is to create and develop tourism that is adapted and controlled by our country - both in terms of the environment, but also in terms of the people who live there. [our translation] (Egede, 2025)

In the Tourism Act, the well-known and often generic wish to foster sustainable tourism development is operationalised with a clear emphasis on the social tenet of the concept, seeking to give priority to Indigenous ownership, reduce leakage and secure a better distribution of revenues from tourism. While the opposition against the law headed by Greenland's Business Association argued that stricter legislation and taxation would stifle tourism growth, the Tourism Act as well as the protests in South Greenland show a commitment to law making that does not prioritise large-scale and growth-optimising development initiatives. In South Greenland, protests and concerns about the land overruled not only prospects for local economic growth, but also a potential national economic boost to support self-sustainability and independence. In both instances, policy to slow down or 'complicate' tourism development for the sake of greater social and environmental goods and longer-term sustainability prevailed.

These examples of policy development supported by citizen protests provide a counter example to challenge other Arctic and Northern mega projects where the idea of the 'left behindness' of both places and people often 'position policy actors as prone to taking higher risks and perceiving fewer opportunities and alternative futures', as argued by Eriksson *et al.* (2025: 2). In such scenarios, 'any promise of a new development path is often seen as a local necessity, regardless of the associated risks [where] lagging regions may agree to establish megaprojects, even when these projects are challenging to oversee, exceed budgets and jeopardize public funds' (Eriksson *et al.*, 2025: 2). In the case of Greenland, this turned out not to be the case.

Demanding climate justice for future generations: The case of tourism protest in Tyrol

In Tyrol, tourism is one of the key sources of economic value creation, with roughly every third euro in value added coming from tourism and tourism-related activity (Land Tirol, 2021). Yet, with the rising temperatures (see Figure 9.2 for impressions of melting landscapes and snow-scarce winters) and global policy shifting towards sustainability, Tyrol's official strategy and tourism development statements include several claims and goals, where sustainable action and change is reflected (Amt der Tiroler Landesregierung, 2021; Land Tirol, 2021). Most points in these governmental and tourism strategies refer to goals such as the United Nations' Sustainable Development Goals and aim to reduce carbon emissions, shift to green and renewable energy resources and focus on green mobility.

These policies only have minor connections to explicit calls for climate justice. For example, the regional climate and sustainability strategy of the regional government in Tyrol (Amt der Tiroler Landesregierung, 2021: 12) mentions 'leave no one behind' (dt. 'niemanden zurücklassen') emphasising the active voice of civil society as a guiding principle, and links this to diversity-related issues (such as limited access to decision making by people with disabilities, people with a history of migration, the youth and the elderly) in its initial document. While 'leaving no one behind' also hints towards regional and geographical issues (e.g. the concentration of wealth, infrastructure and opportunities limited to specific regions), these claims remain situated in the regional bounds of Tyrol and do not discuss broader inequalities, questions of mobility and migration and income inequality in Austria and beyond.

Following up on these guiding principles, there are no explicit and tangible pathways to achieve this call for social justice except for a

Figure 9.2 Warm winter weather and melting snow days in the Tyrolean Alps.
© Monica Nadegger (author's own)

statement on the importance of education in the updated 2025 action plan (Amt der Tiroler Landesregierung, 2024). In the same vein, the regional tourism strategy (Land Tirol, 2021) mentions sustainable tourism as a priority goal for tourism development, positioning Tyrol as a self-perceived pioneering competency leader in developing destination management for sustainable alpine tourism. While climate change indicators (such as reducing CO_2 emissions or increasing the use of clean energy) are prominently featured in Tyrol's tourism strategy, the explicit discussion of issues pertaining to climate justice in tourism are largely absent. The only facets of the strategy reflecting the voice of the local population are social sustainability indicators (with indicators such as tourism intensity and attitude towards tourism) and guiding principles such as 'actions across past and future generations'. The absence of such explicit mentions of climate justice pathways in official documents and policies does not mean that there is no fertile soil for climate justice, even if here it emerges in different places, such as civil society.

Focusing on the power of social movements and the role of affective solidarity in such struggles, we now turn to an illustrative example of collective protests around tourism development in Tyrol. Tourism development, especially in relation to the impact of the climate crisis on winter tourism, has faced criticism from local grassroots movements in Tyrol (Wegerer & Nadegger, 2023). For example, in 2020, a petition was raised against the planned Pitztal-Ötztal glacier ski area, demanding a stop to the proposed tourism development project and reform of the Tyrolean legislation that regulates the number of cable cars and ski areas (Estermann, 2019). Importantly, it was not tourism per se that was under critique, but specific aspects and mechanisms nested within tourism. Wegerer and Nadegger (2023) showed how activists criticised the dominant capitalist growth logic in ski tourism, arguing that greed for profit overshadows environmental and social wellbeing. The activists' criticism focused on a powerful elite that lacks democratic accountability and disregards the long-term consequences of tourism development on future generations. The comments of protesters on the online petition platform emphasised the need to develop alternative forms of winter tourism to safeguard natural resources, demanding justice for future generations who cannot yet participate in democratic decision-making processes and thus have no voice.

These social movements also demonstrate the struggles around climate justice. With more than 170,000 signatures, alliances between the local population and broader public formed (Wegerer & Nadegger, 2023), linked by the dissonance between the status quo (i.e. development and the destruction of nature for the wealth of a few) and worries for how this affects the less powerful groups (i.e. the youth and the future generations yet to come). We see similarities of protesters' claims to some indicators in the public strategies, such as 'leaving no one behind' referring to the youth or inter-generational action plans. However, in contrast to the policy and strategy documents, these grassroots movements explicitly connect such struggles to issues around climate justice, such as wealth distribution, exploitation of nature for profit and calls for alliances across different groups (as in this case, between locals and visitors) and new alternative visions for tourism futures. The dissonance is felt not only by individuals but moves collective protest in standing together in affective solidarity. Such a civil engagement for climate justice can mobilise enough attention and pressure for local governments to rethink and stop development projects, as was the case of this halted development project.

Discussion: Stepping Forward and Standing with Climate Solidarity in Melting Destinations

Examples of social movements in South Greenland and Tyrol show that affective solidarity can emerge along different axes, spanning local and visitors, for and against tourism and with various goals in melting destinations. They demonstrate how civil society steps forward and forms new alliances. Such connections can enhance the demand for climate justice in melting destinations. So how do our examples illustrate the building of affective solidarity? We suggest that it happens by *'stepping forward'* and *'standing with'* (Baxter, 2021; Walters & Butterwick, 2017) in demanding climate justice in melting destinations.

The precondition for *stepping forward* (Baxter, 2021; Walters & Butterwick, 2017) in both destinations was to reflect upon current injustices and articulate this dissonance and the resulting critique against dominant promises of economic development, growth and a 'better' life for the local population. In Tyrol, such dissonance developed from the critique and histories of development that served only a few while marginalising others' needs (such as those of future generations to enjoy natural and preserved alpine terrain). In South Greenland, the promises of jobs and revenue in mining stood against the environmental consequences and the living conditions of present and future inhabitants. The realisation of such dissonances, of how life could be and how it *really* is, is the first step towards an articulation of injustices (Baxter, 2021; Hemmings, 2012). However, nourishing climate justice as a social movement (Sultana, 2022) requires stepping forward and 'bearing witness' in solidarity (Baxter, 2021: 902; see also Hemmings, 2012). Such a bearing witness might not lead to direct solutions in the moment, but may serve to legitimise internalised struggles of feeling like claims for a better life are 'wanting too much' and transform them into broader collective alliances (Hemmings, 2012).

Moving things in the literal sense requires forming such alliances: It requires *standing with* others (Baxter, 2021; Walters & Butterwick,

2017). In the Tyrolean example, the protests against the tourism development project led to a broader alliance between two groups whose interests are usually portrayed as conflicting: visitors and locals (Wegerer & Nadegger, 2023). Their broader alliance formed a critical mass for gaining traction, voicing a more climate-just vision of the Alps for future generations. In Greenland, local farmers and solidarity-driven Greenlanders stood with each other and pushed back against uranium extraction, voicing visions for more just futures for Greenland through tourism. As we see with these examples, climate justice here can advocate against and for tourism, while the underlying practice of forming alliances remains vital for both quests for justice in melting destinations. In the Greenlandic case, local politics became part of the movement, pushing for futures that look towards tourism and against mining. While such involvement in struggles for climate justice by local and national governments is crucial, these protests also show that climate justice will not be realised by nation states and politics alone. The affective solidarity of these diverse groups of people in standing with each other is vital for initiating a shift that promotes the voices of the local population. The understanding of climate justice then becomes especially important in times when actions by nation states deteriorate prior agreements, with, for example, the US exiting the Paris Climate Agreement or cutting funds for climate-change-related research projects at the beginning of 2025.

Conclusion: Staying Connected Through Careful Comparisons

As we have exemplified with the Tyrolean and Greenlandic examples of affective solidarity, *stepping forward* and *standing with* are crucial practices of nurturing climate justice as a collective movement. While we have discussed these examples within each melting destination, we want to end this chapter by speculating about the shared potential of *staying connected* as an ongoing commitment to affective solidarity (Walters & Butterwick, 2017) for climate justice in melting destinations. How can we find and nourish a language that connects such similar yet different destinations without generalising their struggles? How can we mobilise an ongoing commitment of care and support across such melting destinations? Listening to shared struggles (such as the melting landscapes), they are also not equally entangled in systems of oppression (as Greenland and the Tyrolean Alps with their very different tourism histories show). Yet, such an affective solidarity must recognise a shared feeling and shared struggles without diminishing localised differences.

In writing this chapter, we recognised the difficulty in doing so. How can we not fall back into patterns of generalising, speaking for and about others and making assumptions that are not ours to make (as researchers, as people (not) living in or visiting these regions)? We highlight that activists as well as researchers are nested within these tensions that underpin current struggles for climate justice. However, we argue that connections and solidarity between melting destinations can be achieved through situated and careful comparison (Ren & Markussen, 2024; Ren & Nadegger, 2026). Such an approach to comparison requires us to be moved by affect, engage in embodied and localised rather than distanced knowledge making and form partial connections, for example, between local farmers and a broader public (as shown in the illustrative example of Greenland) or across locals and visitors (as shown in the illustrative example of Tyrol).

Such connections are not total but acknowledge ambiguity and friction while standing with each other for a common cause. We speculate that *staying connected* through such careful comparisons then allows affective solidarity to nourish climate justice as a social movement through an ongoing commitment across melting destinations. As our examples show, such connections can happen in alliances and solidarity movements in everyday life in Greenland and the Alps. Citizens, visitors, governments and others step forward and stand with others, even though not all of their goals might always be aligned or come from precisely the same place. By engaging in careful comparison for climate justice, such partial connections then allow us to cultivate a response-ability (Haraway, 2016). It allows us to stay accountable for the worlds we are in and the worlds we make with others while at the same time developing a capacity to respond and form connections and new alliances with 'many others with whom we share this world' (Singh, 2017: 767).

References

Amt der Tiroler Landesregierung (2021) Leben mit Zukunft: Tiroler Nachhaltigkeits- und Klimastrategie. See https://www.tirol.gv.at/fileadmin/themen/landesentwicklung/raumordnung/Nachhaltigkeit/Nachhaltigkeits-_und_Klimakoordination/Publikationen/Nachhaltigkeits-und-Klimastrategie_2021.pdf (accessed February 2025).

Amt der Tiroler Landesregierung (2024) Leben mit Zukunft: Tiroler Nachhaltigkeitsund Klimastrategie. Maßnahmenprogramm 2025-2027; Entwurf vom 16. Dezember 2024. See https://www.tirol.gv.at/fileadmin/themen/landesentwicklung/raumordnung/Nachhaltigkeit/Nachhaltigkeits-_und_Klimakoordination/Publikationen/Entwurf__2__Massnahmenprogramm.pdf (accessed February 2025).

Andersen, M. (2015) Uran eller spisekammer? (Uranium or pantry?) *Weekendavisen* 23, 4 June 2025.

Bazzani, G. (2024) From organic to climate solidarity: Challenges for climate change mitigation. *Globalizations* 21 (8), 1632–1641. https://doi.org/10.1080/14747731.2024.2386138.

Baxter, L.F. (2021) The importance of vibrant materialities in transforming affective dissonance into affective solidarity: How the Countess Ablaze organized the Tits Out Collective. *Gender, Work & Organization* 28 (3), 898–916. https://doi.org/10.1111/gwao.12676.

Bjørst, L.R. (2016) Saving or destroying the local community? Conflicting spatial storylines in the Greenlandic debate on uranium. *The Extractive Industries and Society* 3 (1), 34–40.

Blanco-Romero, A., Blázquez-Salom, M. and Fletcher, R. (2023) Fair vs. fake touristic degrowth. *Tourism Recreation Research* 50 (2), 435–439. https://doi.org/10.1080/02508281.2023.2248578.

Chen, C. (2023) The (ecologically) imperial mode of sport at the exterminist stage of capitalism: Counter stories of Dakar Rally's ride in South America (2009–2019). *International Review for the Sociology of Sport* 58 (8), 1241–1262. https://doi.org/10.1177/10126902231163062.

Egede, Muté B. (2025) New Year Speech 2025. See https://naalakkersuisut.gl/Nyheder/2025/01/0201_Nytaarstale-2025?sc_lang=da (accessed March 2025).

Eriksson, M., Lundgren, A.S. and Eriksson, R.H. (2025) The heroes and killjoys of green megaprojects: A feminist critique. *Cambridge Journal of Regions, Economy and Society* 18 (2), 341–357. https://doi.org/10.1093/cjres/rsaf002.

Estermann, G. (2019) Nein zur Gletscherverbauung Pitztal-Ötztal!: An: LH Günther Platter; LH-Stv. Ingrid Felipe; LR Johannes Tratter. Aufstehn.at - Verein zur Förderung zivilgesellschaftlicher Partizipation. See https://mein.aufstehn.at/petitions/nein-zur-gletscherverbauung-pitztal-otztal (accessed February 2025).

Gobiet, A., Kotlarski, S., Beniston, M., Heinrich, G., Rajczak, J. and Stoffel, M. (2014) 21st century climate change in the European Alps – a review. *The Science of the Total Environment* 493, 1138–1151. https://doi.org/10.1016/j.scitotenv.2013.07.050.

Gössling, S., Hall, C.M., Peeters, P. and Scott, D. (2010) The future of tourism: Can tourism growth and climate policy be reconciled? A mitigation perspective. *Tourism Recreation Research* 35 (2), 119–130. https://doi.org/10.1080/02508281.2010.11081628.

Guia, J. (2021) Conceptualizing justice tourism and the promise of posthumanism. In T. Jamal and J. Higham (eds) *Justice and Tourism* (pp. 370–387). Routledge.

Haraway, D.J. (2016) *Staying with the Trouble: Making Kin in the Chthulucene. Experimental Futures Technological Lives, Scientific Arts, Anthropological Voices*. Duke University Press.

Harbor, L.C. and Hunt, C.A. (2021) Indigenous tourism and cultural justice in a Tz'utujil Maya community, Guatemala. *Journal of Sustainable Tourism* 29 (2-3), 214–233. https://doi.org/10.1080/09669582.2020.1770771.

Hegelund, L.E. (2009) *Tikeraaq*. Kalaallit Nunaanni.

Hemmings, C. (2012) Affective solidarity: Feminist reflexivity and political transformation. *Feminist Theory* 13 (2), 147–161. https://doi.org/10.1177/1464700112442643.

Higgins-Desbiolles, F. (2022) The ongoingness of imperialism: The problem of tourism dependency and the promise of radical equality. *Annals of Tourism Research* 94, 103382.

Higgins-Desbiolles, F. (2024) The end of tourism? Contemplations of collapse. *Journal of Tourism Futures* 10 (3), 476–485. https://doi.org/10.1108/JTF-11-2023-0259.

Intergovernmental Panel on Climate Change (2023) Climate Change 2023: Synthesis Report. See https://www.ipcc.ch/report/ar6/syr/downloads/report/IPCC_AR6_SYR_LongerReport.pdf (accessed March 2025).

Juskus, R. (2023) Sacrifice zones. *Environmental Humanities* 15 (1), 3–24. https://doi.org/10.1215/22011919-10216129.

Land Tirol (2021) Der Tiroler Weg: Perspektiven für eine verantwortungsvolle Tourismusentwicklung. See https://www.tirolwerbung.at/_Resources/Persistent/f/a/b/7/fab7b0004fcf117a589b4bd8622d57568ecc7d23/Tiroler%20Weg%20NEU.pdf (accessed February 2025).

Lazic, S. and Della Lucia, M. (2024) A holistic and pluralistic perspective for justice through tourism: A regenerative approach. *Tourism Geographies* 1–18. https://doi.org/10.1080/14616688.2024.2372114.

Leoni, M. (2019) From colonialism to tourism: An analysis of cruise ship tourism in Ittoqqortoormiit, East Greenland. Unpublished Master's thesis, University of Iceland, Reykjavik.

Markussen, U. and Ren, C. (2023) A just destination? Exploring local hopes, fears, and power asymmetries in East Greenlandic (Tunu) tourism development. *Études/Inuit Studies* 47 (1–2), 253–273.

Mathieu, J. (2022) Alpine tourism: Refashioning a model. In E.G.E. Zuelow and K.J. James (eds) *The Oxford Handbook of Tourism History* (pp. 425–446). Oxford University Press.

Mazza, F. (2023) Tourism and marginalization in the Alps. The case of Media-Alta Valtellina Region. *International Conference on Tourism Research* 6 (1), 440–448. https://doi.org/10.34190/ictr.6.1.1100.

Qian, Z. (2022) Territorial governance, market integration and Indigenous citizens in China's state-led eco-tourism: Developing the Xixi National Wetland Park. *Journal of China Tourism Research* 18 (5), 991–1010. https://doi.org/10.1080/19388160.2021.1973932.

Rambøll Consulting (2014) Hvor kan udviklingen komme fra? Potentialer og faldgruber i de grønlandske erhvervssektorer frem mod 2025. (Where can development come from? Potentials and pitfalls in Greenland's business sectors towards 2025). Report. See https://www.kamikposten.dk/lokal/last/container/da/hvadermeningen/pdf/hvor_kan_udviklingen_komme_fra.pdf (accessed October 2025).

Rantanen, M., Karpechko, A.Y., Lipponen, A., Nordling, K., Hyvärinen, O., Ruosteenoja, K., Vihma, T. and Laaksonen, A. (2022) The Arctic has warmed nearly four times faster than the globe since 1979. *Communications Earth & Environment* 3 (1), 168.

Rastegar, R. (2022) Towards a just sustainability transition in tourism: A multispecies justice perspective. *Journal of Hospitality and Tourism Management* 52, 113–122. https://doi.org/10.1016/j.jhtm.2022.06.008.

Rastegar, R. and Becken, S. (2024) Embedding justice into climate policy and practice relevant to tourism. *Journal of Sustainable Tourism* 33 (10), 2011–2028. https://doi.org/10.1080/09669582.2024.2377720.

Rastegar, R. and Ruhanen, L. (2023a) A safe space for local knowledge sharing in sustainable tourism: An organisational justice perspective. *Journal of Sustainable Tourism* 31 (4), 997–1013. https://doi.org/10.1080/09669582.2021.1929261.

Rastegar, R. and Ruhanen, L. (2023b) Climate change and tourism transition: From cosmopolitan to local justice. *Annals of Tourism Research* 100, 103565. https://doi.org/10.1016/j.annals.2023.103565.

Rastegar, R., Higgins-Desbiolles, F. and Ruhanen, L. (2023) Tourism, global crises and justice: Rethinking, redefining and reorienting tourism futures. *Journal of Sustainable Tourism* 31 (12), 2613–2627. https://doi.org/10.1080/09669582.2023.2219037.

Ren, C. and Bjørst, L.R. (2016) Composing Greenlandic tourism futures: An integrated political ecology and actor-network theory approach. In M. Mostafanezhad, R. Norum, E.J. Shelton and A. Thompson Carr (eds) *Political Ecology of Tourism* (pp. 302–319). Routledge.

Ren, C. and Markussen, U. (2024) A comparative advantage? Using situated comparison for collaborative ways of knowing in Greenlandic tourism. In O. Rantala and D.K. Müller (eds) *A Research Agenda for Arctic Tourism* (pp. 109–124). Edward Elgar Publishing.

Ren, C. and Nadegger, M. (2026) Tinkering with affective, frictuous, and speculative comparison: Comparing 'data' on tourism encounters with climate changes in Greenland and the Alps. In P. Vannini (ed.) *Non-Representational and More-Than-Human Research: Vitalist Methodologies for the End of Data* (pp. 110–126). Routledge.

Ren, C., Jóhannesson, G.T., Ásgeirsson, M.H., Woodall, S. and Reigner, N. (2024) Rethinking connectivity in Arctic tourism development. *Annals of Tourism Research* 105, 1–12. https://doi.org/10.1016/j.annals.2023.103705.

Rose, J. (2014) Ontologies of socioenvironmental justice. *Journal of Leisure Research* 46 (3), 252–271. https://doi.org/10.1080/00222216.2014.11950325.

Scott, D. and Gössling, S. (2022) From Djerba to Glasgow: Have declarations on tourism and climate change brought us any closer to meaningful climate action? *Journal of Sustainable Tourism* 30 (1), 199–222. https://doi.org/10.1080/09669582.2021.2009488.

Sealey-Huggins, L. (2018) The climate crisis is a racist crisis: Structural racism, inequality and climate change. In A. Johnson, R. Joseph-Salisbury, B. Kamunge and C.E. Sharpe (eds) *The Fire Now: Anti-Racist Scholarship in Times of Explicit Racial Violence* (pp. 99–113). Zed Books.

Sheller, M. (2021) Reconstructing tourism in the Caribbean: Connecting pandemic recovery, climate resilience and sustainable tourism through mobility justice. *Journal of Sustainable Tourism* 29 (9), 1436–1449. https://doi.org/10.1080/09669582.2020.1791141.

Singh, N. (2017) Becoming a commoner: The commons as sites for affective socio-nature encounters and co-becomings. *Ephemera: Theory and Politics in Organization*, 17(4), 751–776.

Steiger, R. and Scott, D. (2020) Ski tourism in a warmer world: Increased adaptation and regional economic impacts in Austria. *Tourism Management* 77, 1–10. https://doi.org/10.1016/j.tourman.2019.104032.

Steiger, R., Demiroglu, O.C., Pons, M. and Salim, E. (2024) Climate and carbon risk of tourism in Europe. *Journal of Sustainable Tourism* 32 (9), 1893–1923. https://doi.org/10.1080/09669582.2022.2163653.

Steiger, R., Scott, D., Abegg, B., Pons, M. and Aall, C. (2019) A critical review of climate change risk for ski tourism. *Current Issues in Tourism* 22 (11), 1343–1379. https://doi.org/10.1080/13683500.2017.1410110.

Steiger, R. and Stötter, J. (2013) Climate change impact assessment of ski tourism in Tyrol. *Tourism Geographies* 15 (4), 577–600. https://doi.org/10.1080/14616688.2012.762539.

Sultana, F. (2022) Critical climate justice. *The Geographical Journal* 188 (1), 118–124. https://doi.org/10.1111/geoj.12417.

Vachhani, S.J. and Pullen, A. (2019) Ethics, politics and feminist organizing: Writing feminist infrapolitics and affective solidarity into everyday sexism. *Human Relations* 72 (1), 23–47. https://doi.org/10.1177/0018726718780988.

Voller, L, Obbekjær, M and Tybjerg, J. (2021) Ny grønlandsk leder om Kvanefjeld: 'Man skal lytte til folket, og folket har talt, det bliver ikke til noget' (New Greenlandic leader on Kvanefjeld: 'You have to listen to the people, and the people have spoken, it won't come to anything'). *Danwatch*. See https://danwatch.dk/ny-groenlandsk-leder-om-kvanefjeld-man-skal-lytte-til-folket-og-folket-har-talt-det-bliver-ikke-til-noget/ (accessed March 2025).

Walters, S. and Butterwick, S. (2017) Moves to decolonise solidarity through feminist popular education. In A. von Kotze and S. Walters (eds) *Forging Solidarity* (pp. 27–38). SensePublishers.

Wegerer, P.K. and Nadegger, M. (2023) It's time to act! Understanding online resistance against tourism development projects. *Journal of Sustainable Tourism* 31 (2), 425–441. https://doi.org/10.1080/09669582.2020.1853761.

Wennecke, C., Jacobsen, R.B. and Ren, C. (2019) Motivations for Indigenous island entrepreneurship: Entrepreneurs and behavioral economics in Greenland. *Island Studies Journal* 14 (2), 43–60.

Zhou, W., Faturay, F., Driml, S. and Sun, Y.-Y. (2024) Meta-analysis of the climate change-tourism demand relationship. *Journal of Sustainable Tourism* 32 (9), 1762–1783. https://doi.org/10.1080/09669582.2024.2354882.

10 The Regenerative Vanua Stewardship Framework: *Ol Vanuas blong yumi oli no ol showgrounds blong ol turis!* (Our Vanuas are not Destinations)

Kehana Andrews, Jerry Spooner, Laurana Rakau-Tokataake, Eva Addinsall and Cherise Addinsall

In the face of the global polycrisis and the negative effects of both climate change and many of the policies taken to mitigate it, self-determined access to Vanua is of paramount importance to the resilience of Pacific Islanders and should, therefore, be the main focus of climate justice advocacy and implementation. However, industrial models of tourism which rely on economic growth and the commercialisation of Vanuas as destinations alienate Indigenous peoples from their Vanua – from their own land, oceans, food systems and culture.

Emerging from Critical Participatory Action Research (CPAR) conducted in Vanuatu (2019 to 2024), the Regenerative Vanua Stewardship Framework (RVSF) proposes a model of 'Regenerative Intercultural Hosting' based on self-determined and governed stewardship of Vanua. The RVSF enables Indigenous peoples to maintain deep connection to and stewardship of their Vanuas, and to share this connection with visitors through regenerative and transformational experiences. The RVSF counteracts the domination of colonial narratives in tourism and supports the deconstruction of power and privilege in the co-creation of regenerative practice. The RVSF is a tool for decolonisation and climate justice in ways that tourism cannot be as it aims to disentangle Vanuas from tourism's industrialised patterns of resettlement, reoccupation and reinhabitation.

Introduction: Vanua Tu – Our Land Standing Up

> *'Justice' itself is a problematic modern(ist) concept based on notions of grievance, redress, detachment, dispassion, objectivity, impartiality, rationality, hierarchy, binary balancing of scales, abstractions, judgement, redress and punishment…perhaps justice can grow into earthly healing, synergistic responsibility, critical consciousness and self-determined realities, as a commitment to right relationships between humans, so-called 'nature' and the spiritual realm, encompassing ethical praxis such as seeking truth, harmony, balance and healing.* (Paradies, 2024: 87)

Vanuatu translates roughly as 'land standing up' (Tusitala Marsh, 2014: 6). The name is made up of 'Vanua', meaning land (as a relational concept), and 'Tu', which Walter Hadye Liñi, the first Prime Minister of the Independent Vanuatu, described as embodying 'all the following meanings – to be, to be existing, to stand upright, to have existed, to stand firm, to be well founded', and 'to exist in time to come' (Liñi, 1980a). The word was chosen to represent the new nation of Vanuatu – the act of standing up across time connotes Independence and self-determination (Tusitala Marsh, 2014: 6) and affirms the reverence of Vanua to the people of our Islands. It is clear that an understanding of 'justice' can only be achieved in a Pasifika context through an understanding of Vanua – as it is where our culture, language and communal values come from. The notion of standing up, of standing firm, of existing in time to come is woven into the story of Vanuatu. We will face the challenges of the future with our stories of survival.

Located in the South Pacific Ocean, the Vanuatu archipelago is made up of 83 islands spread over almost 13,000 square kilometres, each with a rich and diverse cultural heritage. Indigenous lives were significantly impacted by colonisation – which began in the 1800s through European missionary settlements and 'blackbirding' practices. As Tusitala Marsh writes:

> Labour shortage for sugar plantations in Queensland, Fiji and New Caledonia, along with the growing trade in sandalwood and beche de mer, saw the rise of 'blackbirding' (Mortensen, 2000) – the kidnapping and enforced labour of Islanders. With no formal colonial power ruling, foreign investors and businessmen viewed the Islands as open territory for their labour requirements…. (Tusitala Marsh, 2014: 1)

However, eventually both British and French colonial powers became interested in establishing a formal colonial Vanuatu, and in 1906

Britain and France took joint colonial control through an Anglo-French Condominium (Brown & Nolan, 2008). Tusitala Marsh writes:

> Although the archipelago remained intact, it produced systems of government that created further divisions amongst Vanuatu's multiple geographical, linguistic, social and cultural identities. There were not only two separate (and often opposing) administrative colonial systems, but also competing Anglophone and Francophone education and religious systems (French Catholicism versus English Protestantism) and policing and health systems which played on intra-village rivalries, creating 'disunity and polarization' ... This made the creation of a unified Indigenous movement for Independence a difficult, but not impossible task.... (2014: 1)

After 73 years of the Anglo-French Joint Condominium, Independence and formal statehood was achieved in July 1980 (Lini, 1980b: 26). The Independence movement (under the Nasional Pati, later Vanua'aku Pati) of the 1960s and 1970s took the reclamation of alienated land as the national, unifying program for Independence (Tusitala Marsh, 2014).

Pasifika conceptualisations of 'justice' are relational, truth seeking, balanced and healing – and informed by and embedded in ancestral land (Teariki & Leau, 2023). As Tuvaluan scholar Maina Talia writes: 'in reality, there is no such thing called "individual rights", all we know is "communal rights" that exist and are scribed in our being' (Talia, 2009: 41). We therefore want to start our conversation about climate justice in tourism by honouring and tracing these relational concepts of land and justice from a Pasifika worldview.

There is a significant volume of literature emerging from Indigenous authors which reorients the pervasive Western imperial framing of land as a commodity and affirms Land as sacred (Gordon, 2017; Lilomaiava-Doktor, 2020; McDonnell & Regenvanu, 2022; Paradies, 2024; Talia, 2021; Tynan, 2020). In Pacific Island countries, Land 'tends to have meanings to those who "belong" to or are "part of it" that are often difficult to encapsulate in English or other colonial languages' (Campbell, 2010: 60). Sa'iliemanu Lilomaiava-Doktor takes this further, arguing that the 'difficulties that arise in interactions between people who have very different conceptualizations of land and land tenure are probably nowhere better illustrated than in scholarship of the Pacific or Oceanic region' (2020: 122). Sa'iliemanu Lilomaiava-Doktor's statement reiterates the importance of specificity and understanding when it comes to providing definitions of land, place and community in the Pacific. Vanua is a term which is felt and understood across Oceania by Indigenous communities, however, it evades adequate translation into English and is difficult to understand from a Western perspective. Our use of the Indigenous term Vanua throughout this chapter is a deliberate acknowledgement of this

Pasifika worldview and we assert that an understanding of justice, one which allows for right relationships between humans, nature and the spiritual realm, can only be achieved in a Pasifika context through an understanding of Vanua.

Vanua is a holistic term for place used in Pacific countries (with regional linguistic variations such as *Belau, Fanua, Fenua, Fonua, Hanua, Henua, Honua* and *Whenua*) and refers to the relationship between custodians of the land and the land itself, to the energetic aspect of this relationship which creates a sense of belonging to place. Vanua is the relational principle referring to life in all its various manifestations – land, waterways, oceans, air, flora, fauna, weather, people, language, song, dance, stories and so on. Vanua extends across time – it is informed by ancestral knowledge and embodies the needs of the future generations. It is dynamic and evolves to enable harmony and balance. In the Pacific, we understand that the role of being a custodian involves taking care of and stewarding Vanua, so that its full potential – environmentally, socially and spiritually – can be actualised.

The Pasifika region has immense linguistic diversity: there are over 138 Indigenous Oceanic languages spoken in Vanuatu alone – however, every place understands the concept of Vanua and its relational aspects. For example, in West Ambae (an Island in Vanuatu's Penama Province) a person who conducts themselves in a way that shows that they are taking care of and acting within the best interests of the Vanua (a steward or custodian) is known as *Kai Tangahuri*. In the same dialect, *Tangaroa* refers to that spirit one carries with them as an essence of their Vanua and which influences how they live within their custom system and belief. *Tangaroa* is the part of the Vanua that moves with you, reflected in the way you conduct yourself on other Lands. There is also another spirit, *Kangwaleva,* the spirit that is embedded in and looks after Vanua. In cases of movement, migration or displacement, when a person returns to their ancestral Vanua for the first time or after a very long time, the Elders will ask the *Kangwaleva* to 'smell' or recognise them, and to accept them and offer protection. This acceptance can be extended to visitors and applies equally to all, whether Indigenous or non-Indigenous, but to become a part of and be protected by *Kangwaleva* is to enter into a relational network of care and responsibility with the Vanua – one cannot simply 'pass through'.[1] So, 'justice' for West Ambaens, and for Oceanic peoples more broadly, relies on the spiritually informed 'ongoing enactment of their constitutive relationships' (Paradies, 2024: 87; see also Campion *et al.*, 2023) to Vanua – a conceptualisation which directs us to carefully consider (balance, harmonise) the effects of travel on Vanuas as hosts and as visitors. The industrial, professional, commercial model of tourism does not fit and is not just – we must move forward as enactors of our 'constitutive relationships' and strive for a model of hosting which allows Indigenous peoples to maintain deep connection to and stewardship of their Vanuas (to be *Kai Tangahuri*) and to share this connection with visitors (*Kangwaleva*) through regenerative and transformational experiences.

Tourism as 'Resettlement, Reoccupation and Reinhabitation' – Applying a Regenerative and Decolonial Praxis

Within recent tourism and development discourse, there is increased reference to decolonial praxis, methodology and projects (Bellato *et al.*, 2023; Chambers & Buzinde, 2015; Grimwood *et al.*, 2019; Hollinshead & Suleman, 2017; Lee, 2017; Mura & Wijesinghe, 2023; Yang & Ong, 2020). The term 'decolonial' is liberally mobilised by academics and industry workers within the tourism and development sector. While critical methodologies and approaches which decentre settler colonial perspectives are essential, Tuck & Yang (2012: 1) argue that referring to these activities as decolonial appropriates the term, as many of the objectives of tourism are incommensurable with decolonisation. For example, analytical studies of the socioeconomic climate in the Pacific demonstrate that tourism and large-scale agriculture are the main ongoing contributors to the dispossession of lands and alienation of Indigenous peoples from their Vanua (Addinsall *et al.*, 2016; De Burlo, 2003; Hosman, 2019). Devyn Holliday highlights the effects of this alienation for Pasifika populations, stating that relocation, displacement or alienation causes 'a break in ties to a sense of place and identity, self-efficacy, rights to land and culture, and bridging and bonding capital that is often derived from physical places and losing access to common property resources' (Holliday, 2020: 2–3).

Tuck and Yang argue:

> Decolonisation in the settler colonial context must involve the repatriation of land simultaneous to the recognition of how land and relations to land have always already been differently understood and enacted; that is, all of the land, and not just symbolically. (2012: 7)

In an attempt to market Vanuas as destinations to global consumers, the industrial model of tourism has impacted the accessibility that Indigenous populations have to their own land, waters, food and culture, either directly (blocking off access or over-developing leased land) or indirectly (priced out of areas, gentrification) while largely still being able to make claims to being 'sustainable', or even, in some cases, 'regenerative' (Loehr *et al.*, 2019). When this paper uses the term decolonise or decolonisation, it is doing so within the context of the repatriation of Indigenous land and life and it is used as an active verb, not as a metaphor. A just transition in tourism involves the project of disentangling tourism from its industrialised patterns of 'resettlement, reoccupation, and reinhabitation' (Tuck & Yang, 2012: 1). As Jason Hickel writes, while we are ultimately aiming to recentre societies around 'human wellbeing and ecological regeneration, rather than around the interests of international capital...economic sovereignty and self-reliance [a]re essential to real decolonization' (2021: n.p.).

It is undeniable that tourism as an industry is involved in the 'resettlement, reoccupation, and reinhabitation' of Indigenous lands (Tuck & Yang, 2012: 1). The idea that we must be critical and wary of the tourism industry and its effects in Vanuatu and across the Pacific is not new – our ancestors have been cautious of the industrial model since it began. In 1973, the Melanesian Council of Churches organised a two-week South Pacific Action for Development Strategy (SPADES) Conference. One of the resolutions to come out of the SPADES Conference was that the 'tourist industry in Oceania needed to be controlled' and that a 'Pacific Pact should be established for regional cooperation' (Swan, 2022: 244). In 1977, Epeli Hau'ofa wrote his beloved essay *Our Crowded Islands*, where he describes the neocolonial influence of Western-centric tourism and wrote 'we are losing the best in us while adopting the cheapest, the most superficial, and often dangerous aspects of other civilisations' (Hau'ofa in Swan, 2022: 249).

Almost 50 years later, this critique of industrial models of tourism which rely on constant economic growth and the commercialisation of Vanua is making its way into the mainstream – with a growing number of authors in multiple disciplines now shifting, in part, towards the regenerative paradigm (Bellato & Pollock, 2023; Lim *et al.*, 2018; Stafford-Smith *et al.*, 2017). Regenerative practice views the world as a complex whole living system, with the ability to self-organise (Bellato *et al.*, 2022). While the application of regeneration is a relatively new concept within academic literature, the foundational principles and concepts echo Indigenous Vanua stewardship practices which have always played a vital role in achieving food security, social security, biodiversity protection and ecological stability, in addition to providing natural medicines and cultural continuity (Addinsall *et al.*, 2024; Weir *et al.*, 2023). Prior to European colonisation in the Pacific Islands, regenerative values and practices were an integral part of custom and culture. Once again, we reiterate the importance of Vanua – one of the key components of the traditional economy is equitable access to our customary land, which provides food security, housing, widespread employment, social security, biodiversity protection and ecological stability (Regenvanu, 2010).

Regenerative principles and practices address goals of wellbeing and health and issues such as food security, community engagement in decision making, restoration and enhancement of the natural environment, social cohesion, cultural diversity and vitality and social justice. Most importantly, regenerative principles are positioned as a maturation, if not a replacement for, sustainable practices designed to shrink footprints and negative impacts (Pollock, 2019). Regenerative tourism literature centres social-ecological systems and encourages development practices to operate in harmony with these systems (Bellato & Pollock, 2023). In acknowledging tourism systems as inseparable from the places and communities in which they operate, Bellato et al. (2022: 2) argue that the 'theory and practice of regenerative approaches also address climate change, urbanisation, justice and inequality'. Owen (2007) first introduced the term regenerative tourism within architectural design literature, specifically regarding ecotourism facilities (Bellato et al., 2022). The development of the concept as well as the integration of Indigenous perspectives and worldviews emerged in the following years, seen in publications such as Becken (2019), Dwyer (2018), Mang and Haggard (2016), Pollock (2015) and Zivoder et al. (2015). Bellato et al. (2022: 7) note that, during the COVID-19 pandemic, 'calls to rethink tourism saw tourism scholars beginning to consider regenerative tourism as an alternative'. Alongside Bellato et al. (2022), many authors since COVID 19 have been influential in developing the conscious travel and regenerative tourism concepts within the tourism literature (Ateljevic, 2020; Cave & Dredge, 2020; Dredge, 2022; Duxbury et al., 2021; Sheller, 2021). It is clear within both the tourism literature and practice that a regenerative approach must be implemented according to locally defined values and outcomes.

An attempt to strictly define regenerative tourism is at odds with its fundamental principles, as such attempts reinforce reductive thinking and universality (Bellato et al., 2022). While the ability to bring about change is often conceptualised at the global, political and scientific level – meaningful change actually requires an integrated and pluralistic approach (Vanhulst & Beling, 2014). To 'think global but act local' connects widespread, complex issues to personal and meaningful action and experience (Jasanoff, 2010). Bellato et al. (2022: 9) argue that pluriversality asserts that knowledge 'cannot be universal due to different cultural contexts and varying impacts of processes such as colonisation and modernity in different places and communities'. Bellato et al. (2022) offer a draft working definition of regenerative tourism based on the principles of pluriversality and informed by intersectional scholarship:

> Regenerative tourism is a transformational approach that aims to fulfil the potential of tourism places to flourish and create net positive effects through increasing the regenerative capacity of human societies and ecosystems. Derived from the ecological worldview, it weaves Indigenous and Western science perspectives and knowledges. Tourism systems are regarded as inseparable from nature and obligated to respect Earth's principles and laws. In addition, regenerative tourism approaches evolve and vary across places over the long term, thereby harmonising practices with the regeneration of nested living systems. (Bellato et al., 2022: 9).

We have paid close attention to language in this chapter so far, and tourism is certainly a loaded term wrought with colonial baggage which we are inclined to do away with entirely. Drawing on concepts of epistemological decolonisation articulated by Chambers and Buzinde (2015), we acknowledge the valuable work from authors seeking to expose the dominant discourses and practices in tourism. However, in order to be truly regenerative, and to be in right relationship, the tourism sector and academia need to engage in deep listening and contemplation, radical action and complete transformation. This radical action requires privileging the voices of Indigenous people's epistemologies in the co-creation of tourism knowledge when applying regenerative principles (Butcher, 2021; Chambers & Buzinde, 2015; Higgins-Desbiolles & Everingham, 2022; Higgins-Desbiolles & Powys Whyte, 2013; Higgins-Desbiolles *et al*., 2019). We advise caution to academics entering the regenerative paradigm to ensure they are not speaking on behalf of others and to address the predominantly colonial narrative entering the regenerative space which can further lead to 'epistemic violence' (Spivak, 1988: 280). As Higgins-Desbiolles (2018) states, the co-creation of knowledge and deconstruction of power and privilege is essential for critical tourism to achieve just outcomes.

Developing an Intercultural Model of Hosting: Using CPAR and Storian to Address Epistemic Injustice in Tourism Literature And Practice

Without reciprocity, research is just another form of exploitation masked as inquiry and knowledge production. Ruth De Souza (2024: n.p)

Designing and implementing a 'just' methodology

The case study presented in this chapter tells a story of Critical Participatory Action Research (CPAR) conducted in Vanuatu (2019–2024) funded by the Australian Centre for International Agricultural Research (ACIAR). Through CPAR, the Indigenous-led, not-for-profit organisation Regenerative Vanua has been engaging in collaborative research across the archipelago to understand the local attitudes to tourism in Vanuatu, investigating the effectiveness of policies and strategies in supporting meaningful participation of Indigenous Ni Vanuatu in the tourism sector. The CPAR process was conducted as part of two ACIAR projects and facilitated through Regenerative Vanua in collaboration with the Vanuatu Government from 2019 to 2024. The methodology consisted of collective discussions and interaction through 'storian' – the Bislama term for 'chatting, yarning or swapping stories' (Crowley, 1995: 235) through a CPAR process.

Our aim as an inter-cultural research team is not to achieve conventional success (scaling out), but to scale deep by building meaningful connections and providing opportunities for Vanuas to lead their own research and development agendas and support long lasting systemic change. Scaling deep prioritises building connections and deepening relationships, recognising the significance of context and prioritising inner work and healing. Through this process, the Vanuas become best placed to scale out by becoming the stewards and mentors for regenerating Vanuas through Regenerative Intercultural Hosting.

This book chapter was written using storian as a way to ensure all co-authors were represented and had their voices heard. A true collaboration, the authors came together to review the CPAR and propose conclusions, and key literature was read collectively and discussed as a group. The result is a book chapter which has been authored by Indigenous Pacific Islanders using qualitative data collected from Vanuas in Vanuatu in collaboration with Australian settler authors and allies who act as invited co-facilitators, co-researchers and editors. This research and writing is guided by the knowledge systems of Indigenous Pacific Islanders and has emerged from these detailed and critical storian sessions.

As critical researchers, the authors of settler decent saw their role in the CPAR and in co-authoring this case study as a process of mindfully listening to and taking the lead from their Indigenous colleagues, providing research, resources and guidance only where needed, and when invited. The CPAR and authoring this chapter has been a regenerative process and is in itself an example of how we can share inter-cultural research, knowledge and conversation as an enactment of epistemic justice. According to Rastegar, Higgins-Desbiolles and Ruhanen:

> Epistemic injustice occurs when a wrong or harm is 'done to someone in their capacity as a knower' ... and justice is secured by counteracting practices of silencing or devaluing alternative forms of knowing and living that do not conform with assumptions about the 'authority' of scientific knowledge. (2023: 2619)

By combining extensive literature analysis of regenerative systems thinking and Pacific epistemologies together with comprehensive dialogue between the participants and Vanuas in which they are located, the research team sought to co-develop an alternative approach to the way tourism had been developed and led in Vanuatu while also applying regenerative thinking. In total, 127 storian sessions were conducted: 27 with agritourism operators, 20 with government representatives and 80 with representatives from project sites. Small group storian sessions worked exceptionally well and suited Ni-Vanuatu participants' preferred communication styles. The central feature of storian in a research context is relationship building between participants and the researcher, which enables members of the community to play an important role in providing a collaborative environment to address the research problem (Warrick, 2009). Implementing these methods enabled this research to be interactive rather than extractive (Beeton, 2006; Warrick, 2009), which was key to ensuring the CPAR was Vanua led.

In applying epistemic justice to tourism, Rastegar *et al.* advise that:

> We can identify cases of epistemic justice in tourism in the projects that ground tourism in Indigenous knowledges and lifeways…[and] we can identify epistemic injustices in tourism when tourism undermines the lifeways of local and Indigenous communities, requiring they change to suit tourists and tourism industry interests. (2023: 2619)

Results and findings of the CPAR process

The CPAR process revealed many of the negative impacts of the colonised tourism and agricultural development on Vanuas as a consequence of the dispossession of land and access to Vanua. This includes: increased economic inequalities and leakage; increased land disputes and distrust in customary governance (NSDP, 2021), dependence on global trade and markets; vegetation clearance and dredging; unethical compensation below living wage; and coastal pollution, which further supports the findings of research conducted in Vanuatu by Loehr *et al.* (2019). Jerry Spooner, Executive Director of Regenerative Vanua, explains the systematic injustices caused through the industrial model of tourism:

> The current industrial colonial tourism model that we seem to continue to aspire for in the Pacific largely excludes the local Indigenous population from meaningful participation as hosts and guests. This model is based on an introduced world view and values system that undermines our custom governance and values system and cuts off access to our Vanuas and increases the costs of living for our people. Yet, it's the only model that our young people are encouraged to learn, through university scholarships and internships funded by foreign countries. Regenerative Vanua is addressing the injustices caused from this industrial model of tourism and providing another model of hosting by validating our knowledge systems, our worldviews and creating opportunities and market access for those Vanuas that compliments their world views and supports them to thrive on their terms. (Personal communication, 9 December 2023)

Chief Rasa Ure describes his experience of the commodification of cultural knowledge caused from over-commercialisation:

> Tourism has been the catalyst for unlawfully claiming Indigenous knowledge, without addressing it, it's been supported that its ok to do this for the purpose of entertaining tourists (in tourism enclaves). Now our young people are learning inappropriate use of cultural knowledge and claiming knowledge that isn't their own. Through intercultural hosting we can support a healing journey where the original custodians of these knowledge systems are recognised and are the only ones that are able to share these knowledge systems. If tourists want to experience these knowledge systems, they need to come to the rightful custodians on the Vanua's terms. These are not performances to be made for tourists; they are rituals that have a time, place, purpose and reason for being. (Personal communication, 10 November 2023)

Kehana Andrews, Project lead for Our Vanua Project, was also one of the participants in many of the storians, and describes the breaking down of the governance systems within the Vanua caused from the shifting focus from Vanua to individual and how this increases vulnerability to climate change:

> Indigenous people, especially those that are located near the coastal areas, are seeking housing on other people's Vanuas due to custom landholders selling ground to tourism investors that then cut off access to members of that Vanua. Our values influence us to be non-confrontational and not question our elders, even when we feel injustice. Therefore, our custom governance needs to be strengthened so that our elders are speaking on behalf of everyone again and listening to their Vanua [*Kangwaleva*]. If we don't, the land disputes and breakdown of Vanuas will increase. Without strong governance and access to our Vanuas we will become more vulnerable to the increasing impacts of the climate crises. (Personal communication, 9 October 2022)

Votausi Mackenzie Reur, Chair of Regenerative Vanua, states:

> It is about rekindling the connection and regeneration of Vanua by enabling Indigenous Pacific people to be meaningfully engaged in hosting visitors, while raising their pride in their food and farming heritage and ensuring land sovereignty. We believe that regenerative food systems can achieve social justice, healing and revival for the Vanuas and countries they operate in. Therefore, it is vital that we have storian with our Vanuas in dynamic ways to restore pride and value to our local food, culture and traditions and to safeguard these for our future generations. (Personal communication, 9 October 2022)

Laurana Rakau-Tokataake, Operations Manager for Regenerative Vanua, describes her personal decolonisation journey and how that has enabled her to see the impacts of colonisation and the mass industrial tourism model on her Vanua:

> Through my role with Regenerative Vanua my eyes feel so open, I can see things I didn't see before and now I can't unsee them. My family and I now talk about critical issues and the impacts of colonisation and the industrial model of tourism on our Vanua which has caused considerable tension within our Vanua. I now feel so determined to support my family and our Vanua in repairing the damage that has been caused. (Personal communication, 10 December 2023)

In Paradies' essay titled 'Saturated Strands of (In/Re)Surgent Solidarity', he writes that Indigenous lifeways (or Vanua, as we would say) 'cannot be abstracted, quantified, rationalized, determined, instrumentalized or standardized, without simultaneously extinguishing it' (Paradies, 2024: 87). The Regenerative Vanua Stewardship Framework (RVSF) aims to resist these abstractions and quantifications, which typically try to 'integrate' Indigenous knowledge into western science.

The Framework instead prioritises Indigenous research leadership and the flourishing of Vanua. There is not one universal model, no specific instruction for implementation, no guaranteed methodological process, that can be applied in every context, in every Vanua. There must be constant collaboration, invention, review and adaptation in order to develop and design Regenerative Intercultural Hosting. It is this interdisciplinary, decolonial, creative and cross-cultural dialogue between Vanuas that is key to finding hope and potential in our increasingly uncertain future. It is this dialogue which informs and builds the RVSF which seeks to not only strengthen land and food systems sovereignty, but also supports Indigenous worldviews. Through Vanua-led truth telling processes, the Framework aims to create opportunities for the strengthening of ties to custom and lore, and regenerate language and other forms of knowledge which have been lost or suppressed through historic and contemporary colonial/neocolonial agendas. The RVSF is interdisciplinary, decolonial, creative and cross-cultural, and paves the way forward for regenerative praxis (see Figure 10.1).

Figure 10.1 Regenerative Vanua Stewardship Framework (RVSF) (Regenerative Vanua, 2025). Used with permission

Purpose: Stewarding Vanua (Kai Tangahuri)

Throughout Oceania, Vanua is the relational principle referring to life in all its various manifestations – land, waterways, oceans, air, flora, fauna, weather, people, language, song, dance, stories and so on. Vanua extends across time – it is informed by ancestral knowledge and embodies the needs of the future generations. It is dynamic and evolves to enable harmony and balance. In the Pacific, we understand that the role of being a custodian involves taking care of and stewarding Vanua, so that its full potential – environmentally, socially and spiritually – can be actualised. The role of humans in general (and hosts in this case) is to honour *Kangwaleva* (as the spirit of Vanua is referred to in West Ambae) by taking care of and stewarding Vanua. A person who conducts themselves in a way that shows that they are taking care of and acting within the best interests of the Vanua (a steward or custodian) is known as *Kai Tangahuri* (again, using terms with permission from West Ambae).

The four categories of self, others, place and nature are attempts at describing a complex concept which moves 'against the prevailing and pervasive Eurocentric story of separation, rather than framed isolated insulated distinct things, Indigenous worlds are constituted by relationship, connectedness and the space between' (Paradies, 2024: 89; Poelina *et al.*, 2024). The categories are not fixed – and could be reorganised, unorganised, have multiple other categories added or be reduced to the one category: Vanua. For the purpose of developing a guiding framework for stewardship, the authors have separated the aspects of self, others, place and nature to describe the values, principles, practices and actions needed to support stewardship.

Practice: Regenerative Stewardship

The RVSF is developed on a set of values, principles, practices and actions to ensure Vanua flourishes and thrives. The principles of the RVSF align with agroecology, which advocates for Indigenous knowledge systems, land rights and sovereignty, and supports small-scale food producers. These principles are designed to challenge neoliberalism, corporate dominance and the globalised food system. Adopting a political ecology approach to agroecology is action oriented and critical in that it confronts the economic and political power structures that have driven the colonist takeover of Vanua and food systems. This also enables a clear understanding of the purpose of regeneration that can ensure the dominant industrial agricultural and tourism players do not misappropriate the use of the term in the Pacific. As to truly be regenerative, land reform must recognise sovereignty and the individual or collective rights of Vanua. The objective should be a 'giving back' of custodianship of land from big agribusiness and tourism, and ensuring the rights of the people to produce food on their own terms is paramount (La Via Campesina, 2003).

Stewardship of place (connecting to Kangwaleva)

Many Western authors have attempted to focus on the experience and meaning of 'place' in Oceania, emphasising the concept of 'sense of place' to explain attachment to ancestral lands while ignoring the unobservable features that apply to Indigenous interpretations of land and space. Indigenous Pacific concepts of Vanua encompass a holistic and spiritual connection to place which is felt through the *Kangwaleva* of that Vanua. Connecting to *Kangwaleva* is a connection to the unobservable entities: cosmological, genealogical, ancestral and spiritual that inform understandings of place (Wilson, 2005). To understand the Vanua as a cultural and spiritual space that allows for genealogical connection, it is important to realise the significance that Pacific Island cultures place on intuition. This intuition can interpret beyond Western notions of intellect, and is a source of truth. In many Pacific Island Cultures, everything has its own intelligence, and Islanders can feel the spirit of the land, which in turn shapes their hearts and mind (Meyer, 2003).

> Many Pacific Islanders and Indigenous Peoples are experiencing a time of healing, cultural renaissance and recovery. We are re-discovering ourselves through the revival of our cultural spaces – our histories, languages and our connections to place. Yet, our very existence as unique people, with unique cultural spaces and understandings, remains threatened by a new form of colonialism, that of globalisation and cultural homogenisation. (Wilson, 2005: 43)

A key feature of the regenerative approach involves developing a shared sense of the uniqueness of place, i.e. co-discerning, creating its own story of place (Mang & Haggard, 2016; Teruel, 2018) based on its own geography–ecology and the composite mix of personalities, strengths and weaknesses of people within the Vanua. This encourages the development of unique solutions to place-based issues, based on commonly shared aspirations (health, safety, equity, fairness, etc.).

Stewardship of nature (nurturing Kangwaleva)

Walter Liñi, Vanuatu's 'Father of Independence', stated that:

> The peoples of the Pacific have always respected and been at peace with our environment. To us, our land, our skies and our ocean have always been a source of spiritual guidance as well as our means of sustaining life. (Liñi in Swan, 2022: 262)

While the land and sea provide an abundance of food, the people must in return be guardians and stewards who nurture the land and ocean as the land and ocean are genealogically connected to the people, like we are to one another. Leonid Vusilai, Local Cuisine Advisor for Regenerative Vanua, describes this realignment with ancestral custodian practices:

> There is a difference between need and want: need is focusing on what is important for your soul, because you're feeding the spirit (*Kangwaleva*). When you take taro from the earth, you are taking the ancestors. If you take what you need, your spirits communicate, and you are at peace. When you take too much from the land, you are not honouring your ancestors. For me regenerative stewardship is about reconnecting with your land, with your Vanua, with your nature, and with your ancestors. (Personal communication, 10 June 2024)

Regenerative stewardship of nature sees agricultural and land management practices restoring damage done to the land and ocean of the Vanua; building soil health, enhancing biodiversity, sequestering carbon, minimising emissions, producing healthy nutritious food, revitalising traditional-Indigenous land and ocean management practices and building a sense of pride, capability and willingness to innovate. Regenerative stewardship of nature becomes the upmost priority and is behind all decision making. Stewardship of nature sees humans not as having ownership of the Vanua, but as being stewards of the natural environment for the future generations. This sometimes means resurrecting practices and rituals that have been forgotten or rejected to support the natural environment to regenerate (Batibasaga *et al.*, 1999; Movono, 2017; Tuwere, 2002).

Stewardship of others (feeling Kangwaleva)

Through the concept of Regenerative Intercultural Hosting, hosts are identified as healers and agents of transformational change through the stewardship of others – helping to revitalise cultural and natural landscapes. Countering the dominant narrative of the tourist as exclusively Western, Regenerative Intercultural Hosting focuses on local and regional visitors, building knowledge exchange, intercultural dialogue and capacity building. Speaking to ancestral migration, trade and exploration activities throughout Oceania pre-colonisation, Epeli Hau'ofa writes in his essay *Our Sea of Islands*:

> We are the sea, we are the ocean, we must wake up to this ancient truth and together use it to overturn all hegemonic views that aim ultimately to confine us again, physically and psychologically…we must not allow anyone to belittle us again, and take away our freedom. (Hau'ofa, 1994: 160)

Vanuas are not destinations, they are our homes. Vanuatu is Our Land Standing Up. Regenerative transformational experiences move beyond the superficial exchange of a payment for services, and instead seek to exchange knowledge, connection, hospitality, stories, ideas and ways of being. It is through this exchange that visitors can be invited to feel *Kangwaleva* by the Elders and knowledge holders, who will ask the *Kangwaleva* to accept them and offer protection. As stated earlier, to become a part of and be protected by *Kangwaleva* is to enter into a relational network of care and responsibility with the Vanua – one cannot simply 'pass through'. As a consequence, visitors can be left transformed, fulfilled and enriched by the experience of deepening into a connection to place and of being invited stewards of that place. The overall aim of the transformative experience is to leave visitors with the desire to connect, heal, unlearn – and to regenerate their own Vanua once returning.

Stewardship of self (remembering Kangwaleva)

Stewardship of self requires a process of unlearning and remembering. Unlearning for Pacific Islanders is directly linked to decolonisation. The RVSF and working definition of Regenerative Intercultural Hosting actively seek to not only strengthen land and food systems sovereignty, but also support Indigenous worldviews, strengthen ties to custom and lore and regenerate language and other forms of knowledge which have been lost or suppressed through historic and contemporary colonial/neocolonial agendas. When we use the term decolonise or decolonisation, we are doing so within the context of the repatriation of Indigenous land and life and it is used as an active verb, not as a metaphor. Within the process of repatriation of Vanua, decolonisation requires an opening of the eyes and the facing of the trauma that was inflicted through colonisation and the introduction of mindsets and ways of being that have disconnected people from stewarding their Vanua. As part of our own healing journey, there is a process of re-examining the assumptions underpinning our behaviour, a need to self-organise and take responsibility for our own healing and then further support members of our Vanuas to unlearn and remember.

Conclusion and Implications

The term climate justice refers not only to the unequal social and geographical causes and impacts of climate change, but also to 'principles of democratic accountability and participation, ecological sustainability and social justice and their combined ability to provide solutions to climate change' (Chatterton *et al.*, 2013: 606). Similar moral considerations arise in relation to efforts to support cultural sustainability. 'Extensive threats to the vitality and viability of cultural expressions across the world, particularly those of Indigenous peoples, are closely interrelated with social injustices' (Grant, 2019: 47, including socioeconomic inequalities, the effects of colonialism, war and displacement of people, Western capitalism and environmental degradation and destruction.

Top-down or outsider-driven strategies attempting to mitigate cultural loss – whether those strategies emanate from the local, regional or international levels – risk perpetuating the very global imbalances of power that led to them in the first place. Time and again, such strategies elicit disengagement or disapproval from Indigenous peoples, who instead call for greater agency in making their own decisions about the future of their cultural heritage (Schippers & Grant, 2016). Within the CPAR, the retention of traditional knowledge was found to be characteristic of thriving Vanuas and individuals. In this sense, cultural retention is intimately interconnected with notions of wellbeing, cultural rights (Weintraub & Yung, 2009) and cultural justice (Niezen, 2009), and resonates closely with the concepts of climate justice.

Importantly, the RVSF does not seek out or imply an ancient, 'original' state of Indigenous and traditional cultures separated and uninfluenced by contemporary sociopolitical life, the marketing of which is prolific in the tourism industry (Taylor, 2001: 9; West & Carrier, 2004: 485). It instead refers to a remembering and re-engagement with Vanua and custodial practices that are meaningful, intentional, culturally appropriate and that protect and honour *Kangwaleva*. Shifting Vanuas from a mass industrial extractive model of tourism to one based on meaningful connection and knowledge exchange through Regenerative Intercultural Hosting can support a bridging of world views addressing the 'difficulties' referred to by Sa'iliemanu Lilomaiava-Doktor that arise in interactions between people who have very different conceptualisations of land and land tenure. As Rastegar *et al.* state:

> In such a context, the intercultural and peace-building capacities of tourism may act as a catalyst to greater solidarity. Peace through tourism analyses suggest that tourism may contribute to peace between peoples, communities, and nations. While the intercultural contact theory of tourism has been challenged because tourism is too superficial to induce conflict reduction and peace …, discussions of solidarity through tourism may advance the possibilities. However, for peace through tourism and solidarity through tourism to be effective and have real meaning, justice underpinnings are essential. (2023: 2620)

The underpinnings of the RVSF enable a realignment with the goals and visions of the Independence movement and with ancestral custodial practices across Oceania, as Dr Lesley Rameka states: *kia whakatōmuri te haere whakamua* – 'I walk backwards into the future with my eyes fixed on my past' (2016). Through Regenerative Intercultural Hosting, this remembering can be further supported through meaningful and self-determined knowledge exchange. We are invited to co-create aspirational futures based on 'responsibility, critical consciousness and self-determined realities, as a commitment to right relationships between humans, so-called "nature" and the spiritual realm, encompassing ethical praxis such as seeking truth, harmony, balance, and healing' (Paradies, 2024: 87). It is our vision and our belief that all visitors who are invited to our Vanuas will leave feeling this: touched by our hospitality, generosity and the strength of culture and connection to Vanua, that they will leave having felt and embodied *Kangwaleva*. As much as we hope to regenerate the spirit and vitality of those who stay with us, we know that being hosts, when done in way that honours *Kangwaleva*, can also ignite in us a pride and a passion to continue this journey to revolutionise, regenerate and reconnect to our Vanuas, to be *Kai Tangahuri*.

Acknowledgements

We would firstly like to thank the Vanuas for leading this important research and believing and trusting in us to tell their stories. We would also like to thank the Vanuatu Cultural Centre and Malvatumauri National Council of Chiefs (MNCC) for their verification and guidance in the sharing of cultural knowledge. We are grateful to Anna Pollock and Dianne Dredge for their mentorship and expertise in developing the RVSF and their contribution to the regenerative movement, and also to the Regenerative Vanua Board members who have been unwavering in their support of the Regenerative Vanua team.

The CPAR research documented in this book chapter spanned 3 Australian Centre for International Agriculture Research projects (ACIAR, 2020; ACIAR, 2021; ACIAR, 2024). Without this financial support, and support from our project partners, we would not be able to conduct the kind of research we believe in, which prioritises building connections and deepening relationships, recognising the significance of context and prioritising inner work and healing.

Note

(1) These terms are specific to West Ambae and are used in this chapter to create a depth of understanding of these belief systems which are not easily translated into English. Each Vanua throughout the Pacific region has unique beliefs and terms – therefore it is important to note that *Kai Tangahuri, Tangaroa and Kangwaleva are not representative of all Vanuas*. We ask that these terms are not reproduced or used in any other context without permission.

References

ACIAR (2020) *Development of a third party verified voluntary sustainable certification program for beef and other key commodities in Vanuatu LS/2020/125*, Canberra, Australia. See https://www.aciar.gov.au/project/ls-2020-155 (accessed April 2025).

ACIAR (2021) *Climate-smart regenerative ridge to reef landscapes for sustaining livelihoods of communities on custom land and food security in Vanuatu (Our Vanua Project) SSS/2021/120*, Canberra, Australia. See https://www.aciar.gov.au/project/sss-2021-120 (accessed April 2025).

ACIAR (2024) *Regenerative Agritourism Vanuatu SSS/2024/137*. Canberra, Australia, See https://www.aciar.gov.au/project/sss-2024-137 (accessed March 2025).

Addinsall, C., Scherrer, P., Weiler, B. and Glencross, K., (2016) Agroecological tourism: Bridging conservation, food security and tourism goals to enhance smallholders' livelihoods on South Pentecost, Vanuatu. *Journal of Sustainable Tourism* 25 (8), 1100–1116. 10.1080/09669582.2016.1254221.

Addinsall, C., Spooner, J., Pollock, A., Rakau-Tokataake, L., Mackenzie-Reur, V., Barrett, J., Bibi, P. and Addinsall, E. (2024) The regenerative Vanua journey: Rekindling the connection and regeneration of Vanua through regenerative agritourism experiences. *Tourism Cases: CABI, Special Issue: Small Islands Developing States. (SIDS) – Tourism in the past, current and future* https://doi.org/10.1079/tourism.2024.0073.

Ateljevic, I. (2020) Transforming the (tourism) world for good and (re)generating the potential 'new normal'. *Tourism Geographies* 22 (3), 467–475. https://doi.org/10.1080/14616688.2020.1759134.

Batibasaga, K., Overton, J. and Horsley, P. (1999) Vanua: Land people and culture in Fiji. In J. Overton and R. Scheyvens (eds) *Strategies for Sustainable Development: Experiences from the Pacific* (pp. 100–108). Zed Books.

Becken, S. (2019) Decarbonising tourism: mission impossible? *Tourism Recreation Research* 44 (4), 419–433. https://doi.org/10.1080/02508281.2019.1598042.

Beeton, S. (2006) *Community development through tourism*. Landlinks Press.

Bellato, L. and Pollock, A. (2023). Regenerative tourism: A state-of-the-art review. *Tourism Geographies* 27 (3–4), 558–567. https://doi.org/10.1080/14616688.2023.2294366.

Bellato, L., Frantzeskaki, N., Lee, E., Cheer, J. and Peters, A. (2023) Transformative epistemologies for regenerative tourism: Towards a decolonial paradigm in science and practice? *Journal of Sustainable Tourism* 32 (6), 1161–1181. https://doi.org/10.1080/09669582.2023.2208310.

Bellato, L., Frantzeskaki, N. and Nygaard, C.A. (2022) Regenerative tourism: A conceptual framework leveraging theory and practice. *Tourism Geographies* 25 (1), 1–21. https://doi.org/10.1080/14616688.2022.2044376.

Brown, M.A. and Nolan, A. (2008) Towards effective and legitimate governance: States emerging from hybrid political orders. *Project Output: Reports, Australian Centre for Peace and Conflict Studies*, University of Queensland, Australia. See https://www.sprep.org/publications/towards-effective-and-legitimate-governance-state-emerging-from-hybrid-political-orders-vanuatu-report (accessed January 2025).

Butcher, J. (2021) Covid-19, tourism and the advocacy of degrowth. *Tourism Recreation Research* 48 (5), 633–642. https://doi.org/10.1080/02508281.2021.1953306.

Campbell, J. (2010) Climate-induced community relocation in the Pacific: The meaning and importance of land. In J. McAdam (ed.) *Climate Change and Displacement: Multidisciplinary Perspectives* (p. 60). Hart Publishing.

Campion, O.B., West, S., Degnian, K., Djarrbal, M., Ignjic, E., Ramandjarri, C., Malibirr, G.W., Guwankil, M., Djigirr, P., Biridjala, F., O'Ryan, S. and Austin, B.J. (2023) Balpara: A practical approach to working with ontological difference in Indigenous land and sea management. *Society and Natural Resources: An International Journal* 37 (5), 695–715. 10.1080/ 08941920.2023.2199690.

Chambers, D. and Buzinde, C. (2015) Tourism and decolonisation: Locating research and self. *Annals of Tourism Research* 51, 1–16. https://doi.org/10.1016/j.annals.2014.12.002.

Chatterton, P., Featherstone, D. and Routledge, P. (2013). Articulating climate justice in Copenhagen: Antagonism, the commons, and solidarity. *Antipode* 45 (3), 602–620.

Cave, J. and Dredge, D. (2020) Regenerative tourism needs diverse economic practices. *Tourism Geographies* 22 (3), 503–511. https://doi.org/10.1080/14616688.2020.1768434.

Crowley, T. (1995) *A New Bislama Dictionary*. University of the South Pacific.

De Burlo, C.R. (2003) Tourism, conservation, and the cultural environment in rural Vanuatu, in D. Harrison (ed.) *Pacific Island Tourism* (pp. 69–81). Cognizant Communication.

De Souza, R. (2024) Linkedin post. See bit.ly/reimagining-Research (accessed February 2025).

Dredge, D. (2022) Regenerative tourism: Transforming mindsets, systems and practices, *Journal of Tourism Futures* 8 (3), 269–281.

Duxbury, N., Bakas, F.E., de Castro, T.V. and Silva, S. (2021) Creative tourism development models towards sustainable and regenerative tourism. *Sustainability* 13 (1), 2. https://doi.org/10.3390/su13010002.

Dwyer, L. (2018) Saluting while the ship sinks: The necessity for tourism paradigm change. *Journal of Sustainable Tourism* 26 (1), 29–48. https://doi.org/10.1080/09669582.2017.1308372.

Gordon, N. (2017) A critical ethnography of dispossession, indigenous sovereignty and knowledge production in resistance in Samoa by Naomi Gordon. A thesis submitted in partial fulfilment of the requirements for the degree of Master of Education in Theoretical, Cultural and International Studies in Education Department of Educational Policy Studies, University of Alberta.

Grant, C. (2019) Climate justice and cultural sustainability: The case of Etëtung (Vanuatu Women's Water Music). *The Asia Pacific Journal of Anthropology* 20 (1), 42–56. https://doi.org/10.1080/14442213.2018.1529194.

Grimwood, B.S.R., Muldoon, M. and Stevens, Z.M. (2019) Settler colonialism, Indigenous cultures and the promotional landscape of tourism in Ontario, Canada's 'near North'. *Journal of Heritage Tourism* 15 (1), 1–16.

Hau'ofa, E. (1994) Our sea of islands. *The Contemporary Pacific* 6 (1), 148–61.

Hickel, J. (2021) How to achieve full decolonization. *The New Internationalist*. See https://newint.org/features/2021/08/09/money-ultimate-decolonizer-fjf (accessed October 2022).

Higgins-Desbiolles, F. (2018) The potential for justice through tourism. *Tourism Review* 13. https://doi.org/10.4000/viatourism.2469.

Higgins-Desbiolles, F. and Everingham, P. (2022) Degrowth in tourism: Advocacy for thriving not diminishment. *Tourism Recreation Research* 49 (1), 215–219. https://doi.org/10.1080/02508281.2022.2079841

Higgins-Desbiolles, F. and Powys Whyte, K. (2013) No high hopes for hopeful tourism: A critical comment. *Annals of Tourism Research* 40, 428–433.

Higgins-Desbiolles, F., Carnicelli, S., Krolikowski, C., Wijesinghe, G. and Boluk, K. (2019) Degrowing tourism: Rethinking tourism. *Journal of Sustainable Tourism* 27 (12), 1926–1944. https://doi.org/10.1080/09669582.2019.1601732.

Holliday, D. (2020) Pacific small island developing states and climate change migration. *Small States Matters* 1, 2–3.

Hollinshead, K. and Suleman, R. (2017) Time for fluid acumen: A call for improved tourism studies dialogue with the decolonizing world. *Tourism, Culture and Communication* 17 (1), 61–74. https://doi.org/10.3727/109830417X14837314056933.

Hosman, T. (2019) Subsistence in Samoa: influences of the capitalist global economy on conceptions of wealth and well-being. *Independent Study Project (ISP) Collection*. See https://digitalcollections.sit.edu/isp_collection/3045 (accessed March 2025).

Jasanoff, S. (2010) A new climate for society. *Theory, Culture & Society* 27 (2-3), 233–253.

La Via Campesina (2003) Towards food sovereignty: Constructive alternatives to the World Trade Organization's agreement on agriculture. Institute for Agriculture and Trade Policy. See https://www.iatp.org/sites/default/files/Towards_Food_Sovereignty_Constructive_Alternat.pdf (accessed March 2025).

Lee, E. (2017) Performing colonisation: The manufacture of Black female bodies in tourism research. *Annals of Tourism Research* 66, 95–104. https://doi.org/10.1016/j.annals.2017.06.001.

Lim, M.M., Jørgensen, P.S. and Wyborn, C.A. (2018) Reframing the sustainable development goals to achieve sustainable development in the Anthropocene—a systems approach. *Ecology and Society* 23 (3). See https://papers.ssrn.com/sol3/papers.cfm?abstract_id=3518972 (accessed October 2025).

Liñi, W. (1980a) 5 March, Letter to GD Turner Esq, Managing Director Melanesia International Trust Company Limited, Port Vila, Vanuatu.

Liñi, W. (ed.) (1980b) *Beyond Pandemonium: From the New Hebrides to Vanuatu*. Wellington, in association with the Institute of Pacific Studies of the University of the South Pacific, Suva, Asia Pacific Books.

Loehr, J., Addinsall, C. and Weiler, B. (2019) Understanding how context affects resilience and its consequences for sustainable tourism. In S.F. McCool and K. Bosak (eds) *A Research Agenda for Sustainable Tourism* (pp. 39–52). Edward Elgar Publishing.

Lilomaiava-Doktor, S. (2020) Oral traditions, cultural significance of storytelling, and Samoan understandings of place or fauna. *Native American and Indigenous Studies* 7 (1), 121–151. https://doi.org/10.5749/natiindistudj.7.1.0121.

Mang, P. and Haggard, B. (2016) *Regenerative Development and Design: A Framework for Evolving Sustainability*. Wiley.

Meyer, M.A. (2003) *Hoʻoulu: Our Time of Becoming*. ʻAi Pōhaku Press.

McDonnell, S. and Regenvanu, R. (2022) Decolonization as practice: Returning land to Indigenous control. *An International Journal of Indigenous Peoples* 18 (2), 235–244.

Mortensen, R, (2000) Slaving in Australian Courts: Blackbirding Cases, 1869-1871. *Journal of South Pacific Law* 4. See https://www.paclii.org/journals/fJSPL/vol04/7.shtml (accessed October 2025).

Movono, A. (2017) Conceptualising destinations as a Vanua: An examination of the evolution and resilience of a Fijian social and ecological system. In J. Lew and J.M. Cheer (eds) *Tourism Resilience and Adaptation to Environmental Change* (pp. 286–302). Routledge.

Mura, P. and Wijesinghe, S.N.R. (2023) Critical theories in tourism - a systematic literature review. *Tourism Geographies* 25 (2–3), 487–507. https://doi.org/10.1080/14616688.2021.1925733.

Niezen, R. (2009) *Rediscovered Self: Indigenous Identity and Cultural Justice*. McGill-Queen's Press-MQUP.

NSDP (2021) NSDP (2019-2020) Baseline survey, well-being in Vanuatu 2021. Vanuatu National Statistics Office. See https://vnso.gov.vu/sites/default/files/Wellbeing_report.pdf (accessed January 2025).

Owen, C. (2007) Regenerative tourism–Re-placing the design of ecotourism facilities. *The International Journal of Environmental, Cultural, Economic and Social Sustainability* 3 (2), 175–181.

Paradies, Y. (2024) Saturated strands of (in/re) surgent solidarity. In *Planetary Justice* (pp. 87–94). Bristol University Press.

Poelina, A., Webb, B., Wooltorton, S. and Godden, N.J. (2024) Waking up the snake: Ancient wisdom for regeneration. In M. Lobo, E. Mayes and L. Bedford (eds) *Planetary Justice* (pp. 25–38). Bristol University Press.

Pollock, A. (2015) Social entrepreneurship in tourism: The conscious travel approach. TIPSE–Tourism Innovation Partnership for Social Entrepreneurship. See https://www.conscious.travel/wp-content/uploads/2018/03/Conscious-Tourism-TIPSE-2016-1.pdf (accessed February 2025).

Pollock, A. (2019) Flourishing Beyond Sustainability. ETC Workshop in Krakow. See https://etc-corporate.org/uploads/2019/02/06022019_Anna_Pollock_ETCKrakow_Keynote.pdf (accessed February 2025).

Rameka, L. (2016) Kia whakatōmuri te haere whakamua: 'I walk backwards into the future with my eyes fixed on my past'. *Contemporary Issues in Early Childhood* 17 (4), 387–398.

Rastegar, R., Higgins-Desbiolles, F. and Ruhanen, L. (2023) Tourism, global crises and justice: Rethinking, redefining and reorienting tourism futures. *Journal of Sustainable Tourism* 31 (12), 2613–2627.

Regenerative Vanua (2025) Vanuatu. See https://www.regenerativevanua.org/ (accessed May 2025).

Regenvanu, R. (2010) The traditional economy as a source of resilience in Vanuatu. In T. Anderson and G Lee (eds) *In Defence of Melanesian Customary Land* (pp. 30–34). AID/WATCH.

Schippers, H. and Grant, C. (eds) (2016) *Sustainable Futures for Music Cultures: An Ecological Perspective*. Oxford University Press.

Sheller, M. (2021) Reconstructing tourism in the Caribbean: connecting pandemic recovery, climate resilience and sustainable tourism through mobility justice. *Journal of Sustainable Tourism* 29 (9), 1436–1449. https://doi.org/10.1080/09669582.2020.1791141.

Spivak, G.C. (1988) In other worlds: On the politics of research by 'first world' geographers in the 'third world'. *Area* 25 (4), 403–408.

Stafford-Smith, M., Griggs, D., Gaffney, O., Ullah, F., Reyers, B., Kanie, N. and O'Connell, D. (2017) Integration: The key to implementing the Sustainable Development Goals, *Sustainability Science* 12, 911–919.

Swan, Q. (2022) *Pasifika Black: Oceania, Anti-colonialism, and the African World*. New York University Press.

Talia, M. (2009) Towards Fatele theology: A contextual theological response in addressing threats of global warming in Tuvalu. Unpublished MTh thesis, Tainan Theological College and Seminary.

Talia, M. (2021) Cultural identity in the face of the climate crisis – the case of Tuvalu. *Errant Journal* 2, 15–24.

Taylor, J.P. (2001) Authenticity and sincerity in tourism. *Annals of Tourism Research* 28 (1), 7–26.

Teariki, M.A. and Leau, E. (2023) Understanding Pacific worldviews: Principles and connections for research. *Kōtuitui: New Zealand Journal of Social Sciences Online* 19 (2), 132–151. https://doi.org/10.1080/1177083X.2023.2292268.

Teruel, S.A. (2018) *Analysis and Approach to the Definition of Regenerative Tourism Paradigm*. Universidad Para La Cooperacion Internacional (UCI).

Tuck, E. and Wayne Yang, K. (2012) Decolonization is not a metaphor. *Decolonization: Indigeneity, Education and Society* 1 (1), 1–40.

Tusitala Marsh, S. (2014) Black Stone Poetry: Vanuatu's Grace Mera Molisa. *Cordite Poetry Review*. See http://cordite.org.au/essays/black-stone-poetry-vanuatus-grace-mera-molisa/5/ (accessed May 2025).

Tuwere, I.S. (2002) *Vanua: Towards a Fijian theology of place*. Institute of Pacific Studies, University of the South Pacific/College of St. John the Evangelist.

Tynan, L. (2020) Thesis as kin: Living relationality with research. *Alternative: An International Journal of Indigenous Peoples* 16 (3), 163–170.

Vanhulst, J. and Beling, A.E. (2014) Buen vivir: Emergent discourse within or beyond sustainable development? *Ecological Economics* 101, 54–63.

Warrick, O. (2009) Ethics and methods in research for community-based adaptation: reflections from rural Vanuatu. In H. Ashley, N. Kenton and A. Milligan (eds) *Participatory Learning and Action, Community-based Adaptation to Climate Change* (pp. 76–87). The International Institute for Environment and Development (IIED).

Weir, T., Dovey, L. and Okrupa, M. (2023) Ecosystem-based adaptation in the Pacific Islands. In W. Filho, G.J. Nagy and D.Y. Ayal (eds) *Handbook of Nature-based Solutions to Mitigation and Adaptation to Climate Change* (pp. 1–24). Springer.

Weintraub, A.N. and Yung, B. (eds) (2009) *Music and Cultural Rights*. Board of Trustees of the University of Illinois.

West, P. and Carrier, J.G. (2004) Ecotourism and authenticity. *Current Anthropology* 45 (4), 483–498.

Wilson, K.L.N. (2005) View from the Mountian, Moloka'i Nui a Hina, *Junctures*. See https://junctures.org/junctures/index.php/junctures/article/view/130/134 (accessed March 2025).

Yang, E.C.L. and Ong, F. (2020) Redefining Asian tourism [editorial]. *Tourism Management Perspectives* 34, 100667. https://doi.org/10.1016/j.tmp.2020.100667.

Zivoder, S.B., Ateljevic, I. and Corak, S. (2015) Conscious travel and critical social theory meets destination marketing and management studies: Lessons learned from Croatia. *Journal of Destination Marketing & Management* 4 (1), 68–77. https://doi.org/10.1016/j.jdmm.2014.12.002.

11 'No Climate Justice without Gender Justice' and Racial Justice: A Critical Feminist Analysis

Freya Higgins-Desbiolles

The phrase 'no climate justice without gender justice' (Terry, 2009) has been invoked for more than a decade. It refers to more than just the differential impacts of climate change on diverse genders; it spotlights issues of structural injustices. This chapter applies a critical feminist lens to better understand climate (in)justices in tourism. A critical feminist lens assists in analysing the oppressions, extractive premises and inequities of tourism, as well as the omissions in tourism management, mitigation and adaptation policies intended to address the threat of climate change. It gives attention to critical feminist voices on tourism to underscore the dangers of a reformist approach when faced with powers of violence from extractivism, which is what the climate crisis represents. Engaging with Angela Davis' (2018) critique of 'glass-ceiling feminism', this analysis reveals the intertwined nature of gender justice, racial justice and climate justice in tourism, calling for a more radical, structural justice-oriented response.

Introduction

> *Our civilisation is being sacrificed for the opportunity of a very small number of people to continue to make enormous amounts of money. Our biosphere is being sacrificed so that rich people in countries like mine can live in luxury. It is the sufferings of the many which pay for the luxuries of the few.* Greta Thunberg (2018)

In opening this chapter, we might concentrate our focus by asking: What is it we would want from a gender lens when thinking about climate justice in tourism? Is it: better outcomes for women; the benefit of women's insights and participation in our planning and action; and greater equality in the experience of climate change impacts and reactions? Or would we want something more – a goal of better ensuring that no one is subject to abuse, inequity, neglect and oppression. Taking an anti-oppression and structural justice perspective derived from critical feminist approaches, this chapter offers insights into the vital need to integrate climate justice, gender justice and racial justice analyses to better secure holistic justice in our responses to climate change in tourism.

While climate change concerns are now firmly on the tourism agenda, climate justice in tourism is only recently being discussed. Approaching climate change with a justice lens compels us to interrogate the differential impacts of climate change on populations made vulnerable and marginal and to spotlight their exclusion from decision making. Justice thinking is increasingly coming to the fore, as exemplified in the Sixth Assessment Report of the International Panel on Climate Change (IPCC) (2021), which for the first time highlighted colonialism as one of the ongoing forces of inequity that act as drivers of climate vulnerability for some.

There are multiple epistemologies, disciplines and lenses we could use to uncover the multilayered oppressions and injustices of tourism's interface with climate change (for example, queer and Indigenist – see Aung, 2023; Bhandar & Ziadah, 2020; Watene, 2016; Whyte, 2016). Despite women's and gender diverse people's leadership being at the forefront in climate change activism and policy dialogues – thinking Vandana Shiva, Greta Thunberg, Mia Motley, Hindou Oumarou Ibrahim, Wanjira Mathai and Nemonte Nenquimo – feminist and gender analyses in tourism discussions of climate change are rarely evident to date. This analysis, prompted by the now common mantra 'no climate justice without gender justice' (Terry, 2009), proposes to employ a critical feminist lens to explore the full meaning of this insight, particularly in the realm of tourism's engagement with the threat of climate change.

Significantly, the United Nations Framework Convention on Climate Change (UNFCCC) initially overlooked the gender dimensions of climate change. This changed with the 2007 UNFCCC Bali Action Plan and the establishment of the Global Gender and Climate Alliance (GGCA). This was followed with the Women and Gender Constituency to the UNFCCC in 2009. A Gender Action Plan was adopted at the 23rd Conference of the Parties (COP23) in 2017, advocating that gender equality must be part of climate change policy discussions. Following these milestones, gender responsiveness has become a standing agenda item in the Convention and through references in the text of the Paris Agreement of 2015 (for a chronology of gender in the Intergovernmental Process, see UNFCCC, 2024).

This chapter engages a critical feminist approach to offer a preliminary analysis of gender justice in tourism. It follows the premise that the climate crisis is the result of wider structural injustices occurring for many decades, and in fact, for centuries. As the Women's Environmental Leadership Australia argued: 'Uneven climate and environmental impacts upon women and gender diverse people are a direct result of structural factors including the gendered power relations produced by the overlapping forces of patriarchy, colonialism, racism and unfettered capitalism' (n.d.: 1). The ideology of supremacy, of some groups over others, including cis men over women, corresponds to an ideology of human supremacy over nature. Domination, oppression and exploitation are the results of such structural systems, causing gender, racial and climate injustices.

Sultana stated:

> [...] critical feminist analyses help to reveal oft-overlooked or buried concerns, exposing interconnected inequities and harms. Indeed, it was at the 2007 COP where 'No climate justice without gender justice' was a rallying cry from feminist climate activists and social movements. Women have been at the forefront of climate justice movements [...] They have raised awareness of the importance of paying attention to the differential gendered burdens and harms worldwide [...]. (2022: 120)

A full exposition on the evolution of feminist thought and action is beyond the remit of this chapter. Instead, this work engages the recent analysis by Angela Davis (2018) as a framework for evaluating the intersections between gender justice, racial justice and climate justice in tourism responses to the climate crisis. Davis discussed 'mainstream' liberal feminism as a white, bourgeois, 'glass-ceiling feminism' that remains comfortable with hierarchies and has little relevance for those women that subsist far below this metaphorical glass ceiling. This contrasts to critical revolutionary feminisms that refuse the collaborations with capitalism that 'glass-ceiling feminists' make as they seek greater 'inclusion' through 'women empowerment' strategies; the critical feminists instead demand liberation for all. This analysis can be applied to current tourism, revealing that current mainstream tourism feminist analyses tend to focus on 'gender alone' (Christoffersen & Emejulu, 2023: 633), rather than the wider and shared struggles against oppression. These ideas will be explored in the sections that follow.

Centring Women in Gender Justice Analyses?

Gender is commonly understood as the 'socially constructed roles, behaviours, activities and attributes that a given society considers appropriate for men and women' (World Health Organization cited in Morgan *et al.*, 2016). This raises a critical question for gender justice analysis: should the focus be on centring women, and if so, in what ways? In their work on 'queer thinking' in climate action, Aung (2023: n.p.) reported the insight that 'LGBTQI+ communities are systematically excluded from receiving government support because legal documents rely on a gender binary system and do not recognise diversity of genders and expressions'. This participant added: 'when [governments] design climate plans and policies, they only consider gender as women and girls. We are completely forgotten' (Aung, 2023: n.p.). In light of these insights, work such as this must support the inclusivity of all women – cisgender, transgender, femme/feminine-identifying, genderqueer and nonbinary individuals – all of whom bring unique experiences and perspectives to conversations in climate justice. This position is reinforced by the insights of Vinyeta *et al*.:

> Gender is a socially constructed concept that is fluid and ever-changing, the meanings and accepted norms of which vary across cultures and across individuals. Despite its fluidity, gender has concrete, socially determined implications in people's lived experiences. For genders most frequently subject to oppression in the United States, such as women, and transgender and other gender nonconforming peoples, climate change may have more significant impacts. (2016: 1–2)

This work specifically focuses on gender, rather than women (and girls) exclusively. To focus gender analysis only on women is both harmful and myopic as: it may exclude the experiences of non-binary and gender non-conforming people; it may place the burden of responsibility for climate adaptation on women in development processes; and it may overlook the other intersectional factors shaping women's experiences of marginalisation (e.g. racialisation, class, age and ability). Instead of an exclusive focus on women's equality, a gender perspective should look at the power relationships between different genders and the oppressions and disadvantages that these may bring. Most 'gender' analyses to date may give rhetorical acknowledgement to 'women and gender diverse' people, but still remain largely focused on women and girls. The Women's Environmental Leadership Australia group explained why this may be the case:

> […] we note that a lack of research on the experiences and perspectives of gender diverse people in relation to environmental change limits what we can say about trans and non-binary people. This research is urgently needed if we are to continue to deepen our understanding of the intersections between all genders and the environment. (n.d.: 6)

In addition to a non-binary and wide-ranging inclusivity in our approach to gender, foundations for this analysis must also critically interrogate what form of feminist analysis supports the goals of climate just futures.

Moving on from White Liberal Feminism

The modern feminist movement developed out of advocacy and action for women's rights beginning in the mid-19th century, starting with the struggle of the suffragettes in the United States for the right to vote. The movement has sought political, economic and social equality between the genders. It has also illuminated how patriarchal power has oppressed people, enculturated sexism and resulted in lasting injustices and inequities.

Since its first emergence, there have been waves of feminism that have held different foci. One of these, liberal mainstream feminism, focuses specifically on achieving equal rights and opportunities for women within the legal and political systems that exist. The main concerns of liberal feminism are the enjoyment of individual liberties and equality before the law. It was at this point in the second wave of feminism that feminist analysis and theorising flourished, often in gender studies programmes, to help understand the dynamics of gender oppressions and to counter them.

However, there were simultaneous criticisms and searches for alternative pathways for feminism, including from 'radical feminists', who spotlighted patriarchy and the violence of men, and Marxist feminists, who emphasised the ways women's oppression resulted from both patriarchy and capitalism. A sub strand of radical feminism emerged from Black, Chicana and decolonial feminist critiques of privileged, mainstream, white feminism (Icaza & Vázquez, 2025) and has inspired the recent emergence of critical revolutionary feminism.

Liberal mainstream feminist advocacy was critiqued for its narrow focus on (certain) women only. For example, Vinyeta *et al.* explained: '[…] mainstream feminism fails to address the intersectional oppressions faced by contemporary indigenous women and other women of color' (2016: 13). However, the critiques have also come from cultures that recognise diverse genders (such as some Indigenous nations), who are respected and valued for their knowledge, sociocultural roles and contributions; this stands in contrast to the binary gender cultural constructs of much of the Global North (see Simpson, 2020: 144). Additionally, there is a history of race privilege that adheres to liberal feminists that has been called out. For example, First Nations scholar from the Michi Saagiig Nishnaabe Nation Leanne Betasamosake Simpson has noted the history of white women's complicity in violent colonialism and argued that, in contemporary times, '[…] white mainstream feminist analyses… are too simplistic and racist to be of use in our world-building projects…' (2020: 144).

Angela Davis described this liberal mainstream feminism as a '… "Glass ceiling feminism" [that] necessarily privileges those who are already nearest the ceiling—white, middleclass women, while erasing the collective nature of social struggle' (2018: n.p.). However, this critique may apply to individuals who are not in fact 'white', as there are many 'collaborators' who have made the bargain to purchase privilege through subservience to the hegemonic system. Lila Abu-Lughod has critiqued the universalising and patronising tendencies of liberal feminism as a: '…worldview that drives the much-documented arrogance of Western feminists in their attempts to "save" colonized women, not just from patriarchal violence, but from culture itself' (Abu-Lughod, 2013: 31–32). Also relevant is Nancy Fraser's analysis explaining how feminism has become 'capitalism's handmaiden' by supporting the neoliberal market agenda of individualism, competition and 'careerism' (2013: n.p.).

Intersectionality is central to discussions of feminism and its role in addressing oppression(s). The concept of intersectionality was first described by Kimberlé Crenshaw (1991) and has become a foundational pillar of Black women's theorising of racialised oppressions and unmasking of whiteness in feminist discourses. Christoffersen and Emejulu described intersectionality:

> […] core to its meaning is that systems of inequality, including capitalism/class, sexism, racism and white supremacy, heterosexism, cisgenderism, ableism, and borders, constitute one another, meaning that they construct one another and interact to create institutions and differential social positions … Social institutions and positions are therefore shaped by multiple, mutually constituting, divisions operating simultaneously. Applying intersectionality, in both theory and practice, therefore means engagement with the interrelationship of these systems of inequality. This engagement is in turn predicated on acknowledgment of and reckoning with the ontology of each of these structures themselves. (2023: 633)

The intersectional lens allows us to unpack the different positionalities and privileges that diverse women may embody in the 'feminist movement'. As Lorde explained:

> As women, we have been taught to either ignore our differences or to view them as causes for separation and suspicion rather than as forces for change. Without community, there is no liberation, only the most vulnerable and temporary armistice between an individual and her oppression. But community must not mean a shedding of our differences, nor the pathetic pretense that these differences do not exist.
>
> Those of us who stand outside the circle of this society's definition of acceptable women; those of us who have been forged in the crucibles of difference; those of us who are poor, who are lesbians, who are black, who are older […]. (1979: 98)

These critiques demand that we be discerning of our feminism to understand that those forms focused on individual 'empowerment' of women are defective in securing goals for justice. Collective struggle for freedom, human dignity and flourishing is what is needed to build a world where all can thrive through dismantling the structures of privilege and exploitation. As Davis noted, we can use the concept of 'intersectionality', a central concept of contemporary feminism, to not only talk about the 'interelating character of identities, but…intersectionality is most helpful when we think about the intersectionality of social justice struggles' (2018: n.p.).

This critical feminist analysis is also inspired by Southern feminisms which privilege the voices, experiences and practices of the Global South. This enables what Mohanty called 'noncolonizing feminist solidarity across borders' (2002: 503) and genuine anti-capitalist and anti-imperialist agendas that are able to address the structural injustices. Byrne and Imma explained that the term '…Global South' refers to a politically useful grouping of 'spaces and peoples negatively impacted by contemporary capitalist globalization' (Mahler, 2017: n.p. cited in Byrne & Imma, 2019: 2). However, the applications of this term are not literal, rather they are complex and deserving of nuanced application; for example, as Mahler (2017: n.p.) noted, there exists 'Souths' in the 'geographic North'.

Additionally, queer thinking and practice should also inform our critical feminist approach to climate justice in tourism. As Aung (2023: n.p.) explained, 'climate justice and queer rights movements share critiques of structural power asymmetry and in their activism to champion the rights of marginalised groups'. The critical interrogations and challenges of queer analysis question the received wisdoms and norms in important ways:

> Incorporating queer thinking and practice into climate justice could shed light on different approaches to gender and intersectionality in climate justice. Queer theory challenges norms around gender, identity and sexuality. It challenges concepts of what is 'normal' within our identities and what is 'normal' within society. It challenges economic models and accepted forms of knowledge in society such as capitalism and colonisation so often embedded in global North thinking and practice. (Aung, 2023: n.p.)

Icaza and Vázquez analysed the 'coloniality of gender' bringing to the fore '… the historical positioning of subjects crossed by or on the other side of the colonial difference' (2025: 1). This acts as a challenge to privileged feminist positionalities by uncovering hidden dividing lines and as a result suppressed epistemologies of those subjected to centuries of colonial violence. They note:

> Under coloniality, people are actively excluded from the discursive field that sets the condition of recognition and action of being political,

of political belonging. Coloniality can be understood as a form of exclusion from the field of discourse that grants access to humanity, to gender …, to the political. The enslaved people are outside gender. (Icaza & Vázquez, 2025: 8–9)

These brief forays into critical feminisms contrast the mild, reformist approaches which seek amelioration of (mostly) women's 'inequality' (Davis' 'glass-ceiling feminism') to those focused on collective struggles to overturn the structural injustices through building movements for liberation.

Gender Justice in Current Tourism Climate Change Policies

It is firstly important to note that gender just and critical feminist approaches to climate change are not well established in the wider sphere before we move to specifically discuss it in current tourism planning and in tourism academia. Mainstream climate strategies for mitigation and adaptation are often technically focused and piecemeal failing to get to the root causes of the climate crisis. Such failures are illuminated in the submission made by the global network called Gender CC – Women for Climate Justice to the COP 13 held in 2007 in Bali, Indonesia addressing plans to use forestry for carbon abatement:

> Gender CC wishes to highlight that forestry is not only about trees and their carbon content, but also about the ecosystem in a broader sense, and the people who live in and from the forest [...] rather than only focussing on technical and methodological issues [...] The real direct and underlying causes of deforestation must be addressed, such as overconsumption, agrofuel expansion, fossil fuel extraction, the replacement of natural forests by monoculture tree plantations, and the lack of respect for indigenous peoples' rights. (Gender CC, 2007: n.p.)

Learning from this intervention, we could substitute tourism for this focus on forestry, to challenge the current focus of many tourism strategies. That is, in examining tourism strategies focused on climate change, we should be seeking responses that address the root causes of the crisis rather than the symptoms. Thus, are such strategies presenting ideas of terminating tourism's emissions in order to be in alignment with governmental commitments to emissions reductions, noting tourism is a non-essential activity, even a luxury, when we are facing the threat of 'near-term collapse' from climate change (Jones & Steffen, 2019: n.p.)? Are they directly and effectively addressing tourism's capitalistic overconsumption or are they presenting voluntary measures for consumers to engage in 'responsible' consumption of tourism and technological fixes for the industry such as reduction of plastic (an issue, but not the most important climate change source), electrification of transport and 'sustainable' aviation fuels. A gender justice lens enables illuminating and distinguishing mild, ineffective and often greenwashing responses versus those that are focused on undoing structural injustices so that climate action is climate just in its orientation.

Yet, in tourism stakeholder responses to climate change, currently evident is only a preliminary recognition that women and other 'marginalised' segments of our communities are likely to be most negatively impacted. For example, the Glasgow Declaration on Climate Action in Tourism noted:

> Climate change impacts are most severely felt by under-represented and vulnerable groups such as women, Indigenous communities, people living with disabilities, and small island states. A just and inclusive transformation of tourism must prioritise their voices and needs, as well as those of younger generations who will otherwise pay the full price of our inaction. (One Planet Sustainable Tourism Programme, 2021)

The UN Climate Change COP29 Action Agenda put tourism on the radar for the first time, resulting in the Baku Declaration on Enhanced Climate Action in Tourism. It stated: 'We underscore our dedication to advancing climate action in tourism through global partnerships and dialogue, considering the perspectives of local communities and indigenous peoples, migrants, women, children, youth, persons with disabilities, and people in vulnerable situations' (UN Tourism, 2024a).

The Stocktake Report of the Tourism Panel on Climate Change (TPCC) addressed tourism as a luxury good and the inequities that arise from this fact. However, the report engaged with climate justice and the social costs of carbon emissions only briefly, noting:

> The social cost of tourism carbon emissions is increasing and is likely to equal or exceed its direct contribution to the global economy by as much as US$2 trillion in 2030. The climate justice implications of travel emissions predominantly from high-income countries, and the disproportionate burden of the social costs of tourism emissions in highly vulnerable countries, compels greater consideration in tourism sector climate responses. (TPCC, 2023: 19)

As one possible solution to such inequity, the report made a brief reference to a proposed levy on air travel by the UN Special Rapporteur on Human Rights and the Environment, an intervention that underscores that justice and human rights groundings are critical foundations for climate justice.

The only tourism industry report to date focused on climate justice, entitled Climate Justice in Tourism, went further in recognising the need for women's and Indigenous people's leadership and ownership rather than merely noting they are impacted by climate change, and that we should hear their voices. But this report only engaged indirectly with a gender analysis when it recommended as a priority action:

> Ensure marginalised communities – who are more impacted by climate change – have ownership and leadership of experience design and delivery. Prioritising experiences led by women, Indigenous-owned businesses and local youth for example, provides needed empowerment and gives visitors access to less readily accessible voices and viewpoints. (Bigby *et al*., 2024: 18)

One interviewee quoted in the report, Judy Kepher Gona, Founder of Sustainable Travel and Tourism Africa, showed insight into the interweaving of gender, climate and racial justice, rather than setting them apart in silos: 'Inclusive tourism must be about inclusive policies, inclusive governance, inclusive assets, inclusive opportunities – it must go beyond just talking about gender and race' (cited in Bigby *et al*., 2024: 41).

A particularly noteworthy actor in this space is Equality in Tourism (EIT), a UK-based non-profit established in 2018, which positions itself as an advocate for 'gender equality in tourism'. EIT epitomises the liberal feminist approach focused on gender equality or 'glass-ceiling feminism'. The choice of the word 'equality' rather than 'equity' signals this liberal feminist agenda and lack of focus on structural injustices and intersectional oppressions (see Minow, 2021). In their own words: 'EIT works to transform the lives of women in tourism by advocating for gender equality across the industry by providing practical solutions and best practices that inspire global action towards a more equitable and responsible tourism industry' (EIT, 2024: 4). An example of its focus is the concern with surveying female representation in tourism and hospitality boardrooms and providing recommendations for ameliorating the poor showing (EIT, 2018). Such an approach remains tourism myopic rather than concerned with the wider contexts of structural injustices in tourism. It has also offered 'women's empowerment projects' such as the 2018 'Empowering Women Farmers through Tourism' in Tanzania which was presented as addressing 'poverty, women's rights and climate change':

> The project has now trained 120 impoverished, marginalised women from three villages to farm well and to produce quality produce, to understand farming as a business, in entrepreneurship, to understand their rights as women and to save by managing their own micro-finance group and to work together as a co-operative. (EIT, n.d.)

The description 'impoverished' signals the concerns Abu-Lughod expressed for the agenda of Western feminists to 'save' (quoted above) and the training described offers no clear respect for these women's already hard-won knowledge and the structural contexts which may situate them as 'impoverished' and 'marginalised' rather than any deficits in their capacities.

Turning to tourism academia, it is also important to note that good deal of tourism scholarship to date conforms to Davis' 'glass-ceiling feminism', as it focuses on progressing women's equality in tourism (notably speaking very little about diverse women). For example, Pritchard explained:

> Feminist tourism scholarship is espousal of a commitment to addressing gender inequality in tourism and to improving women's experiences and conditions as tourism workers and consumers. Feminists seek to define, establish, and defend equal political, economic, and social rights for women; as such feminism is largely focused on women's issues…. (2014: 314)

Similarly, this glass-ceiling feminism is also well represented in a spate of work that came out particularly focused on gender equality in the tourism academy. For example, the report 'The Gender Gap in the Tourism Academy' (Munar *et al.*, 2015) focused on mapping 'gender equality in the tourism academy' through statistics and indicators which included, for example, women's occupation of editorial positions on journals, positions on conference committees and keynote speaker roles. Soon thereafter, Chambers *et al.* (2017) led a special issue of the journal of Anatolia applying feminist epistemologies to analyse the place of women in the tourism academy.

Other work has focused on building the conceptual foundations and approaches to gender scholarship in tourism. For example, Munar (2017) explored liberal, critical and postmodern forms of feminism with a key concern of how to be a feminist tourism academic. Kalisch and Cole (2022: 2706) have proposed 'a Feminist Alternative Tourism Economics (FATE)' based on approaches including a Feminist Ethic of Care, Social Solidarity Economy and Human Rights Based Economy together with 'a transnational feminist, decolonising lens'. This, they argue, 'could offer an alternative pathway to achieving a more gender just society and, by extension, tourism system' (2022: 2706). This approach seemingly recommends creating an alternative sub-system running in parallel to the wider and destructive capitalist economy, because these authors are mute on this structural justice concern (on such reformism versus radical overturning, see Higgins-Desbiolles, 2008). While this feminist alternative tourism economy might foster change on the margins, capitalist tourism continues to expand, driving social, environmental and climate injustices, and perpetuating largely unimpeded destruction. This analysis also suffers the weakness of a 'problematic lacunae around race' that the proponents of diverse economies have been accused of (see Bledsoe *et al.*, 2022: 282).

In the main, we have reformists and co-opting agendas of implementing gender budgeting, 'women empowerment' and giving voice to women and others in tourism strategies. However, these act as trojan horses for the imposition of universalising modernity, bringing women into capitalistic tourism to struggle to make a living from it while it continues to contribute to the wrecking of the planet through climate change. A critical feminist approach would advocate for a radical, anti-capitalist, anti-oppression, anti-hierarchy and collectivist approach as the essential foundation to intertwining gender justice, racial justice and climate justice so that all can thrive with dignity and hope for their children.

Critical Feminist Climate Justice

> *[…] these systems are so deeply flawed – instead of commons, we have sacrifice zones. And how climate change is really a byproduct of this systematic world of winners and losers. And then we talk about the ways we really need to flip this on its head.* Jacqueline Patterson, the director of the NAACP Environmental and Climate Justice Program (2017)

From this critical analysis, it is possible to derive a conceptualisation of critical feminist climate justice. Critical feminist climate justice spotlights the unequal impacts of both climate change and the responses to climate change on women, diverse genders, peoples made marginalised and vulnerable, future generations and our more than human relations. It employs an intersectional approach that unpacks, makes visible and responds to the multiple, interwoven and compounded injustices, oppressions and marginalisations that occur as climate change intersects with and reinforces ongoing racism, classism, ableism, sexism and other forms of discrimination and oppression. It also takes a structural justice approach understanding climate injustice is wrapped up with other forces of social and environmental injustices that shape our world today. Finally, the solutions and responses it offers are not solely focused on empowering women, but rather advancing the collective struggle for justice. It is therefore an essential pillar to the development of any 'critical climate justice praxis' in tourism (see the Introduction).

There are many voices building this understanding of climate justice as interwoven with gender justice. For example, some voices of the Pacific have insisted that gender justice in climate justice movements must have a wider focus than women, with some advocating 'queering climate' because:

> Fa'afafine [those biologically assigned male at birth, who express their gender in a feminine way], fa'atama [those assigned female at birth, who express their gender in a masculine way], trans and queer peoples are disproportionately affected by natural disasters and the wider effects of climate change. This talanoa [discussion] explores the overlap between the climate and queer movements, and their interconnected struggles within small island environments. (Talanoa Forum, 2023)

But we must also move beyond the limited understanding that women and gender diverse people should be understood to be disproportionately impacted and, in response, have their voices heard as we mitigate and adapt to climate change. A critical feminist lens challenges us to get to the structural injustices of tourism that see some exploited, oppressed and treated with violence under its capitalistic power formation. This empowers us to rethink the very basis of tourism as a phenomenon.

In pursuit of such a programme, there are critical feminist voices in tourism's past that we should return to for guidance. Professor Haunani-Kay Trask could be classed among the most prominent when she characterised tourism to her homeland, Hawai'i, as both 'prostitution' and a form of 'environmental racism'. She co-wrote with her sister, Mililani Trask:

> Tourism is not here to sell haole (white) culture. It is here because we are the native people of this aina (land). It is our culture the tourists come to see. It is our land the tourists come to pollute. Without beautiful Hawaiian women dancing, there would be no tourism.
>
> Tourism deforms culture to the point of cultural prostitution [...] Hawaii is itself the female object of degradation. Our aina are no longer the source of food and shelter, but rather the source of money. Land is now called real estate, rather than papa, our word for mother. Beautiful areas, once sacred to my people, are now expensive resorts. Now, even access to beaches and near hotels is strictly regulated or denied to the local people altogether. (Trask & Trask, 2010)

The concept of environmental racism describes the 'deliberate targeting of communities of color for toxic waste disposal and the siting of polluting industries. It is racial discrimination in the official sanctioning of the life-threatening presence of poisons and pollutants in communities of color' (Chavis, 1993: 3). Trask's work demands that we analyse the ways in which tourism acts as a form of environmental racism. The truth of this assertion can be uncovered in its gated resorts, golf courses, privatised beaches, luxury swimming pools and debasement of cultures all marking forms of pollution to ecology and community, in places converted to tourism destinations. These were places where commons once existed, communities exercised impactful stewardship and complex cultures nourished relations between people, ecologies and future generations; such people and places are the ones we frequently refer to as 'least responsible for climate change'. For Trask, Hawai'i is emblematic of the structural injustices of rapacious tourism and she warned tourists: 'if you are thinking of visiting my homeland, please don't. We don't want or need any more tourists, and we certainly don't like them. If you want to help our cause, pass this message on to your friends' (1993: 196). This marks a clear refusal of the imposition of capitalist tourism on a community.

As another critical vignette for perspective, we might turn to the materials released for International Women's Day 2024 by UN Tourism and a response from the Tourism Alert and Action Forum (TAAF), a global network advocating for communities impacted by tourism. UN Tourism (2024b) selected the theme 'Invest in women' and was very much in line with liberal feminism's focus of promoting 'gender equality' (only focused on women). UN Tourism (2024b: n.p.) took an econometric view arguing: 'The benefits of increased investment in women could be huge, with evidence showing that closing gender gaps could boost GDP per capita by 20% and create almost 300 million jobs in the global economy by 2035'.

TAAF's International Women's Day statement reacted with a critical feminist response:

> March 2024 International Women's Day comes around again. This year's theme [declared by UN Tourism] is 'Invest in women: Accelerate progress' [...] TAAF counters 'Build solidarity, letting women lead the way' [...] Women and girls are part of a collective society. The Western feminist interpretation of the goals of International Women's Day pushes a materialist, individualist and apolitical worldview that does not resonate with the Global South. Justice for the Global South will come from an expansive approach based on solidarity. It necessitates resistance to the imposition of Western universalistic notions of 'progress' and 'development' that really are covert means for ongoing colonial oppressions and wealth extractions. It demands dignity for women, girls, men and boys as they are all people of value collectively. (TAAF, 2024)

Too often the approach taken for addressing climate change is a managerial and technical approach rather than addressing structural injustices. There is also a marked ignoring of the violence against those who speak out. In a growing number of countries, environmental defenders and climate activists, including Indigenous leaders, Elders, youth and women, who face dangerous repression for demanding environmental rights and justice. They have been jailed, threatened with or subjected to violence, forcibly disappeared or even murdered (including an affiliate of TAAF, see https://aplaneta.org/11173-2/). There is also a growing trend to equate environmental activism with terrorism (see Salter, 2011).

Truth telling is vital in our efforts for climate justice. An analysis from the insurance sector offering a 'planetary solvency' assessment of climate change estimated that if we reach 2°C of warming or more, GDP losses of 25% or more and two billion deaths or greater could be expected (Trust *et al*., 2025: 32). In such a context, how can the consumption of tourism with its climate emissions be justified, when it is such a non-essential and luxury indulgence?

In this era, when lives have different value according to their proximity to privilege, it is imperative that critical feminist interrogations of power and privilege are foregrounded and sustained. These words from Icaza and

Vázquez are very relevant for our considerations of tourism futures in the face of critical climate injustices:

> To what extent is the political subjectivity of the hyper-consumer of today derivative or belongs to this genealogy of a subject that acquires pleasure and understands himself through the consumption of the life of others and the life of earth? (2025: 10)

That is, contemporary tourism's power imbalances, oppressions, wealth extractions and devastations are in part built on the foundations of historical conquest and imperialism and processes of ongoing racial capitalism (Edwards, 2021; Higgins-Desbiolles, 2022). It is a sense of supremacy that allows some to demand the right to consume the lives of others (as cultures in tourism) and the life of earth (as destinations to visit and consume) and that supports the insatiable and ceaseless demand for tourism growth and expansion despite the threat of collapse that emanates from exceeding several planetary boundaries, including climate change (Rockström *et al.*, 2023).

Conclusion

The master's tools will never dismantle the master's house. They may allow us temporarily to beat him at his own game, but they will never enable us to bring about genuine change. Audre Lorde (1984)

This chapter has taken seriously Patterson's assertion that 'Climate justice is racial justice is gender justice' (2017). It has used Angela Davis' critique of liberal 'glass-ceiling' feminism as a tool for this analysis. This has demonstrated the reasoning as to why liberal feminism is a suboptimal pathway for climate justice in tourism and instead has proposed the merits of radical critical feminist analyses to develop insights into the requirements for climate justice in tourism.

Rather than focus on the micro concerns of women's inclusion through mild reforms of current systems, the issues raised by the climate crisis demand we scope out to structural justice dynamics and understand the intersectional nature of current gender, racial and other oppressions. To be clear, the truth of climate change has been known since at least the 1950s and it is the powerful capitalist drive for profits of the fossil fuels industry (Yale Climate Connections, 2025) that has so far prevented meaningful action to get the world to stop this dangerous trajectory. Many tourism stakeholders have been too comfortable with mild reformist approaches that have failed to deliver the radical change (Higgins-Desbiolles, 2008) that we need as we confront the multiple crises presented in this era of polycrisis. Arguably, this is a strategy to maintain business as usual and avoid undermining their capacities to generate profits through growth and continual exploitation of the means of tourism production (the ecologies, the workers, the cultures and in fact all 'resources' pertaining to 'tourism destinations'). Perversely, some in the tourism industry may use 'climate washing' to grow their profits by marketing their 'responsible' response to the threat of climate crisis as a competitive selling point to climate concerned and guilt-ridden travellers. These are the mechanisms that have gotten us into the dire situation in which we find ourselves.

Recommendations arising from this analysis:

- We must have mechanisms for dialogues and decision-making on the choices and trade-offs with which we are faced so that the domination of decision-making by the powerful and privileged ceases (see Patterson, 2017). The COP processes are currently captured by petro-states and fossil fuel interests (Malm, 2016) and are arguably not fit for purpose.
- Climate finance funds must be established and under the control of Global South organisations as reparations for the climate crisis. At the 2009 COP15 convened in Copenhagen, Global North countries agreed to provide $100 billion climate finance per year by 2020 to Global South countries to support climate change mitigation and adaptation. These targets have yet to be realised.
- We must delink from capitalist systems of extraction, oppression and pollution. Considered attention to actions to realise these radical changes is vital (noting the terms 'transformations' and 'transitions' signal too slow and accommodating processes). Possibilities include commoning, collectivising, socialising, refusing and many others (see, for example, Fletcher *et al.*, 2023; Higgins-Desbiolles, 2020).

Critical feminist approaches would prompt us to ask different questions of tourism. How to we create counter hegemonic cultural models for tourism that replace current privatised and profit-extractive forms? How can decolonial practices be embedded in tourism responses to climate change? How do we alter the framing for all tourism stakeholders so that they all come to understand tourism as subservient to the flourishing of people, place and planet? There remains much work to be done, but success is dependent on approaches that take us to the heart of the structural injustices rather than distract us and delay us with focus on mere symptoms.

References

Abu-Lughod, L. (2013) *Do Muslim Women Need Saving?* Harvard University Press.
Aung, M.T. (2023) Queering climate justice – what climate justice can learn from queer groups. International Institute for Environment and Development. See https://www.iied.org/queering-climate-justice-what-climate-justice-can-learn-queer-groups (accessed April 2025).

Bhandar, B. and Ziadah, R. (2020) *Revolutionary Feminisms*. Verso.
Bigby, B.C., Smith, J. and Higgins-Desbiolles, F. (2024) Climate justice in tourism: An introductory guide (Travel Foundation). See https://www.thetravelfoundation.org.uk/climatejustice/ (accessed April 2025).
Bledsoe, A., McCreary, T. and Wright, W. (2022) Theorizing diverse economies in the context of racial capitalism. *Geoforum* 132, 281–290. https://doi.org/10.1016/j.geoforum.2019.07.004.
Byrne, D.C. and Imma, Z. (2019) Why 'southern feminisms'? *Agenda* 33 (3), 2–7. https://doi.org/10.1080/10130950.2019.1697043.
Chambers, D., Munar, A.M., Khoo-Lattimore, C. and Biran, A. (2017) Interrogating gender and the tourism academy through epistemological lens. *Anatolia* 28 (4), 501–513. https://doi.org/10.1080/13032917.2017.1370775.
Chavis, B. (1993) Foreword. In D.R. Bullard (ed.) *Confronting Environmental Racism: Voices from the Grassroots* (pp. 3–5). South End Press.
Christoffersen, A. and Emejulu, A. (2023) 'Diversity within': The problems with 'intersectional' white feminism in practice. *Social Politics: International Studies in Gender, State & Society* 30 (2), 630–653. https://doi.org/10.1093/sp/jxac044.
Crenshaw, K. (1991) Mapping the margins: Intersectionality, identity politics, and violence against women of colour. *Stanford Law Review* 43 (6), 1241–1299. https://doi.org/10.2307/1229039.
Davis, A. (2018) Angela Davis Criticises 'Mainstream Feminism'/'Bourgeoisie Feminism'. Afromarxist. YouTube, 8 January 2018. Video.
Edwards, Z. (2021) Racial capitalism and COVID-19. *Sustainable Human Development*, August. See https://jussemper.org/Resources/Economic%20Data/Resources/ZophiaEdwardsRacialCapitalism.pdf (accessed June 2025).
Equality in Tourism (EIT) (n.d.) Impact projects: Wamboma Co-operative Society Ltd. See https://www.equalityintourism.org/impact-projects/ (accessed May 2025).
Equality in Tourism (EIT) (2018) Sun, sand and ceilings: Women in tourism and hospitality boardrooms 2018. See https://www.equalityintourism.org/wp-content/uploads/2018/11/SUN-SAND-AND-CEILINGS-new.pdf (accessed March 2025).
Equality in Tourism (EIT) (2024) Equality in tourism gender policy. See https://www.equalityintourism.org/wp-content/uploads/2024/05/GEP_EIT.pdf (accessed March 2025).
Fletcher, R., Blanco-Romero, A., Blázquez-Salom, M., Cañada, E., Murray Mas, I. and Sekulova, F. (2023) Pathways to post-capitalist tourism. *Tourism Geographies* 25 (2–3), 707–728. https://doi.org/10.1080/14616688.2021.1965202.
Fraser, N. (2013) How feminism became capitalism's handmaiden – and how to reclaim it. *The Guardian* (Online). See https://www.theguardian.com/commentisfree/2013/oct/14/feminism-capitalist-handmaiden-neoliberal (accessed April 2025).
Gender CC (2007) Protecting tropical forests and gender justice. Position paper. See https://seors.unfccc.int/applications/seors/attachments/get_attachment (accessed April 2025).
Higgins-Desbiolles, F. (2008) Justice tourism and alternative globalisation. *Journal of Sustainable Tourism* 16 (3), 345–364. https://doi.org/10.1080/09669580802154132.
Higgins-Desbiolles, F. (2020) Socialising tourism for social and ecological justice after COVID-19. *Tourism Geographies* 22 (3), 610–623. https://doi.org/10.1080/14616688.2020.1757748.
Higgins-Desbiolles, F. (2022) The ongoingness of imperialism: The problem of tourism dependency and the promise of radical equality. *Annals of Tourism Research* 94, 103382. https://doi.org/10.1016/j.annals.2022.103382.
Icaza, R. and Vázquez, R. (2025) Vulnerability and the coloniality of performativity. *Globalizations*. https://doi.org/10.1080/14747731.2025.2491972.
IPCC (2021) Sixth assessment report. See https://www.ipcc.ch/assessment-report/ar6/ (accessed April 2025).
Jones, A. and Steffen, W. (2019) Our climate is like reckless banking before the crash – it's time to talk about near-term collapse. *The Conversation* (Online). See https://

theconversation.com/our-climate-is-like-reckless-banking-before-the-crash-its-time-to-talk-about-near-term-collapse-128374 (accessed August 2023).

Kalisch, A.B. and Cole, S. (2022) Gender justice in global tourism: Exploring tourism transformation through the lens of feminist alternative economics. *Journal of Sustainable Tourism* 31 (12), 2698–2715. https://doi.org/10.1080/09669582.2022.2108819.

Lorde, A. (1979) The master's tools will never dismantle the master's house. Comments at 'The Personal and the Political' Panel, Second Sex Conference, October 29, New York University. See https://tinyurl.com/44yh928w (accessed May 2025).

Lorde, A. (1984) *Sister/Outsider*. Crossing Press.

Mahler, A.G. (2017) Global South. In E. O'Brien (ed.) *Oxford Bibliographies in Literary and Critical Theory*. Oxford University Press. Excerpt available at: https://globalsouthstudies.as.virginia.edu/what-is-global-south.

Malm, A. (2016) *Fossil Capital: The Rise of Steam Power and the Roots of Global Warming*. Verso.

Minow, M. (2021) Equality vs equity. *American Journal of Law and Equality* 1, 167–193. https://doi.org/10.1162/ajle_a_00019.

Mohanty, C.T. (2002) 'Under Western eyes revisited': Feminist solidarity through anticapitalist struggles. *Signs* 28 (2), 499–535. https://www.journals.uchicago.edu/doi/10.1086/342914.

Morgan, R., George, A., Ssali, S., Hawkins, K., Molyneux, S. and Theobald, S. (2016) How to do (or not to do)… gender analysis in health systems research. *Health Policy Plan* 31 (8), 1069–1078. https://doi.org/10.1093/heapol/czw037.

Munar, A.M. (2017) To be a feminist in (tourism) academia. *Anatolia* 28 (4), 514–529. https://doi.org/10.1080/13032917.2017.1370777.

Munar, A.M., Biran, A., Budeanu, A., Caton, K., Chambers, D., Dredge, D., Gyimothy, S., Jamal, T., Larson, M., Nilsson Lindström, K., Nygaard, L. and Ram, Y. (2015) The gender gap in the tourism academy: Statistics and indicators of gender equality. While waiting for the dawn. See https://research-api.cbs.dk/ws/portalfiles/portal/45082310/ana_maria_munar_the_gender_gap.pdf (accessed April 2025).

One Planet Sustainable Tourism Programme (2021) Glasgow Declaration on Climate Action in Tourism: A commitment to a decade of climate action. See https://www.oneplanetnetwork.org/sites/default/files/2022-02/GlasgowDeclaration_EN_0.pdf (accessed August 2024).

Patterson, J. (2017) Climate justice is racial justice is gender justice. The Just Transition Issue, Fall issue, *Yes Magazine*. See https://www.yesmagazine.org/issue/just-transition/2017/08/18/climate-justice-is-racial-justice-is-gender-justice (accessed March 2025).

Pritchard, A. (2014) Gender and feminist perspectives in tourism research. In A. Lew, C.M. Hall and A. Williams (eds) *The Wiley Blackwell Companion to Tourism* (pp. 314–324). Wiley.

Rockström, J., Gupta, J., Qin, D. et al. (2023) Safe and just Earth system boundaries. *Nature* 619, 102–111. https://doi.org/10.1038/s41586-023-06083-8.

Salter, C. (2011) Activism as terrorism: The green scare, radical environmentalism and governmentality. *Anarchists Developments in Cultural Studies* 1, 211–238.

Simpson, L.B. (2020) Interview. In B. Bhandar and R. Ziadah (eds) *Revolutionary Feminisms* (pp. 139–148). Verso.

Sultana, F. (2022) Critical climate justice. *The Geographical Journal* 188, 118–124. https://doi.org/10.1111/geoj.12417.

Talanoa Forum (2023) Queering Climate. Powerhouse Museum, Sydney, 9-12 October. See https://powerhouse.com.au/stories/queering-climate (accessed April 2025).

Terry, G. (2009) No climate justice without gender justice: An overview of the issues. *Gender and Development* 17 (1), 5–18. http://www.jstor.org/stable/27809203.

Thunberg, G. (2018) High-level Segment Statement COP 24. Katowice Climate Change Conference – December. See https://unfccc.int/documents/187780 (accessed August 2024).

Tourism Alert and Action Forum (TAAF) (2024) TAAF counters 'Build solidarity, letting women lead the way'. See https://www.facebook.com/share/p/16HzoMTn2b/ (accessed September 2024).

Tourism Panel on Climate Change (TPCC) (2023) Tourism and Climate Change Stocktake 2023. [Eds Becken, S. and Scott, D.]. See https://tpcc.info/ (accessed March 2025).

Trask, H. (1993) *From a Native Daughter: Colonialism and Sovereignty in Hawai'i*. University of Hawai'i Press.

Trask, H.K. and Trask, M. (2010) The Aloha Industry: For Hawaiian women, tourism is not a neutral industry. *Cultural Survival*. See https://www.culturalsurvival.org/publications/cultural-survival-quarterly/aloha-industry-hawaiian-women-tourism-not-neutral-industry (accessed April 2025).

Trust, S., Saye, L., Bettis, O., Bedenham, G., Hampshire, O., Lenton, T.M. and Abrams, J.F. (2025) Planetary solvency – finding our balance with nature. Collaborative Insights in the Public Interest. See https://actuaries.org.uk/document-library/thought-leadership/thought-leadership-campaigns/climate-papers/planetary-solvency-finding-our-balance-with-nature/ (accessed April 2025).

UNFCCC (2024) Chronology of gender in the intergovernmental process. See https://unfccc.int/topics/gender/workstreams/chronology-of-gender-in-the-intergovernmental-process (accessed March 2025).

UN Tourism (2024a) UN Climate Change COP29 thematic day on tourism. See https://www.unwto.org/events/un-climate-change-cop29-thematic-day-on-tourism (accessed March 2025).

UN Tourism (2024b) UN Tourism Calls on Sector to #InvestInWomen. See https://www.unwto.org/news/un-tourism-calls-on-sector-to-invest-in-women (accessed September 2024).

Vinyeta, K., Whyte, K. and Lynn, K. (2016) Climate change through an intersectional lens: Gendered vulnerability and resilience in Indigenous communities in the United States. Gen. Tech. Rep. PNW-GTR-923. Portland, OR: U.S. Department of Agriculture, Forest Service, Pacific Northwest.

Watene, K. (2016) Valuing nature: Māori philosophy and the capability approach. *Oxford Development Studies* 44 (3), 287–296. doi:10.1080/13600818.2015.1124077.

Whyte, K.P. (2016) Is it colonial déjà vu? Indigenous Peoples and climate injustice. In J. Adamson and M. Davis (eds) *Humanities for the Environment: Integrating Knowledge, Forging New Constellations of Practice* (pp. 88–105). Routledge.

Women's Environmental Leadership Australia (n.d.) Gender, climate and environmental justice in Australia. See https://genderclimatetracker.org/sites/default/files/Resources/Full-report-Gender-Climate-and-Environmental-Justice-in-Australia-WELA.pdf (accessed March 2025).

Yale Climate Connections (2025) Scientists knew about global warming in the 1950s. 27 April 2025. See https://yaleclimateconnections.org/2025/04/scientists-knew-about-global-warming in the 1950s/ (accessed March 2025).

12 Deep Adaptation, Climate Justice and Tourism Futures

Freya Higgins-Desbiolles

Climate scientists stated in 2019 that it is 'time to talk about near-term collapse'. This is due to the lack of action to prevent climate change and the crossing of a number of other planetary boundaries. This threatens to cause widespread societal collapse in many countries around the world and threatens the globalised way of life many have come to expect. However, the threat of collapse is experienced differentially, thus bringing concerns with climate justice to the fore. The analysis will engage with the framework of Jem Bendell's 'deep adaptation' analysis. Proponents of deep adaptation argue that societal collapse is either likely, inevitable or already underway. Bendell's framework included five aspects of response to this recognition: resilience, relinquishment, restoration, reconciliation and reclamation. In this chapter, I use this framework to analyse what a deep adaptation approach to tourism might offer. It considers how we might revise tourism practices, tourism education and tourism scholarship in a context of the possible collapse of the industrial model of tourism in a climate just way.

Prologue

> We need stories. And not just stories about the stakes, which we know are high, but stories about the places we call home. Stories about our own small corners of the Earth as we know them. As we love them.
> Julian Aguon, 'To Hell With Drowning' (2021)

I grew up on an island off the coast of North Carolina, unaware of its experiences of collapse. Historical collapse was evident in the aftermath of the dispossession and erasure of the Native Americans who would have thrived along its shores, leaving traces such as the 'Indian trail tree' and some place names. A form of collapse was also present during my youth in terms of the neo-conquest of outsiders, mostly invading as 'holiday homeowners' and beach holiday tourists. I grew up with stories about when the island was still wild, hosting black bears and cougars and marshes abounding with alligators and wading birds.

This was a little foretaste of the collapse we may now be facing. A moment of knowing of what stands to be lost, nostalgia in advance of this loss and realisation of how strong the odds are against salvage, much less salvation. The island was developed, many locals displaced and the coastal ecology irreparably changed. This chapter is written with these experiences in mind.

It is also crafted with mindfulness to the critical insights gleaned from Potawatomi scholar Kyle Whyte's work which alerts us to the dystopias colonised people have already confronted and survived:

> Some Indigenous perspectives on climate change can situate the present time as already dystopian. Instead of dread of an impending crisis, Indigenous approaches to climate change are motivated through dialogic narratives with descendants and ancestors. (2018: 224)

Whyte's insights might awaken privileged people to the fortitude that can be summoned when we reckon with our histories. Privileged communities are only recently experiencing some of the devastating impacts of climate reckonings. Glenn Albrecht offered a word for this new experience of facing loss of what we have known – 'solastalgia' – defined '… as an emplaced or existential melancholia experienced with the negative transformation (desolation) of a loved home environment' (2012: n.p.).

These insights invite us to look at present difficulties with some humility, sense of responsibility and attention to relationalities across time and space. For it is in owning responsibility for the present situation we confront, while also acknowledging the pathologies of others (fossil fuel executives, capitalists and/or the one per cent), that we may adapt ourselves to the collapse of our unviable 'civilisation' and the unviable form of tourism associated with it. It is through articulating dialogic narratives with our ancestors and our descendants that we may confront what we have done and define what we might take forward.

Introduction

> *Courage is accepting how bad climate and ecological breakdown is without grasping for false hope, while continuing to fight hard for a rapid shift into emergency mode. False hope is an opiate, it reduces pain but also acts to block the rapid change we need to save lives.* Peter Kalmus, NASA scientist @climatehuman, tweet 8 February 2023

Climate scientists stated in 2019 that it is 'time to talk about near-term collapse' (Jones & Steffen, 2019). Collapse threatens due to a lack of action to prevent climate change and the crossing of a number of other planetary boundaries. These events not only imperil communities with potential widespread societal collapse, but also threaten the globalised way of life we have come to expect.

In 2023, media reported a number of climate-induced emergencies, leading to United Nations Secretary Antonio Guterres to assert that 'the era of global boiling has arrived' (Guterres, 2023). Soon thereafter, the nation of Iran announced a national shut down for two days as temperatures in some parts exceeded 50°C (Reuters, 2023). Globally, there were travel warnings to countries around the world as fires, smoke pollution, heat waves, floods and storms struck Europe, Asia and the Americas. Clearly, tourism will have to change as these irreversible dynamics from climate heating will continue and strengthen. But, as this chapter explores, are we facing the end of tourism as we have known it? And if so, how might we adapt in the most optimum way to ensure that we act to salvage as much as possible for current humans, future generations and more-than-human kin, preferably through cooperation and consensus and with attention to just transitions and outcomes, as well as intragenerational and intergenerational equity?

The analysis will engage with the framework of Jem Bendell's (2018, 2020, 2021) 'deep adaptation' analysis in order to contemplate just transitions in tourism in the context of possibilities of collapse. Proponents of deep adaptation argue that societal collapse is either likely, inevitable or already underway. Bendell's framework included five aspects of response to this recognition: resilience, relinquishment, restoration, reconciliation and reclamation. This framework is used here to analyse what a deep adaptation approach to tourism might offer and how might we revise tourism practices and tourism scholarship with attention to justice in a context of the possible collapse of the industrial model of tourism.

The Situation

Today, we find ourselves confronted with a climate emergency as greenhouse gases trapped in the Earth's atmosphere are causing a rise in the average global temperature. The predominant gas is carbon dioxide, making up some two thirds, and is particularly the result of burning fossil fuels, which drives our industries, transport and homes. This global heating is causing crises and catastrophes, killing people, threatening economies and societies and undermining public health and food security.

The climate emergency is wrapped up with other crises resulting from human pressures on the Earth system. Recent reporting revealed humanity has exceeded seven of eight Earth Systems Boundaries (ESBs).

ESBs are scientifically measured limits for human impacts on climate, biosphere, water and nutrient cycles and aerosols (Rockström *et al.*, 2023). Operating within these limits, can help maintain a stable and resilient planet. This exceeding of the ESBs marks profound change and damage to the climate, biosphere and biodiversity. As a result:

> Human pressures have put the Earth system on a trajectory moving rapidly away from the stable Holocene state of the past 12,000 years, which is the only state of the Earth system we have evidence of being able to support the world as we know it. (Rockström *et al.*, 2023: n.p.)

We are not reducing greenhouse gas emissions quickly enough, impactful action is lacking and false leads are evident to both divert meaningful efforts and profit from a climate-washing emergent industry (Schuijers, 2023). Misplaced hope in technology and human ingenuity is used to mislead us, including carbon dioxide removal and carbon storage, carbon offset schemes and 'sustainable' aviation fuels, thus allowing continued emissions despite the warnings. As Ho noted, 'Humanity has never removed an atmospheric pollutant at a global, continental or, even, regional scale – we have only ever shut down the source and let nature do the clearing up' (Ho, 2023: 9). Drastic emissions cuts are essential, which would require rapidly changing our growth-based economic model. There is little evidence that such an idea is under consideration by those in power and the window of opportunity to prevent dangerous feedback loops and run-away warming is closing. In fact, Lamb *et al.* outlined powerful discourses of 'climate delay' that work to 'justify inaction or inadequate efforts' (2020: 1).

Additionally, the power of 'fossil capital' (Malm, 2016) is evident in the firm refusal of wealthy and/or powerful states such as Australia, the UK, the US and China to turn away from policies of infinite growth fuelled by polluting fossil fuels. Additionally, fossil fuel interests have captured the key climate action forum, the Convention of the Parties (COP) for the United Nations Framework Convention on Climate Change, possibly in an effort to obstruct measures to phase out fossil fuels. The two most recent Climate Summits, COP 28 and 29, have been held in petro states, the UAE and Azerbaijan. Former UN climate chief Christiana Figueres said of the 2023 UAE summit: 'fossil fuel lobbyists outnumbered representatives of scientific institutions, Indigenous communities and vulnerable nations. We cannot hope to achieve a just transition without significant reforms to the Cop process that ensure fair representation of those most affected' (Harvey *et al.*, 2024: n.p.).

New vocabulary has entered our lexicon as examples of 'climate washing', 'climate profiteering' and 'climate neo-colonialism' are growing (e.g. it was reported that Liberia has agreed to grant 10% of its territory to an Emirati company for carbon credit production (Caramel,

2023)). These manoeuvrings for profit and power are not uncontested, however. Individuals and jurisdictions (such as Maui County, Hawai'i – which recently suffered one of the deadliest wildfires in US history) are suing 'for decades of climate deception and delay' through the courts as part of an effort to break this power dominance of fossil fuel capital (Saunders, 2023: n.p.).

Added to these complex difficulties is the need to account for social and ecological justice as we seek ways to address these dire circumstances. '…Disparity in harm and benefits has been a central tension at international climate negotiations, particularly when trying to allocate responsibility and financial compensation between developed and developing countries' (Starr *et al.*, 2023: 1). The terms 'climate justice', 'climate debt' and 'just transitions' are meant to underscore the crises we face are historically grounded in colonialism, imperialism, capitalistic exploitation, uneven development and notions of Western/White supremacy. 'Uneven responsibilities and vulnerabilities continue to challenge the fairness and effectiveness of global environmental governance' (Sultana, 2023: 2).

Returning to the Maui County wildfire as a case in point, commentators noted the ways settler colonisation with its plantation economy laid the grounds for the disaster through dismantling Indigenous Kanaka Ma'oli land care (e.g. Klein & Sproat, 2023; Peters, 2023). As Mei-Singh *et al.* explained: 'The tourism industry is deeply interwoven with settler colonial capitalist development and has profited by appropriating land, water, infrastructure, and other resources as large-scale industrial plantation enterprises declined' (2024: 324). The predatory behaviour of land speculators in the aftermath of the disaster has been labelled 'plantation disaster capitalism', describing both the neo-colonialism and climate profiteering evident in their actions to try to profit from Kanaka Ma'oli vulnerabilities (Klein & Sproat, 2023). In addition to this aspect of intragenerational justice, we must contend with interspecies justice and intergenerational justice issues as well (Rockström *et al.*, 2023). Attending to such justice issues is vital if we are to secure needed collaboration to successfully respond and adapt to a climate-changed future.

While adaptation is now widely agreed as vital, efforts of some sectors such as tourism are focused on maintaining 'business as usual' for as long as possible, supported by discourses of 'delayed climate action' resulting in only incremental adaptation measures (Lamb *et al.*, 2020). Historically, tourism has been touted to be a 'benign' development pathway in comparison to other industries. However, recent research by Sun *et al.* (2024) provides detailed insights into tourism's climate impacts, exacerbated by its phenomenal growth trajectory. They found 'global tourism in 2019 was responsible for 8.8% of global anthropogenic warming' and in the decade between 2009 and 2019, the 'tourism carbon footprint expanded at a rate (3.5% p.a.) twice that of the global economy (1.5% p.a.) during this period, highlighting its status as a sector that is challenging to decarbonize' (Sun *et al.*, 2024: 2). As we contemplate just transitions, we clearly must include rethinking of tourism in this work.

The situation in which we find ourselves is dire. Needed action is not happening rapidly enough, while 'ongoing climate change is simultaneously continuing to reduce the range of viable adaptation options' (Lawrence *et al.*, 2023: 2). This is a context prompting some experts to turn to thoughts of 'deep adaptation'.

Deep Adaptation

Proponents of 'deep adaptation' come together on the premise that 'societal collapse is now likely, inevitable or already occurring' (Bendell, 2020: 22). Whether we face a 'great unravelling' (Gergis, 2020), 'societal disruption' from climate change and associated crises or eventual collapse of society as we have known it, 'deep adaptation' seems worthy of exploration as a way to come to terms with the difficult futures we face (Bendell, 2021). Deep adaptation advocates urge us to consider what we want to try to save and how we can face collapse together with kindness and compassion. Its unique value is in banishing our tendencies to self-delusion through 'hopium', hope as unhelpful distraction, forcing us to confront our true circumstances and to act accordingly.

Collapse refers to 'the breakdown of socioecological systems characterized by the loss of complexity, structure, and order' (Just Collapse, 2023: 2). However, these more localised experiences of collapse may not remain isolated and limited; 'cascading tipping points in the Earth system – such as melting ice sheets and forest collapse – may be existential long-term threats' (Jones & Steffen, 2019: n.p.). Jones and Steffen explained the global ramifications of these new realities:

> ... multiple stresses across human-made systems lead to catastrophic collapses in their functioning. These collapses, given how interconnected our global system is, can affect one country directly but lead to the failure of our finance systems or global supply chains in many others. To paraphrase English poet John Donne, no country is an island when it comes to shielding itself from collapses in other countries. (2019: n.p.)

This realisation of the growing likelihood of collapse has catalysed a new interdisciplinary field of study called 'collapsology' in France and it has gained some traction in French politics and society (Servigne *et al.*, 2021).

Tourism thought leaders have largely failed to date to engage with the possibility of collapse and challenge the logic of tourism in an era when human-induced climate change and overshoot of planetary boundaries present an existential threat. In fact, a recent study from the European

Commission projected under a 4°C warming scenario, 'overall impact on European tourism demand is expected to be *positive*, with a projected rise of 1.58% for the highest warming scenario' (Matei *et al.*, 2023: 29, emphasis added). While this report did note there would be heterogenous experiences across Europe (e.g. Wales would be likely to fare better than Corsica), it seems unmoored from the reality of the resulting chaotic and unpredictable weather (which, indeed, we are already experiencing at 1.5°C), as we cross multiple thresholds of planetary boundaries.

Through engaging with Bendell's deep adaptation analysis, we join the dialogue he started on how to engage with the threat of collapse; focused on what role tourism might play in a just transition by analysing what a deep adaptation approach to tourism might offer. The next section considers the interface between tourism and collapse to date.

Tourism and Collapse

Burns and Bibbings (2009) were prescient in their contemplation of the 'end of tourism' as tourism authorities began to face the significance of climate change. Generally, the engagement with the possibilities of collapse in tourism has been narrowly focused on the collapse of tourism, usually due to some specific event. For instance, McKercher and Chon (2004) critiqued the management of the 2003 Severe Acute Respiratory Syndrome outbreak which they alleged led to the 'collapse' of Asian tourism. In another example, Tomczewska-Popowycz and Quirini-Popławski (2021) analysed the impact on tourism visitation to the Ukraine from 2013 due to political instability. Most recently, the COVID-19 global pandemic reinvigorated recognition of tourism system vulnerability, as shutdown of global tourism was implemented in an effort to halt disease spread (e.g. de Bellaigue, 2020; Higgins-Desbiolles, 2020a; Hall *et al.*, 2020).

In this work, I am considering the more macro sense of collapse and asking how we might understand tourism's future in such a context and soften the blow. This might seem to be an unnecessary and discouraging endeavour to those immersed in tourism concerns. However, in 2019, scientists Jones and Steffen stated that it is 'time to talk about near-term collapse': 'These are not distant existential issues raised by uncertain and abstract models of future climatic risk. They are urgent questions that humanity has been ducking for decades, but now demand urgent answers' (2019: n.p.).

The field of tourism studies has a long engagement with agendas of 'sustainability' and the United Nations' (UN) Sustainable Development Goals (SDGs). Despite decades devoted to sustainability, we are no closer to arresting tourism's damages to habitats, consumption of scarce resources, insatiable appetite for new and 'undiscovered' places and threats to biodiversity. In the aftermath of the pandemic crisis and exacerbated by climate change crises, the SDG agenda is currently derailed. The most recent assessment found: '… of the around 140 targets for which trend data is available shows that about half of these targets are moderately or severely off track; and over 30 per cent have either seen no movement or regressed below the 2015 baseline' (UN, 2023: 4). Taking account of the failures of the sustainability and the SDG agendas, it should be clear that we cannot decouple economic growth from the real constraints of finite natural resources and fixed planetary boundaries. The promise of technological fixes and resource substitutions has proven hollow, and the outcomes are resource exhaustion, climate crisis and biodiversity extinctions. Efforts at 'sustainable', 'responsible' and 'regenerative' tourism approaches are unlikely to remedy the fundamentals of unsustainability that arise from a tourism system operated under neoliberal capitalism's premise of continual economic growth.

It is also important to note that sustainability initiatives and more recently climate change action have become the focus of conspiracy theories in our fracturing global community. For example, the urban planning concept of the 15-minute city has been associated with centralising power, authoritarian control and repression of citizen's mobility by such conspiracy adherents; they have likened it to ghettos, concentration camps and the storyline of *The Hunger Games* (Baker & Weedon, 2023). This has become a facet of the cultural wars between those who champion libertarian freedom and those who support social democratic management. In the stress induced by multiple crises, we may not be able to count on citizens/consumers supporting needed change as they are swayed by such conspiracy theories and comforted that they do not need to transform their habits and hopes (e.g. reducing meat consumption, avoiding flying and reducing the use of private automobiles). This may prove very important to the futures we are able to secure – whether a social Darwinian future beset with conflict and violence or a humane one that is compassionate, cooperative and community-minded (see Rand, 2016).

In the following sections, the analysis specifically engages with issues of collapse of global civilisation and the implications for those of us working in tourism. When I use the term 'global civilisation' I mean the systems we have grown to rely on built on the globalisation of the industrialised, marketised, capitalistic system with its consumer-driven and economic-growth oriented economies. Some communities are more vulnerably integrated into this ideological system, while others have maintained their subsistence capacities. Tourism-dependent economies in the Global South may find themselves particularly vulnerable during these transition times. Thus, a collapse analysis spotlights climate justice in tourism planning and advocates for a proactive approach to a just transition as the central foundation for just and equitable futures. The next sections address the five aspects of the deep adaptation framework: resilience, relinquishment, restoration, reconciliation and reclamation.

Applying the Deep Adaptation Framework: Resilience, Relinquishment, Restoration, Reconciliation and Reclamation

Resilience in the face of collapse

As we try to adapt to the enormous and unpredictable changes that the era of polycrisis will bring, we will have decisions to make about what we can try to save and how to go about this. According to Bendell, resilience requires us to consider 'how do we keep what we really want to keep?' (2020: 21). This means identifying what matters, making hard choices and also considering more-than-human kin. It will also necessitate choices in tourism as well.

Resilience is 'the capacity of a system to absorb disturbance and reorganize while undergoing change so as to retain essentially the same function, structure, identity and feedbacks' (Walker *et al.*, 2004: 2). This articulation might be understood in one of two ways: resilience as 'bounce-back' from a shock or resilience as a form of adaptive change. The former implies an anticipated return to normality while the latter does not. Being attentive to such nuances is essential in critically evaluating the use of the terminology of resilience.

The Tourism Panel on Climate Change (TPCC) has called for a 'new era of climate resilient tourism' (TPCC, n.d.: n.p.). This indicates a myopia similar to the European Commission's report mentioned in the introduction that envisioned a mere migration of tourism northwards from the Mediterranean as greater heat sets in from climate change. This focus on 'climate resilient tourism' suggests that the TPCC is more focused on sustaining tourism rather than curtailing tourism to address non-essential emissions. The TPCC efforts to date are congruent with what Lamb *et al.* (2020: 2) identified as pushing 'non-transformative solutions' in order to avoid the necessity of disruptive change; Lamb *et al.* delineate non-transformative solutions as: technological optimism; 'all talk, little action'; fossil fuel solutionism; and 'no sticks, just carrots'. The TPCC's approach also seems incongruent with the reality of unpredictable tipping points and crises we confront, as few places are likely to remain unscathed. Business as usual is going to become increasingly untenable everywhere under these conditions.

In the tourism studies literature, Hall *et al.*(2018) demonstrated that tourism studies' engagement with resilience has been relatively recent, lacking clear conceptualisation and tending to focus on economic and ecological resilience and particular levels such as communities and regions. Dredge argued that the 'default' approach in most tourism studies of resilience is human-centred and 'scientific in its knowledge inputs' (2019: 60). She offered critical interrogations of 'what kind of resilience and for what/whom' and challenged us to situate tourism in a complex, more-than-human world. Resilience should be understood in relation to human vulnerabilities; such vulnerabilities are often not accidental but are instead imposed and built over long periods of time. Clearly, resilience policies and actions require critical interrogations, unmasking power, exploitation and profiteering.

If we follow the deep adaptation premise on collapse, we might anticipate that the global tourism system will also collapse. Those communities and countries that have considerable dependency on tourism will find themselves exceedingly vulnerable, particularly Global South communities and Island states. The theorisation of 'localising tourism' offered by Higgins-Desbiolles and Bigby (2023) offers one model of resilience in tourism futures that might offer a useful alternative. They champion 'place-based leadership for place-based governance' and explain: 'Particularly as we face multiple, complex, compounding and cascading crises, such leadership is called on to defend place and peoples. However, this is not necessarily in prickly isolation but rather in related localisms' (Bigby *et al*., 2023: 35).

Resilience will prove insufficient to cope with the types of changes we are set to experience; particularly when our societies continue to demand resources and pollution sinks from the Earth system at extraordinary levels. We must also relinquish some of the ideologies, practices and lifeways that have led us into this situation.

Relinquishment of that which no longer serves us

Those taking a critical perspective would argue that it is the neoliberal capitalism and its attendant ideology that have led us to the situation we confront with human overshoot of fixed planetary boundaries (e.g. Bianchi, 2009; Fletcher, 2011; Higgins-Desbiolles, 2018). Therefore, in adapting to the future we have created, we will need to change in significant ways including relinquishing some things that are taken for granted. The principle of 'relinquishment' asks us 'what do we need to let go of in order to not make matters worse?' (Bendell, 2020: 21).

Careful attention to the source of the situation rather than the symptoms is essential. Mike Joy noted: 'The climate crisis is seen as a problem requiring a solution rather than a symptom of overshoot. The problem is generally formulated as looking for a way to maintain current lifestyles in the wealthy world, rather than reducing overshoot' (2023: n.p.). This suggests that resorting to technological fixes to sustain 'business as usual' may only exacerbate our difficulties rather than resolve the underpinning unsustainable lifeways enjoyed in the Global North and the gross inequalities in sharing the Earth's finite resources. The metrics of inequalities illuminates just how much of our predicament is an inequality crisis as much as it is a climate crisis (see Chancel *et al.*, 2022). For instance, 'In 2019, fully 40% of total U.S. emissions were associated with income flows to the highest earning 10% of households' (Starr *et al.*, 2023: 1). In terms of travelling by flights, Gössling and Humpe outlined the stark global inequality: 'the share of the world's population travelling by air in 2018 was 11%, with at most 4% taking international flights' (2020: 1). Furthermore, only a minute elite of the most frequent fliers, representing at most 1% of the world population, 'likely accounts for more than half of the total emissions from passenger air travel' (Gössling & Humpe, 2020: 1). These facts require us to centre structural justice in order to get at the root causes.

The proponents of deep adaptation also advise people to reflect on their own circumstances and consider what they might relinquish in these times. Some are leaving their paid employment to take up activities supporting ecological protection/restoration or community building, for example. These have been described as 'holding actions' to counter the damage done by capitalism's 'business as usual' (Kelly & Macy, 2021: 198–199). Others are engaging in activism risking their jobs and possible arrest such as those joining Scientist Rebellion and Extinction Rebellion. Youth are leading legal actions to establish requirements for governments to protect their future (Donger, 2022). In such cases, individuals are increasingly relinquishing their dreams of career progression and normal life pathways in order to play their individual part in stopping further damage and salvaging can be salvaged.

In terms of tourism, the principle of relinquishment would lead us to interrogate tourism's contribution to the crisis and act to arrest that contribution. This might include addressing tourism's contributions to greenhouse gas emissions through activities such as long-haul flights (see Gössling & Humpe, 2020). But taking the wider perspective, it should also be focused on addressing tourism's contribution to the wider drivers of these crises, including contributing to a 'culture-ideology of consumption' that has in part caused our circumstances (Higgins-Desbiolles, 2010). Under capitalist globalisation, the culture-ideology of consumption is the essential pillar for continuous growth as it induces 'wants' (for travel, for luxury items, for prestige and status conveyed through consumer choices) in people through a 'relentless torrent of "brainwashing"' (Higgins-Desbiolles, 2010: 123).

There are growing social movements delineating what we must relinquish in tourism in order to arrest tourism's contribution to non-essential consumption adding to greenhouse gas emissions. For instance, Stay Grounded (see Chapter 4), a global network focused on advocacy for a just, environmentally sound transport system and for a rapid reduction in air travel, has lobbied for the reduction in flying, the ban of private jet use and the termination of frequent flyer programmes (Stay Grounded, n.d.).

If we wish to adapt with social and ecological justice in mind, we might quickly relinquish the most unjustifiable, high-pollution and elitist tourism pursuits. For instance, in current circumstances it is very difficult to justify elite travel by private jet (see Gössling *et al.*, 2024), space tourism development (see Horton, 2025) and 'last chance tourism' to the world's last remaining, relatively intact natural environments, such as Antarctica. This is growing more important as some destinations respond to the pressures for decarbonisation of tourism to reach 'net zero by 2050' by shifting to 'low volume, high-yield' tourists – another way of characterising elite tourism (for an example, see Schaller & Vickers, 2020).

In addition to relinquishment of our present practices of self-destruction and harm, we might also pause to consider what we could and should restore. As the prologue indicated, reflexive consideration focused on insights from our collective past may help to inform our present times and help us navigate the future more wisely.

Restoration of that which we mistakenly left behind

The pillar of restoration asks us: 'what can we bring back to help us with the coming difficulties and tragedies?' (Bendell, 2020: 21). We might consider the transformations societies have undertaken to access the finance and goods of capitalism, the change in values that has occurred with marketisation and assess the real costs of these decisions and question who benefits. This might result in rethinking decisions leading to the privatisation of the commons (in a tourism context, parks and protected areas, beaches and waters). We might consider how we rebuild community connections that have been sacrificed in development decision making; for instance, overturning decisions to allow corporates such as Airbnb access to housing that has damaged urban neighbourhoods in old European cities such as Venice and Barcelona (Araya López, 2021). As Bendell (2023, n.p.) has explained, his view is that collapse of the current civilisation will lead us to 'collapsing into communities' and we need to now act to ensure that communities have what they need for a just transition.

Rewilding is a type of restoration that may play an important role in restoring ecologies, biodiversity and human-nature relationships. As Hall (2019) explained, there are many definitions and views on rewilding, including an approach that seeks to rewild urban areas. Rewilding in brief refers to allowing natural processes to recover and return to places previously impacted by human interventions. Rewilding has also been described as a wider social movement (Bekoff, 2014).

Restoration of pluriversal economies is another promising opportunity, as the global capitalist system has been pressing monocultures on Global South communities through debt and development traps (including dependency on continued revenues from fossil fuel extraction to pay off debt – see Woolfenden, 2023). Restoring or reviving such pluriversal approaches will reduce vulnerabilities and restore resilience (see one example in Little, 2023). The invaluable opportunities that restoration of pluriversal economies and lifeways provide are well articulated by Akomolafe and Benavides:

> We discern that the urge of the times is not to fix a broken system, but to acknowledge our inherent power to summon other worlds.
>
> We know a way. Our cultures teach us that in turning to each other, we become disruptive to old realities and hospitable to new ones. Because we will not co-create the world by proxy, we need to turn to ourselves again, and rekindle the realness we have lost. (2023: n.p.)

Similarly, there are enduring forms of more humanistic tourism that have been overshadowed in the corporatisation of tourism. This includes niches such as social tourism, solidarity tourism, educational tourism and pilgrimages. Higgins-Desbiolles (2020b) has called for 'socialising tourism' to revive public good forms of tourism such as these. Additionally, we could support social enterprises, non-governmental organisations and workers' cooperatives rather than accepting governments subsidising and supporting the large multinational corporations of tourism (see Fletcher *et al.*, 2023).

Regenerative tourism is emerging as the newest contribution to ensuring tourism serves wider purposes by shaping tourism according to a more holistic, ecological worldview. Regenerative tourism is focused on tourism's role in contributing to the transition to a regenerative economy. As Bellato *et al.* state: 'regenerative tourism is a transformational approach that aims to fulfil the potential of tourism places to flourish and create net positive effects through increasing the regenerative capacity of human societies and ecosystems' (2023: 1034). However, current regenerative tourism analyses remain stubbornly tourism-centric, with some exceptions, and thus fail to get to the roots of structural injustice (see Higgins-Desbiolles, 2025). The relationships between ecological, sociocultural, economic and spiritual restorations and the ways past knowledges and lifeways can be drawn on to address our present difficulties and to shape future visionings are vital tasks we should advance as part of the restorative agenda.

Reconciliation with all whom we owe care

The penultimate pillar of Bendell's deep adaptation framework is 'reconciliation'. This addresses the question with what/whom must we make peace as we recognise our mutual vulnerabilities and mortality (Bendell, 2020: 21). This is arguably the more open-ended pillar of the framework, as the possibilities are many.

One form of reconciliation we might consider is the human relationship to 'nature' (noting the artificial distancing that occurs with this phrasing and framing; see Fletcher, 2017). The 'biophilia hypothesis' (Kellert & Wilson, 1995) suggests 'that humans have evolved with nature to have an affinity for nature' (Jimenez *et al.*, 2021: 1). However, with processes of modernisation and capitalism many people living in the Global North and in cities have become estranged from nature. Findings that human health benefits from nature are numerous (see Jimenez *et al.*, 2021) and experiences during COVID lockdowns only underscored how important access to nature is. How we may better relate to and care for the natural world may be one of the most significant acts of reconciliation we may make.

One of the paradoxes of tourism is that it can cause environmental devastation and yet, in its 'eco' form, it may also be a tool to learn about, reconnect with and care for nature. However, Fletcher's (2017) analysis demonstrating that the discourse concerning 'connection with nature' requires critical unpacking for its problematic reinforcement of a false human-nature divide is a vital caveat. What is required, according to Fletcher, is a form of environmental education through tourism that employs a critical political-economy lens and reflexivity that continually works to unmask power, complicity with neoliberal marketisation and to overcome the false human-nature dichotomy that modernisation fosters.

Reconciliation also might be required between communities of people. Stinson *et al.* ask how might we '… design, develop, and manage tourism landscapes and installations that might further encourage the awakening of an ethics of reconciliation' (2022: 10). These considerations invite us to reflect on how we might 'do tourism otherwise' (Everingham *et al.*, 2021). Figure 12.1 shows community participation in a planning activity for an Aboriginal Cultural Place which features a triple-layered activation. It offers a cultural place supportive of the local Aboriginal community, fosters truth telling in a settler-colonial place in order to reshape cross-community relations and restores coastal ecologies through a living shorelines approach. This is an example of multiple layers of reconciliatory action through a cultural heritage tourism facility.

Perhaps the hardest aspect of reconciliation will be with our ourselves, as we gain greater recognition of just how much damage we have done when the heat, the fires and other crises only worsen rather than abate. Collapse may force us to reconcile ourselves to the abandonment of the high consumption and mobile lifestyles which some of the privileged took as their birth right. This might guide us to embracing an ethos of down-shifted, collective sufficiency, voluntary simplicity, frugality, direct democracy and radical localisation (see Higgins-Desbiolles & Bigby, 2023).

Figure 12.1 Community participation in planning for Yitpi Yartapuultiku Aboriginal Cultural Place, Port Adelaide, South Australia, 17 November 2022 (photo credit: author)

Reclamation of our lives, communities and futures from destructive systems

The final 'r' in the deep adaptation framework is reclamation. It is focused on the question of what might we reclaim from what has been lost in our submission to this dominant, extractive system which has led us into our predicament. Bendell explained: 'Reclamation is about taking back our lives, our dignity, our communities, and our connection to nature from the systems that have co-opted them' (2025: n.p.). In identifying the roots of social-ecological devastations resulting from our capitalist, hegemonic systems, we begin to identify lifeways we could reclaim to reinstate our humanity and connections. 'This R is about agency, power and dignity – reclaiming meaning-making, local control, spiritual autonomy, and relational depth from the alienation of modernity and global capitalism' (Bendell, 2025: n.p.). This pillar is similar to restoration but extends to ontological concerns.

The regenerative turn in tourism, discussed above, could be viewed as clear reclamation of diverse ways of 'being' and 'doing' in tourism (see Chapter 10, this volume). Tourism is especially conducive to cultural reclamations that protect autonomous lifeways. One example explored by Renkert from the Añangu of Ecuador explains how they use their Kichwa '[…] vision of *Sumak Kawsay* [or lifeway of 'Good Living' – 'Buen Vivir'] in the management and daily practice of tourism […] to strive towards specific, community-defined goals, including the production of local livelihood opportunities, cultural reclamation, and environmental stewardship' (2019: 1894). Additionally, the agenda of 'localising tourism' as described by Higgins-Desbiolles and Bigby (2023) is relevant to reclamation: 'defining tourism by the local community, local community empowerment, more localised geographies of travel and tourism, localising decision-making to the lowest level (subsidiarity) and the local interrelationships between people, place, ecology and all living things' (2023: 2). Reclamation in tourism in the deep adaptation framing would encourage us to empower the local, foster sociocultural–ecological diversity through tourism and scale back industrialised and extractive forms of tourism.

Implications for Tourism Scholarship, Tourism Education and Practice

The findings from this analysis applying the five aspects of deep adaptation to tourism are potentially profoundly significant. Wise planning would require that we apply the precautionary principle at this juncture, an approach in times of uncertainty recommending caution in planning and action to avoid or reduce harm. How might we act with cautious forethought in our tourism work in such a way that we secure human (with attention to intragenerational and intergenerational justice) and more-than-human welfare to the greatest extent possible in times of polycrisis?

Firstly, tourism academia might require profound rethinking. We need to be attentive to the futures of tourism graduates and ensure that we are not possibly placing them in debt for 'dud degrees' (Ross, 2023). Instead of guiding our students to careers leading an unsustainable tourism industry, might we re-orient their studies to be focused on tourism and recreation for local wellbeing rather than driven by growth and profit? Just as tourism education evolved to be dominated by location in business schools (Tribe, 2003), it might now be transitioned to recreation and leisure schools which are possibly more geared to human and ecological wellbeing (see Coleman, 2023). Tourism studies and tourism research need to be conducted with greater mindfulness of normative approaches and ethical values (Bramwell, 2007), especially in these times of crossing planetary boundaries and confronting multiple and cascading crises. Sound thinking now might guide us to specifically practical initiatives, as well. For example, Higham and Font (2019) advocated a 'low carbon' model for academia that would ensure the sector played its part in needed reductions, which should include rethinking our academic conference and meetings participation, shifting to virtual gatherings. Higgins-Desbiolles (2023) has offered a plan she labelled 'subsidiarity in tourism and travel circuits' for degrowing tourism with attention to climate just transitions.

As discussed in earlier sections, tourism practices must also change. As the Maui case demonstrated, the tensions between local communities and tourists are exacerbated in the disaster situations that are unfolding from these polycrises. Tourists will be torn in their decisions to forego travel when communities are struggling to recover, especially when rebooking and insurance cover may be an issue (Chung & Kircher, 2023). The micro, small and medium enterprises of tourism, which predominate numerically, will find it hard to offer such flexibility when tourism seasons become less predictable due to crises; this may mean that government support for tourism will become increasingly required (for an example, see Little & Ke, 2023). It is a critical question whether such spending can be justified when we can foresee that the climate crisis will be placing exponentially rising demands on limited government resources.

Additionally, we might consider how communities and certain other tourism stakeholders might take proactive control of their futures. Work on 'insurgent planning' may be particularly suited to this era of transitions during polycrisis:

> Insurgent planning does not rely on governments for decision-making and action. Instead, communities and citizens exert and extend power by setting their own agendas and implementing their own actions. They do not wait for elected representatives, or other powerbrokers, to act on their behalf. They act outside of formal processes and structures to achieve more equitable outcomes. By nature, insurgent planning is not sanctioned by government. (Just Collapse, 2023: 12)

Fletcher *et al*. (2023) have proposed 'eroding tourism' through 'pathways to a post-capitalist tourism', as a contribution to degrowth analyses and the 'socialising tourism' agenda. They present the possibilities of post-capitalist tourism futures through case studies: dismantling exploitative tourism through municipal regulation in Barcelona; Serviço Social de Comércio (SESC) Bertioga in Brazil as an example of social tourism; cooperative ownership in the case of Hotel Bauen in Buenos Aires, Argentina; and the use of common property regimes in the case of La Trapa, Mallorca, Spain (see Fletcher *et al*., 2023).

Scoping out to the wider context, it is essential that we transition away from globalising capitalism in the face of these urgencies from polycrisis. Not only does neoliberal, globalising capitalism directly contribute to the

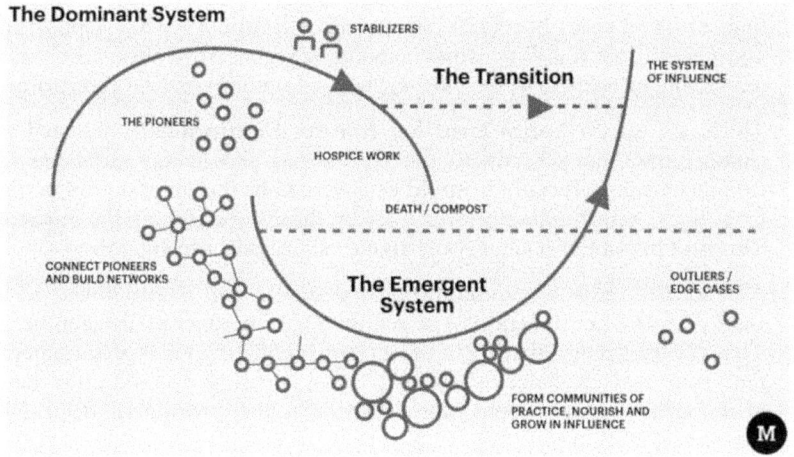

Figure 12.2 Berkana Institute's 'Two Loops Model' of Transition between Systems. Used with permission of the Berkana Institute

crises we face through its demand for endless growth, it also privatises, individualises and commodifies such that it undermines the cooperation we will require to meet the global challenge of the climate emergency. It has also overseen a global system of great inequality between the Global North and South which also works against agreement on action agendas (see Bello, 2008). We will need to overturn this dominant system and embark on a just transition to another way of living (see Figure 12.2).

Tourism dependent communities must transition away from vulnerable tourism dependency as the climate crisis threatens to collapse current tourism systems. As part of a climate just transition, the wealthy states of the Global North should support this transition with funds and resources, proposed as a 'new global climate fund for travel and tourism' (Bigby *et al.*, 2024), similar to the Green Climate Fund proposed as part of the COP processes.

It remains to be seen if the tourism industry's leadership will rise to this challenge, but as this chapter was being finalised, the leadership behind the Glasgow Declaration announced a new consultative agenda they planned in the lead up to COP 30. They shared a proposed four-pronged strategy, which included:

> First, a Global Climate Risk Register for Destinations – an open-access platform providing localised climate risk profiles to help governments, businesses, and investors act on the most urgent threats. This tool would enable transparent decision making and support stronger governance and targeted investment.

Second, a global initiative to sunset harmful tourism models – such as high-emission short breaks or over-touristed hotspots – while supporting a just transition: reskilling workers, building regenerative alternatives, and redirecting investment toward low-carbon, community-owned enterprises.

Third, a Climate Justice Fund for Tourism Destinations – designed to channel resources directly to the places and people most affected by climate change. Crucially, it would be governed by the communities on the front lines, with funding priorities set by those experiencing the impacts. This isn't just about finance – it is about repair, resilience and fairness.

And fourth, ensuring community-led planning and equity-linked KPIs (key performance indicators) a requirement in tourism development. That includes redistributing tax revenues, mandating local procurement, requiring real evidence of benefit, and ensuring communities have meaningful influence over tourism growth and investment decisions. (Sampson, 2025: n.p.)

If effectively realised, such collaborations for a more just transition would build greater resilience in vulnerable communities, reduce tourism's contributions to the polycrisis and help ensure more socially and ecologically just forms of tourism for the future. This is potentially a step towards what the introductory chapter explained as 'critical climate justice praxis' – a critical reading of the foundations of structural injustices and acting to address them in a holistic manner. However, there is a risk this is not the systemic transition needed and the initiative will succumb to climate washing; critical work on deep adaptation remains vital.

This chapter has briefly explored how building community capacities locally for care and nurturance is advocated as a promising path to deep adaptation. Correspondingly in tourism, 'localising tourism' may prove essential for a just transition (Higgins-Desbiolles & Bigby, 2023). This may be one promising pathway to deep adaptation in tourism, addressing polycrisis and fostering just transitions to a network of communities living in new ways that are low-consumption, low-growth, high in equity and conducive to social and ecological justice. The demands of climate justice are many – between humans, both intragenerationally and intergenerationally, and with our more than human kin. The deep adaptation framework offers a tool we can use to rethink tourism in extraordinary times so that it builds communities and collaboration in the face of the threat of collapse.

Conclusion

Collapse followed by transformation is a common way that complex systems evolve. Perhaps collapse of our high consumption, climate-destabilising society can lead to transformation towards a brighter human future. The Deep Adaptation framework ... is a helpful way to seek that transformation. Prof Will Steffen (in Bendell & Read, 2021: n.p.)

It is still not clear when global consciousness of the possibilities of collapse will develop, catalysing concerted action to avert the catastrophe we are facing. Climate disasters are amassing all around the globe and yet we are only at the beginning of the climate-related disasters we will experience in our lifetimes. This chapter has engaged with Bendell's deep adaption framework in order to understand how tourism might navigate a just social and ecological transition in the face of the threat of a great unravelling or collapse of the globalised capitalistic system. Addressing in succession the framework's pillars of resilience, relinquishment, restoration, reconciliation and reclamation, it outlined some pathways for a more climate just approach to tourism.

While thinking about human futures in terms of portending collapse, the possibility to see analyses such as this as pessimistic 'doomism' and conducive to fatalism is clear. Those concerned with collapse in a time of widespread denial may not be viewed well (Fetterman *et al.*, 2019). However, the evolving thinking of deep adaptation advocates actually anticipates some positive outcomes from the transition, as expressed by Will Steffen above and in Bendell's work where he explains he envisions that we will be 'collapsing into community' (2023).

It is both surprising and disappointing that leadership is not acting sufficiently in this era of threat of climate catastrophe. Future research in tourism might examine how cultural lags prevent needed change to transition the industry and academia to address such extraordinary circumstances (see Koutsobinas & Michalopoulou, 2023). However, such lag is not sustainable in the long-term because business as usual in tourism will be up-ended by a crisis challenged world.

It would be an opportune time to rethink tourism practice, education and scholarship with critical thought on how we might deeply adapt in such a way as to ensure best possible outcomes for humans – all humans, not just the elite and privileged – as well as more-than-humans, living now and those still to come. The global pandemic sparked the beginnings of such a rethink and now we should engage with new catalysts, such as the deep adaptation framework explored here, to further our work in transforming tourism for socially and ecologically just futures.

Acknowledgements

This chapter is an expanded version of a journal article entitled 'The end of tourism? Contemplations of collapse' (2024) published in the *Journal of Tourism Futures*. It has been included here with the kind permission of the journal editor, Professor Ian Yeoman. Thanks are due to Professor Jem Bendell, Dr Katie Carr and the Deep Adaptation Forum for the community of thought and practice they have developed to help us in 'breaking together'.

References

Aguon, J. (2021) To hell with drowning. *The Atlantic* (Online). See https://www.theatlantic.com/culture/archive/2021/11/oceania-pacific-climate-change-stories/620570/ (accessed July 2023).

Akomolafe, B. and Benavides, M. (2023) The times are urgent: Let's slow down. See https://www.bayoakomolafe.net/post/the-times-are-urgent-lets-slow-down (accessed August 2023).

Albrecht, G. (2012) The age of solastalgia. *The Conversation*, 7 August. See https://theconversation.com/the-age-of-solastalgia-8337 (accessed April 2023).

Araya López, A. (2021) A summer of phobias: media discourses on 'radical' acts of dissent against 'mass tourism' in Barcelona. Open Research Europe 1 (66). https://doi.org/10.12688/openreseurope.13253.1.

Baker, N. and Weedon, A. (2023, 27 February) What is the '15-minute city' conspiracy theory? ABC News (Online). See https://www.abc.net.au/news/2023-02-27/the-15-minute-city-conspiracy/102015446 (accessed August 2023).

Bellato, L., Frantzeskaki, N. and Nygaard, C.A. (2023) Regenerative tourism: A conceptual framework leveraging theory and practice. *Tourism Geographies* 25 (4), 1026–1046. https://doi.org/10.1080/14616688.2022.2044376.

Bello, W. (2008) Will capitalism survive climate change? *Chain Reaction* 104, 44–45.

Bendell, J. (2018) Deep adaptation: A map for navigating climate tragedy. Initiative for Leadership and Sustainability Occasional Paper 2. See https://www.lifeworth.com/deepadaptation.pdf (accessed January 2025).

Bendell, J. (2020) Deep adaptation: A map for navigating climate tragedy. Institute for Leadership and Sustainability (IFLAS) Occasional Papers Vol. 2. University of Cumbria, Ambleside, UK. (Unpublished).

Bendell, J. (2021) Deep adaptation: A map for navigating climate tragedy. In J. Bendell and R. Read (eds) *Deep Adaptation: Navigating the Realities of Climate Chaos* (pp. 42–86). Polity Press.

Bendell, J. (2023) Let's tell the moodsplainers they're wrong and then get back to work. See https://jembendell.com/2023/08/05/lets-tell-the-moodsplainers-theyre-wrong-and-then-get-back-to-work/ (accessed August 2023).

Bendell, J. (2025) The 5Rs of deep adaptation. See https://jembendell.com/2025/05/25/the-5rs-of-deep-adaptation/ (accessed June 2025).

Bendell, J. and Read, R. (eds) (2021) *Deep Adaptation: Navigating the Realities of Climate Chaos*. Polity Press.

Bekoff, M. (2014) *Rewilding Our Hearts: Building Pathways of Compassion And Coexistence*. New World Library.

Bianchi, R.V. (2009) The 'critical turn' in tourism studies: A radical critique. *Tourism Geographies* 11 (4), 484–504. https://doi.org/10.1080/14616680903262653.

Bigby, B.C., Edgar, J. and Higgins-Desbiolles, F. (2023) Place-based governance in tourism: Placing local communities at the centre of tourism. In F. Higgins-Desbiolles and B.C. Bigby (eds) *The Local Turn in Tourism: Empowering Communities* (pp. 31–53). Channel View Publications.

Bigby, B.C., Smith, J. and Higgins-Desbiolles, F. (2024) *Climate Justice in Tourism: An Introductory Guide*. Travel Foundation. See https://www.thetravelfoundation.org.uk/climatejustice/ (accessed April 2025).

Bramwell, B. (2007) Critical and normative responses to sustainable tourism. *Tourism Recreation Research* 32 (3), 76–78. https://doi.org/10.1080/02508281.2007.11081544.

Burns, P.M. and Bibbings, L.J. (2009) The end of tourism? Climate change and societal changes. *21st Century Society* 4 (1), 31–51.

Caramel, L. (2023) Liberia set to concede 10% of its territory to Emirati company for carbon credit production. *Le Monde* (Online), 2 August. See https://www.lemonde.fr/en/le-monde-africa/article/2023/08/02/liberia-set-to-concede-10-of-its-territory-to-emirati-company-for-carbon-credit-production_6077402_124.html (accessed August 2023).

Chancel, L., Piketty, T., Saez, E. and Zucman, G. (2022) World Inequality Report 2022. See https://wir2022.wid.world/www-site/uploads/2021/12/WorldInequalityReport2022_Full_Report.pdf (accessed August 2023).

Chung, C. and Kircher, M.M. (2023) After Maui wildfires, travelers ask: Would a trip help or hurt? *New York Times* (Online), 10 August. See https://www.nytimes.com/2023/08/17/travel/maui-wildfires-travel-tourism.html (accessed August 2023).

Coleman, C. (2023) Opinion: In case of apocalypse, find the nearest 4-H club. *Washington Post* (Online), 21 August. See https://www.washingtonpost.com/opinions/2023/08/21/4h-apocalypse-survival/ (accessed August 2023).

de Bellaigue, C. (2020) The end of tourism? *The Guardian* (Online), 20 June. See https://www.theguardian.com/travel/2020/jun/18/end-of-tourism-coronavirus-pandemic-travel-industry (accessed August 2023).

Donger, E. (2022) Children and youth in strategic climate litigation: Advancing rights through legal argument and legal mobilization. *Transnational Environmental Law* 11 (2), 263–289. https://doi.org/10.1017/S2047102522000218.

Dredge, D. (2019) Governance, tourism, and resilience: A long way to go? In J. Saarinen and A.M. Gill (eds) *Resilient Destinations and Tourism: Governance Strategies in the Transition towards Sustainability in Tourism* (pp. 48–66). Routledge.

Everingham, P., Peters, A. and Higgins-Desbiolles, F. (2021) The (im)possibilities of doing tourism otherwise: The case of settler colonial Australia and the closure of the climb at Uluru. *Annals of Tourism Research* 88, 103178. https://doi.org/10.1016/j.annals.2021.103178.

Fetterman, A.K., Rutjens, B.T., Landkammer, F. and Wilkowski, B.M. (2019) On post-apocalyptic and doomsday prepping beliefs: A new measure, its correlates, and the motivation to prep. *European Journal of Personality* 33 (4), 506–525. https://doi.org/10.1002/per.2216.

Fletcher, R. (2011) Sustaining tourism, Sustaining capitalism? The tourism industry's role in global capitalist expansion. *Tourism Geographies* 13 (3), 443–461. https://doi.org/10.1080/14616688.2011.570372.

Fletcher, R. (2017) Connection with nature is an oxymoron: A political ecology of 'nature-deficit disorder'. *The Journal of Environmental Education* 48 (4), 226–233. https://doi.org/10.1080/00958964.2016.1139534.

Fletcher, R., Blanco-Romero, A., Blázquez-Salom, M., Cañada, E., Murray Mas, I. and Sekulova, F. (2023) Pathways to post-capitalist tourism. *Tourism Geographies* 25 (2–3), 707–728. https://doi.org/10.1080/14616688.2021.1965202.

Gergis, J. (2020) The great unravelling: 'I never thought I'd live to see the horror of planetary collapse'. *The Guardian* (Online), 15 October. See https://www.theguardian.com/australia-news/2020/oct/15/the-great-unravelling-i-never-thought-id-live-to-see-the-horror-of-planetary-collapse (accessed August 2023).

Gössling, S. and Humpe, A. (2020) The global scale, distribution and growth of aviation: Implications for climate change. *Global Environmental Change* 65, 102194. https://doi.org/10.1016/j.gloenvcha.2020.102194.

Gössling, S., Humpe, A. and Leitão, J.C. (2024) Private aviation is making a growing contribution to climate change. *Communications Earth Environment* 5 (666), 1–11. https://doi.org/10.1038/s43247-024-01775-z.

Guterres, A. (2023) The era of global boiling has arrived. *The Guardian* (Online), 28 July. See https://www.theguardian.com/world/video/2023/jul/27/the-era-of-global-boiling-has-arrived-warns-the-un-video (accessed August 2023).

Hall, C.M. (2019) Tourism and rewilding: An introduction – definition, issues and review. *Journal of Ecotourism* 18 (4), 297–308. https://doi.org/10.1080/14724049.2019.1689988.

Hall, C.M., Prayag, G. and Amore, A. (2018) *Tourism and Resilience: Individual, Organisational and Destination Perspectives.* Channel View Publications.

Hall, C.M., Scott, D. and Gössling, S. (2020) Pandemics, transformations and tourism: Be careful what you wish for. *Tourism Geographies* 22 (3), 577–598. https://doi.org/10.1080/14616688.2020.1759131.

Harvey, F., Noor, D., Carrington, D. and Niranjan, A. (2024) Cop summits 'no longer fit for purpose', say leading climate policy experts. *The Guardian*, 15 November. See https://www.theguardian.com/environment/2024/nov/15/cop-summits-no-longer-fit-for-purpose-say-leading-climate-policy-experts (accessed January 2025).

Higgins-Desbiolles, F. (2010) The elusiveness of sustainability in tourism: The culture-ideology of consumerism and its implications. *Tourism and Hospitality Research* 10 (2), 116–129. https://doi.org/10.1057/thr.2009.31

Higgins-Desbiolles, F. (2018) Sustainable tourism: Sustaining tourism or something more? *Tourism Management Perspectives* 25, 157–160. https://doi.org/10.1016/j.tmp.2017.11.017.

Higgins-Desbiolles, F. (2020a) The end of global travel as we know it: an opportunity for sustainable tourism. *The Conversation* (Online). See https://theconversation.com/the-end-of-global-travel-as-we-know-it-an-opportunity-for-sustainable-tourism-133783 (accessed August 2023).

Higgins-Desbiolles, F. (2020b) Socialising tourism for social and ecological justice after COVID-19. *Tourism Geographies* 22 (3), 610–623. https://doi.org/10.1080/14616688.2020.1757748.

Higgins-Desbiolles, F. (2023) Subsidiarity in tourism and travel circuits in the face of climate crisis. *Current Issues in Tourism* 26 (19), 3091–3101. https://doi.org/10.1080/13683500.2022.2116306.

Higgins-Desbiolles, F. (2024) The end of tourism? Contemplations of collapse. *Journal of Tourism Futures* 10 (3), 476–485. https://doi.org/10.1108/JTF-11-2023-0259.

Higgins-Desbiolles, F. (2025) On what terms might regenerative tourism proponents justly engage with Indigenous knowledges? *Journal of Travel Research* 64 (8), 2047–2055. https://doi.org/10.1177/00472875251341308.

Higgins-Desbiolles, F. and Bigby, B.C. (eds) (2023) *The Local Turn in Tourism: Empowering Communities*. Channel View Publications.

Higham, J. and Font, X. (2019) Decarbonising academia: Confronting our climate hypocrisy. *Journal of Sustainable Tourism* 28, 1–9. https://doi.org/10.1080/09669582.2019.1695132.

Ho, D.T. (2023) Carbon dioxide removal is not a current climate solution — we need to change the narrative, *Nature* 616 (7955), 9. doi: https://doi.org/10.1038/d41586-023-00953-x.

Horton, A. (2025) Celebrities criticize all-female rocket launch: 'This is beyond parody'. *The Guardian*, 16 April (Online). See https://www.theguardian.com/culture/2025/apr/15/blue-origin-space-flight-criticism (accessed April 2025).

Jimenez, M.P., DeVille, N.V., Elliott, E.G., Schiff, J.E., Wilt, G.E., Hart, J.E. and James, P. (2021) Associations between nature exposure and health: A review of the evidence. *International Journal of Environmental Research and Public Health* 18 (9), 4790. https://doi.org/10.3390/ijerph18094790.

Jones, A. and Steffen, W. (2019) Our climate is like reckless banking before the crash – it's time to talk about near-term collapse. *The Conversation* (Online). See https://theconversation.com/our-climate-is-like-reckless-banking-before-the-crash-its-time-to-talk-about-near-term-collapse-128374 (accessed August 2023).

Joy, M. (2023) Critics of 'degrowth' economics say it's unworkable – but from an ecologist's perspective, it's inevitable. *The Conversation* (Online). See https://theconversation.com/critics-of-degrowth-economics-say-its-unworkable-but-from-an-ecologists-perspective-its-inevitable-211496 (accessed August 2023).

Just Collapse (2023) The little book of insurgent planning. See https://justcollapse.org/2023/03/13/a-little-book-of-insurgent-planning/ (accessed July 2023).

Kellert S.R. and Wilson E.O. (1995) *The Biophilia Hypothesis*. Island Press.

Kelly, S. and Macy, J. (2021) The great turning: reconnecting through collapse. In J. Bendell and R. Read (eds) *Deep Adaptation: Navigating the Realities of Climate Chaos* (pp. 197–208). Polity Press.

Klein, N. and Sproat, K. (2023) Why was there no water to fight the fire in Maui? *The Guardian* (Online), 18 August. See https://www.theguardian.com/commentisfree/2023/aug/17/hawaii-fires-maui-water-rights-disaster-capitalism (accessed August 2023).

Koutsobinas, T. and Michalopoulou, P. (2023) Networks of culture creatives in Patras: the relevance of cultural lag. *European Planning Studies* 31 (8), 1651–1672. https://doi.org/10.1080/09654313.2022.2093099.

Lamb, W.F., Mattioli, G., Levi, S., Roberts, J.T., Capstick, S., Creutzig, F., Minx, J.C., Müller-Hansen, F., Culhane, T. and Steinberger, J.K. (2020) Discourses of climate delay. *Global Sustainability* 3 (218), 1–5. https://doi.org/10.1017/sus.2020.13.

Lawrence, J., Wreford, A., Blackett, P., Hall, D., Woodward, A., Awatere, S., Livingston, M.E., Macinnis-Ng, C., Walker, S., Fountain, J., Costello, M.J., Ausseil, A.-G.E., Watt, M.S., Dean, S.M., Cradock-Henry, N.A., Zammit, C. and Milfont, T.L. (2023) Climate change adaptation through an integrative lens in Aotearoa New Zealand. *Journal of the Royal Society of New Zealand* 54 (4), 491–522. https://doi.org/10.1080/03036758.2023.2236033.

Little, M. (2023) Enhanced food security through localised community cryptocurrency: Experiences of a Costa Rican tourism town. In F. Higgins-Desbiolles and B.C. Bigby (eds) *The Local Turn in Tourism: Empowering Communities* (pp. 116–132). Channel View Publications.

Little, S. and Ke, G. (2023) What does B.C.'s brutal wildfire season mean for the future of the tourism sector? *Global News Canada* (Online), 22 August. See https://globalnews.ca/news/9910223/bc-brutal-wildfire-season-future-of-tourism/ (accessed August 2023).

Malm, A. (2016) *Fossil Capital: The Rise of Steam Power and the Roots of Global Warming*. Verso.

Matei, N.A., García-León, D., Dosio, A., Batista e Silva, F., Ribeiro Barranco, R. and Císcar Martínez, J.C. (2023) *Regional Impact of Climate Change on European Tourism Demand*. Publications Office of the European Union.

McKercher, B. and Chon, K. (2004) The over-reaction to SARS and the collapse of Asian tourism. *Annals of Tourism Research* 31 (3), 716–719. https://doi.org/10.1016/j.annals.2003.11.002.

Mei-Singh, L., Mostafanezhad, M. and Jamal, T. (2024) Plantation disaster capitalism in Maui: Reckoning with tourism after the fire. *Tourism Geographies* 26 (3), 311–328. https://doi.org/10.1080/14616688.2024.2337185.

Peters, A. (2023) Maui's Lahaina was once filled with wetlands. Can it be rebuilt differently? See https://www.fastcompany.com/90939766/mauis-lahaina-was-once-filled-with-wetlands-can-it-be-rebuilt-differently (accessed August 2023).

Rand, D.G. (2016) Cooperation, fast and slow: Meta-analytic evidence for a theory of social heuristics and self-interested deliberation. *Psychological Science* 27, 1192–1206. https://doi.org/10.1177/0956797616654455.

Renkert, S.R. (2019) Community-owned tourism and degrowth: A case study in the Kichwa Añangu community. *Journal of Sustainable Tourism* 27 (12), 1893–1908. https://doi.org/10.1080/09669582.2019.1660669

Reuters (2023) Iran shuts down for two days because of 'unprecedented heat'. *Reuters* (Online), 1 August. See https://www.reuters.com/world/middle-east/iran-shuts-down-two-days-because-unprecedented-heat-2023-08-01/ (accessed August 2023).

Rockström, J., Gupta, J., Qin, D. *et al.* (2023) Safe and just Earth system boundaries. *Nature* 619, 102–111. https://doi.org/10.1038/s41586-023-06083-8.

Ross, J. (2023) Refund students for 'dud' degrees, Australian Liberal Party urges. *Times Higher Education* (Online). See https://www.timeshighereducation.com/news/refund-students-dud-degrees-australian-liberal-party-urges (accessed August 2023).

Sampson, J. (2025) Where next? See https://www.linkedin.com/pulse/where-next-jeremy-sampson-eogic/ (accessed June 2024).

Saunders, E. (2023) Maui is suing Big Oil. *ExxonKnews* (Online), 16 August. See https://www.exxonknews.org/p/maui-is-suing-big-oil (accessed August 2023).

Schaller, C. and Vickers, J. (2020) Visitor value versus volume for international tourists to New Zealand. Wellington, N.Z.: Thinkstep Inc. See https://www.thinkstep-anz.com/assets/Whitepapers-Reports/Visitor-Value-vs-Volume-Report-2020-10-30.pdf (accessed August 2023).

Schuijers, L. (2023) Capitalising on climate anxiety: What you need to know about 'climate-washing'. *The Conversation*. See https://theconversation.com/capitalising-on-climate-anxiety-what-you-need-to-know-about-climate-washing-202507 (accessed July 2023).

Servigne, P., Stevens, R., Chappelle, G. and Rodary, D. (2021) Reasons for anticipating societal collapse. In J. Bendell and R. Read (eds) *Deep Adaptation: Navigating the Realities of Climate Chaos* (pp. 87–104). Polity Press.

Starr, J., Nicolson, C., Ash, M., Markowitz, E.M. and Moran, D. (2023) Income-based U.S. household carbon footprints (1990–2019) offer new insights on emissions inequality and climate finance. *PLOS Climate* 2 (8), e0000190, https://doi.org/10.1371/journal.pclm.0000190.

Stay Grounded (n.d.) 13 Steps for a Just Transport System and for Rapidly Reducing Aviation. See https://stay-grounded.org/position-paper/ (accessed August 2023).

Stinson, M.J., Hurst, C.E. and Grimwood, B.S.R. (2022) Tracing the materiality of reconciliation in tourism. *Annals of Tourism Research* 94, 103380. https://doi.org/10.1016/j.annals.2022.103380.

Sultana, F. (2023) Whose growth in whose planetary boundaries? Decolonising planetary justice in the Anthropocene. *Geo: Geography and Environment* 10 (2), e00128. https://doi.org/10.1002/geo2.128.

Sun, Y.-Y., Faturay, F., Lenzen, M., Gossling, S. and Higham, J. (2024) Drivers of global tourism carbon emissions. *Nat Commun* 15 (1), 10384. https://doi.org/10.1038/s41467-024-54582-7.

Tomczewska-Popowycz, N., and Quirini-Popławski, Ł. (2021) Political instability equals the collapse of tourism in Ukraine? *Sustainability* 13 (8), 4126.

Tourism Panel on Climate Change (TPCC) (n.d.) Code red for climate resilient tourism. See https://tpcc.info/code-red/ (accessed August 2023).

Tribe, J. (2003) The RAE-ification of tourism research in the UK. *International Journal of Tourism Research* 5, 225–234. https://doi.org/10.1002/jtr.433.

UN (2023) The Sustainable Development Goals Report 2023: Special Edition. See https://unstats.un.org/sdgs/report/2023/ (accessed August 2023).

Walker, B., Holling, C.S., Carpenter, S.R. and Kinzig, A. (2004) Resilience, adaptability and transformability in social–ecological Systems. *Ecology and Society* 9 (2), 5. http://www.jstor.org/stable/26267673.

Whyte, K.P. (2018) Indigenous science (fiction) for the Anthropocene: Ancestral dystopias and fantasies of climate change crises. *Environment and Planning E: Nature and Space* 1 (1–2), 224–242.

Woolfenden, T. (2023) The debt-fossil fuel trap: Why debt is a barrier to fossil fuel phase-out and what we can do about it. London: Debt Justice. See https://debtjustice.org.uk/wp-content/uploads/2023/08/Debt-Fossil-Fuel-Trap-Report_2023.pdf (accessed August 2023).

Conclusion: Advancing Climate Justice – Pathways Forward for More Just Tourism Futures

Roshis Krishna Shrestha, Raymond Rastegar and Freya Higgins-Desbiolles

This chapter critically synthesises insights from 12 contributions to this edited book, revealing the systemic injustices in tourism and outlining pathways for just transformation, research and policy work. It reveals three overarching themes: the persistence of growth-driven, extractive tourism models; structural governance inequities that marginalise vulnerable voices; and the tension between degrowth imperatives and conventional development. Subsequently, it applies a multidimensional justice framework, encompassing environmental, distributive, procedural, recognitive, restorative, ecological and cosmopolitan justice, to narrate how case studies included in this book advance knowledge on climate justice. It then proposes pathways for radical transformation, advocating post-growth paradigms, justice-centred governance, global solidarities, a just transition for tourism workers and a commitment to address the impacts of conflict and instability, ultimately calling for collective action towards a more equitable tourism future.

Introduction

From the Himalayas of Nepal to the remote island nation of Vanuatu, this book has explored the complex and, often disproportionate, impacts of climate change. It aligns with the growing global consensus that major contributors, particularly wealthy nations and carbon-intensive industries, must take stronger, more responsible climate action. However, despite the early awareness, concerted public and political action has been significantly delayed. International agreements, such as the United Nations Framework Convention on

Climate Change (UNFCCC) and the 2015 Paris Agreement, while establishing frameworks for collective action, have been criticised for their non-binding nature and weak enforcement mechanisms, which hinder the delivery of outcomes that are both adequate and just (Venn, 2023). This is especially problematic given the increasing recognition of the unequal and dynamic impacts of polycrisis on the tourism sector, as well as the politicisation of the climate agenda – factors that further obstruct progress towards just tourism (Warlenius, 2017).

The focus on climate justice, therefore, carries the potential to address the deep-rooted inequities and injustice inherent in the causes and consequences of climate change (Becken & Rastegar, 2025). As demonstrated in this book, climate justice acknowledges that the impacts of a rapidly changing climate are not distributed evenly, nor are the burdens of mitigation and adaptation efforts borne equally. The urgency of this issue is underscored by the authors in this book, who demonstrate that the differential vulnerability resulting from climate (in)justice falls along the lines of class, gender, race and Indigeneity, making it an inherently ecological, decolonising and even spiritual concern. While summarising the key insights from the book, we revisit the multidimensional framework for understanding climate justice in tourism introduced in the introductory chapter and discuss the key ideas that emerged across different chapters. Doing so, this concluding chapter does not offer simple solutions, for there are none. Rather, it synthesises the powerful, and often unsettling, insights from the preceding chapters to chart a pathway, providing future policy and research directions and a call for action for a radical and just transformation.

Emergent Themes from the Chapter

A clear and consistent picture has emerged from the diverse cases studies and critical analyses presented in this book: *climate justice in tourism is not a series of isolated problems but the outcome of deeply flawed and inequitable socioeconomic systems.* Three overarching themes have recurred throughout the chapters, highlighting the structural roots of climate injustice.

First, despite a widespread recognition of the negative consequences of neoliberal approaches to tourism, many countries continue to adopt tourism strategies that reflect the same growth driven, resource-extractive paradigm (Higgins-Desbiolles & Everingham, 2024) – one that prioritises short-term and mid-term economic gains over long-term equity and sustainability. In this regard, in Chapter 1, Sun *et al.* present evidence of asymmetries in the global tourism industry, highlighting the divide between high-emitting, high-consuming nations of the Global North and the climate-vulnerable nations that disproportionately bear the costs. Similarly, in Chapter 3, Broekema *et al.* demonstrate how this dynamic is reinforced by policies that unintentionally penalise peripheral regions, arguing that even well-intended aviation policies can exacerbate inequality, severely impacting remote and climate vulnerable countries. In Chapter 2, He *et al.*, through interviews with global tourism professionals, confirm this reality. They acknowledge the concentration of benefits and the externalisation of social and environmental harms onto tourism-dependent host communities. Similarly, through the case studies from carbon credit in Kenya conservancies (Gona & Atieno, Chapter 7) and entrenched mobility issues in US Virgin Islands (Sheller *et al.*, Chapter 6), this book also adds that extractive economic approaches often manifest into neocolonial resource appropriation in vulnerable tourism destinations, where historical patterns of exploitation continue to shape the benefit and impact of tourism. In this regard, Broekema *et al.* advocate for a forward-thinking governance approach that takes a wider view of justice.

Second, the chapters also show that the prevailing extractive model is sustained by structural inequities in governance. A recurring theme is the procedural and epistemic exclusion of those most affected by climate change, political hegemony and tourism. Local and Indigenous voices are systematically marginalised by both local and global political forces in decision-making processes that often determine their very futures (Higgins-Desbiolles, 2025; Shrestha *et al.*, 2024). The chapters on Vanuatu (Andrews *et al.*, Chapter 10) and Palestine (Qumsiyeh & Bibee, Chapter 8) present powerful counter narratives, showing how Indigenous-led frameworks and alternative tourism models serve as acts of resistance against colonial and/or hegemonic governance structures that deny self-determination. Similarly, in Chapter 5, Dahal and Subedi illustrate how power asymmetries between large operators and small-scale lodges undermine community-based efforts to manage climate-induced water crises in Nepal. These issues, compounded by ethical concerns such as transparency, accountability and meaningful local involvement, are not incidental but rather integral to a system designed to maintain existing power relations.

Third, the chapters reveal a profound tension between degrowth and dominant development aspirations. In Chapter 4, Sulub and Subtil Fialho show that activism-driven non-governmental organisations such as Stay Grounded make an uncompromising case for the pursuit of aviation degrowth as a climate justice imperative – a challenging but necessary feat to mitigate the ethnocidal processes that destroy Indigenous identity and lifeways. Yet, as analysis of emissions data (Sun *et al.*, Chapter 1) and aviation policy (Broekema *et al.*, Chapter 3) shows, the current trajectory forces a cruel choice upon the Global South: *pursue a high-carbon development path for economic prosperity or sacrifice economic aspirations to address a crisis they did not create*. Higgins-Desbiolles (Chapter 11, Chapter 12) dismantles this false dichotomy, arguing that the growth-centric, capitalistic paradigm is the root of the problem, and achieving real justice requires moving beyond reformist solutions, working towards collective solidarity, embracing localised approaches to tourism development and centring wellbeing in focus. In this regard, in Chapter 9, Nadegger and Ren's argument becomes salient, entailing a cohesive and collective effort for affective solidarity through three moves: *stepping forward, standing with* and *staying connected*. These works mark a preliminary engagement with the vital agenda of developing a 'critical climate justice praxis' in tourism.

Returning to the Multi-dimensional Justice Framework for Our Climate Justice Analysis

To fully grasp how these themes are interwoven, the multi-dimensional justice framework from our introduction offers a vital analytical lens. It allows us to see the systemic failures of tourism not as separate issues, but as interconnected facets for a climate injustice.

Environmental justice

Environmental justice, rooted in the equitable access to environmental benefits and meaningful participation of all people, regardless of race, gender, ethnicity or nationality, is required in the development, implementation and enforcement of environmental policies, laws and regulations (Morea, 2021). The example of the debate over uranium mining in South Greenland, as demonstrated in Chapter 9 by Nadegger and Ren, exemplifies how environmental injustice is embedded in colonial exploitation and systemic marginalisation. Through affective solidarity and collection action, local communities in Greenland resisted extractive agendas that posed severe environmental and social risks: contamination of land, disruption of ecosystems and threats to local livelihoods (e.g. farming, hunting and emerging tourism). While Greenland's case illustrates tourism's potential to support environmental justice, in Chapter 6, Sheller *et al.* show that, when governments prioritise tourism in a way that it disrupts the local ecology and livelihoods, increased tourism mobilities intensifies pressures on local environments and livelihoods, thereby making the entire community more vulnerable to climate disruptions. Therefore, this book postulates that environmental justice in tourism requires confronting uncomfortable truths about the industry's complicity in historical and ongoing harm, challenging entrenched power structures and embracing diverse ways of knowing and being.

Distributive justice

Distributive justice, in the context of tourism and climate change, concerns the fair allocation of benefits and burdens (Jamal & Camargo, 2014). This book reveals that climate injustice in tourism stems from a stark asymmetry between those who contribute most to the problem and those who suffer its worst consequences. In this regard, in Chapter 1, Sun et al. provide compelling evidence: *High-income countries and affluent individuals drive tourism-related emissions while low-income countries, small island developing states and marginalised communities contribute minimally yet bear the most severe and immediate consequences of climate change.* The book chapter also shows that the economic benefits from tourism are unequally distributed. For example, in Chapter 2, He et al., giving examples from the Caribbean, show that the profits from luxury resorts often repatriate to investors' home countries, leaving local economies with little gain. In such situations, local populations may find their traditional water access and cultural practices blamed or restricted to accommodate tourist demands. These examples demonstrate that climate change impacts and the burdens of unjust tourism do not fall evenly between the Global North and South, or even within specific communities.

Procedural justice

Procedural justice concerns the fairness, transparency, inclusivity and accountability of the processes through which decisions are made and implemented (Rastegar, 2020). A recurrent theme across the chapters is the systematic exclusion or tokenistic inclusion of local and Indigenous communities in tourism and climate-related decision-making processes. Using the case of carbon offsetting projects in Kenyan conservancies, in Chapter 7, Gona and Atieno illustrate how community engagement can be superficial or constrained. While collaborative mechanisms like community meetings and consultations exist, the actual influence of landowners can be limited by factors such as pre-conditioned non-disclosure agreements or a failure by project developers to seriously consider unsolicited community input. This suggests that formal participation does not always translate into genuine procedural justice when power imbalances persist and community voices are not truly empowered. Geoffrey Lipman's critique of the tourism industry's tendency to make grand promises while continuing 'business as usual' through marketing hype underscores this challenge (cited in He *et al.*, Chapter 2). As Gona and Atieno argue, this effectively silences alternative perspectives, devalues experiential and traditional ecological knowledge and consolidates decision-making power in the hands of a select group of elites, who are often affiliated with corporate interests or institutions in the Global North.

Recognitive justice

Recognitive justice emphasises the imperative to acknowledge, respect and value the diverse identities, cultures, histories, knowledge systems and inherent rights of all peoples (Martini, 2024). In Chapter 4,

Sulub and Subtil Fialho demonstrate that the struggle of the Mayan people in Southeastern Mexico against the tide of touristification and associated mega-projects like the Mayan Train is fundamentally a fight for the recognition of their autonomy, their ancestral connection to the land (embodied in practices like milpa agriculture) and their distinct cultural lifeways. Here, a critical aspect of recognition justice involves challenging policies and practices that ignore, undervalue or appropriate the lived experiences and profound ecological knowledge of Indigenous peoples and local communities. Recognitive justice also requires dismantling the supremacist ideologies – be they racial, gender based, colonial or anthropocentric – that have historically underpinned and continue to fuel exploitation and injustice within tourism and beyond. In this regard, in Chapter 11, Higgins-Desbiolles draws on Trask's critique of tourism in Hawai'i as a form of 'cultural prostitution', where sacred traditions and identities are deformed and sold. Her chapter demonstrates that recognitive justice demands that cultural expressions and knowledge are shared with integrity, respect and on the terms of the communities themselves. This includes the right of individuals and communities to define their own forms of engagement with tourism, and even to reject approaches that do not align with their cultural values, ecological principles or aspirations.

Restorative justice

Restorative justice, within the multidimensional framework, calls for a profound reckoning with historical and ongoing wrongs that have shaped many tourism landscapes and continue to inflict harm on communities and ecosystems (Rastegar, 2025). Several chapters in this book point towards the salience of addressing restorative injustice. The Regenerative Vanua Stewardship Framework (RVSF) in Vanuatu is explicitly framed as a tool for decolonisation (Andrews *et al.*, Chapter 10). It seeks to disentangle their lands, cultures and lifeways from the extractive, industrial tourism approach, aiming instead for the repatriation of Indigenous land, life and sovereignty. Likewise, the Mayan struggle in Mexico against touristification is understood as resistance to a development model that perpetuates colonial forms of dispossession and cultural erasure (Sulub & Subtil Fialho, Chapter 4). In Palestine, the ongoing Israeli occupation represents an active and continuing colonisation through violent settler colonialism where tourism itself is used as a tool of colonisation, through the appropriation of sites and narratives; and also, conversely, tourism is used for decolonial resistance and the assertion of Palestinian identity and rights (Qumsiyeh & Bibee, Chapter 8). Restorative justice, therefore, necessitates a decolonial praxis that actively challenges and dismantles these enduring structures of oppression within tourism.

Ecological justice

Ecological justice expands the purview of justice beyond the human realm, advocating instead for the recognition of the intrinsic value, rights and wellbeing of non-human species, ecosystems and the Earth system as a whole (Rastegar, 2022). The wildfires in Maui county, Hawai'i serve as a stark reminder of how the tourism industry, deeply interwoven with settler colonial capitalist development, has profited by appropriating land, water and other resources. For decades, this approach has fuelled climate deception, and amid polycrisis, intensified tensions between locals and tourists (Higgins-Desbiolles, Chapter 12). In this regard, as Dahal and Subedi note in Chapter 5, even seemingly benign activities like trekking can lead to waste mismanagement and ecological degradation if not properly regulated. Similarly, Gona and Atieno (Chapter 7) highlight that carbon offsetting projects, despite being framed as environmentally beneficial, can produce negative ecological impacts when they involve inappropriate use of land or overlook local ecological dynamics. Importantly, climate change acts as a powerful magnifier of these ecological injustices within tourism. It necessitates a fundamental ethical transformation, a shift in consciousness that recognises the intrinsic value of all life and our profound interdependence with the natural world (Sheppard & Fennell, 2019).

Cosmopolitan justice

In the context of climate justice, cosmopolitan justice necessitates the recognition of our shared planetary existence, the transboundary nature of climate change impacts and the ethical obligations that arise from global interconnectedness (Rastegar & Ruhanen, 2023). A stark manifestation of cosmopolitan injustice in tourism, as noted by Sun *et al*. in Chapter 1, is the global mobility asymmetry: a relatively small, affluent segment of the global population, predominantly residing in the Global North, accounts for the vast majority of international travel and its associated carbon emissions, particularly from aviation. The case of the US Virgin Islands, as a US territory with a long history of colonial influence and current economic dependencies, illustrates how these complex interdependencies and power imbalances shape tourism mobilities and climate vulnerabilities in ways that demand a cosmopolitan justice lens to fully understand and address (Sheller *et al*., Chapter 6). The development of affective solidarity, as explored by Nadegger and Ren in Chapter 9, offers a potent pathway towards cosmopolitan justice. This form of solidarity is not necessarily based on shared identity or geographical proximity, but on a shared emotional and ethical response to perceived injustice and a collective desire for transformation. It involves 'stepping forward' to voice dissent, 'standing with' those who are struggling and 'staying connected' to a common cause across differences.

Future Directions for Climate Justice in Tourism

As this volume has shown, climate justice in tourism demands more than incremental reform or rhetorical inclusion; it calls for radical reimaginings of tourism practice, policy, governance and scholarship. Looking forward, several future directions emerge that build on the work presented here and signal a shift toward transformative praxis grounded in equity, pluralism and planetary responsibility.

Reorienting tourism toward post-growth and post-capitalist paradigms

One of the most urgent tasks is to move tourism beyond the exploitative logics of capitalist accumulation and growth. As multiple chapters demonstrate, growth-centric models are structurally incompatible with both planetary boundaries and justice imperatives (Bianchi & Milano, 2024; Hickel, 2023). Post-growth tourism futures must centre sufficiency rather than surplus, wellbeing rather than GDP and redistribution rather than accumulation. These alternatives should prioritise low-carbon, community-based, slow and regenerative tourism that is democratically governed and ecologically sustainable (Fletcher *et al.*, 2021; Higgins-Desbiolles, 2018, 2022; Higgins-Desbiolles & Everingham, 2024; Rastegar 2025). Pursuing such futures requires new indicators and governance tools to replace growth-focused benchmarks. Instead of tourist arrival numbers and spending, planners should assess the contribution of tourism to community resilience, ecological regeneration and intergenerational equity. Importantly, degrowth tourism frameworks must be applied with caution and reflexivity. As highlighted in the chapters by Sun *et al.* and Sulub and Subtil Fialho, Global South destinations cannot be expected to sacrifice economic opportunity without structural compensation, climate finance and equitable transition plans.

Embedding justice-centred governance and accountability mechanisms

Future tourism governance must place justice, not just sustainability, at its core. This includes rejecting tokenistic stakeholder inclusion and creating transparent, binding and community-led mechanisms that ensure accountability across scales (Becken & Rastegar, 2025; Higgins-Desbiolles & Bigby, 2023; Rastegar & Ruhanen, 2023). Several chapters highlight innovations such as the Regenerative Vanua Stewardship Framework (Andrews *et al.*, Chapter 10) and community-led water governance in Nepal (Dahal & Subedi, Chapter 5), which exemplify what just tourism governance can look like when Indigenous and local actors hold meaningful power. International legal and policy instruments must also evolve to support justice-based governance. The Loss and Damage

Fund, the International Court of Justice's advisory opinion on climate change and human rights and the principle of common but differentiated responsibilities all offer entry points to integrate climate justice into tourism policy. Localised versions of these principles, such as climate justice levies, tourist taxes with redistributive functions and reparative tourism funds, should be explored. The goal is not only to mitigate climate risks but to restore agency, redistribute power and repair historical harms.

Building solidarities through affective, epistemic and multispecies justice

A climate just tourism future must be grounded in affective solidarity across species, spaces and scales. Nadegger and Ren (Chapter 9) offer a powerful framework of 'stepping forward, standing with, and staying connected' as the foundation of cross-border and cross-species solidarities. Tourism must be reframed as a space where solidarities are cultivated, not commodified. This includes solidarities between tourists and host communities, between displaced people and privileged travellers and between humans and more than human worlds (Rastegar, 2022). Tourism research and governance must centre plural epistemologies, particularly Indigenous, feminist and place-based knowledge, and redress the epistemic violence caused by colonial erasure (Higgins-Desbiolles, 2025; Grimwood *et al*., 2024). Future frameworks must support not only the inclusion of diverse knowledge systems but their leadership in setting tourism agendas. Simultaneously, a multispecies justice framework reminds us that tourism must extend beyond human concerns to care for nonhuman kin and ecosystems (Fennell & Sheppard, 2021). This requires practical shifts in how tourism infrastructure is planned, how wildlife is protected and how ethical co-existence is prioritised.

Advancing a just transition for tourism workers and communities

A just transition in tourism is essential, not only for tourists and destinations but also for the millions of workers and communities whose livelihoods depend on the sector. Current decarbonisation strategies risk deepening inequalities if they are not matched with protections for the most vulnerable (Rastegar & Becken, 2024). This book demonstrates how tourism workers, especially in precarious or informal roles in the Global South, remain structurally excluded from decision making and benefits (He *et al*., Chapter 2; Sheller *et al*., Chapter 6). Transitioning to a more just and climate-resilient tourism sector must include investments in education, skills training and the creation of dignified employment opportunities in climate-adaptive roles. Cooperative tourism models, heritage conservation, local food systems and green infrastructure development offer avenues for reimagining labour and ownership.

Likewise, community co-ownership of tourism ventures, as seen in Indigenous-led examples throughout this volume, can offer models of economic autonomy and cultural revival. Climate justice cannot be achieved without socioeconomic justice for those whose labour sustains the industry.

Strengthening climate justice research, pedagogy and advocacy

Finally, tourism scholarship, education and advocacy must continue to evolve with climate justice at the centre. Future research should go beyond emissions metrics to interrogate the structural injustices embedded in tourism systems including racial capitalism, climate colonialism and global mobility hierarchies (Rastegar *et al*., 2023; Sultana, 2022). This requires a shift from extractive to co-productive research practices and the prioritisation of ethical, place-based inquiry (Higgins-Desbiolles & Bigby, 2023). Tourism curricula must prepare students and researchers not only to manage tourism but to transform it to navigate uncertainty, resist climate denial and delay and advocate for equitable futures. As Higgins-Desbiolles (Chapter 12) argues, tourism education must now engage with concepts like deep adaptation, decolonial transition and post-capitalist ethics to respond to polycrisis with integrity and imagination. Partnerships between academia, civil society and policy communities will also be crucial to advance public discourse and collective action on climate justice in tourism.

Addressing the impacts of conflicts and instability on climate justice

Emerging conflicts and intensifying geopolitical instability present a growing and often overlooked threat to the pursuit of climate justice in tourism. The fragility of climate action in conflict-affected regions is starkly exposed by the ongoing genocide in Gaza, civil war in Sudan, the Russia–Ukraine war and escalating tensions involving Israel and Lebanon, Syria, Iran and the broader Middle East. There is little doubt that such conflicts are at least in part evidence of destabilisation and competition for resources in the face of climate change (UNFCCC, 2022). Additionally, these overlapping crises underscore how quickly climate and justice commitments can be sidelined in the face of militarisation, occupation and systemic violence. Despite accounting for nearly 5.5% of global greenhouse gas emissions, military operations remain one of the least scrutinised sectors, routinely excluded from international emissions reporting and relegated in climate policy discussions (Rokke, 2025). War not only redirects financial, political and institutional resources away from climate adaptation and mitigation efforts but also triggers displacement, fuels mobility crises and severely disrupts tourism flows. Moreover, it erodes the foundations of cross-border collaboration and environmental governance, which are essential for addressing transnational climate impacts. In such contexts, the promise of climate justice (grounded in equity, solidarity and collective responsibility) becomes increasingly difficult to realise. A critical challenge for tourism research and policy is to confront these realities and resist the normalisation of climate silence in the face of war and occupation.

As several analysts have noted, tourism has long been promoted as a soft-power tool for peacebuilding (e.g. Guasca *et al.*, 2021; UN Tourism, 2024). Yet, this role is increasingly precarious. Armed conflict and occupation can erase years of progress in community-based tourism, endanger livelihoods, displace populations and deepen pre-existing inequalities. Future climate justice frameworks must address how militarisation, forced migration and environmental degradation under conflict erode local resilience and diminish the agency of affected communities. In regions experiencing or recovering from conflict, justice-oriented tourism must prioritise safeguarding human rights, supporting displaced and marginalised populations and protecting environmental defenders. It must also develop long-term recovery strategies that integrate ecological restoration, cultural healing and durable peace. Without accounting for the structural violence of war and its aftermath, climate justice in tourism will remain incomplete and fragile.

Conclusion

As the first academic book to address the issue of climate justice in tourism, we have covered a wide domain of ideas and concerns but by no means have we exhausted the discussions and action agendas we must progress. This book has been a collaboration, engaging academics from across disciplines, tourism planners and consultants, representatives and founders of non-governmental organisations, leaders of communities and activists all concerned with climate (in)justice. Our authors are based in diverse sites around the world where climate injustice is already real and narrowing futures. Together, we have uncovered and explained many of the complexities and challenges that climate change presents as well as offering some indicative solutions and pathways forward – co-developing a preliminary 'critical climate justice praxis' in tourism.

As we have defined climate justice in tourism, many of us through the chapters comprising this book have illuminated the ways the disparities in power and resources have disadvantaged and endangered those already made vulnerable and marginalised in a global system built on imperialism and extractive, exploitative and supremist ideologies. Noting these facts, we must take aim at the rich and the privileged as a substantial barrier to achieving a more climate just future in tourism and the wider realms of our human society. Their material reality of political and economic power and privilege, their ability to physically separate themselves from the rest of society with gated communities, private jets and planned safe havens and their ideologies of individualism, competition and selfishness work to limit our ability to secure a safe and thriving future for all (Larson, 2024). The relentless demand to grow high emitting forms of tourism – luxury tourism, tourism to the last remote, wild places and space tourism – in the midst of a climate catastrophe is symptomatic of a pathology that is endangering us, our children and the planet. In such a context where wealth and power are intensely consolidating in the hands of a very few (Piketty, 2014), we must interrogate the equity of tourism's carbon emissions (see Gössling & Humpe, 2020) in a time when we must do all we can to drive emissions down.

Thus, this topic of climate justice in tourism is profoundly political and moves us beyond the realm of neutral tourism business planning and management to zones of contestation and political re-ordering. While we must base our analyses on scientifically rigorous assessments, we must also have a normative inclination about what is 'just' in tourism development and the ways tourism can be shaped towards a greater good for humanity and for our 'environment'.

As we close this book, let us be clear that we are by no means making sufficient efforts to address climate change and thereby work towards climate justice. As of June 2025, only 22 countries out of 197 had submitted revised nationally determined contribution targets towards 2035, covering just 21.5% of global emissions and 11.9% of global population (Climate Action Tracker, 2025). Only the UK's planned targets have been assessed as aligned with the Paris Agreement's 1.5°C goal. Exceeding this agreed 1.5°C threshold risks a great deal – extreme and deadly weather events, loss of coral reefs, loss of vital glaciers and sea ice, deteriorating public health and triggering of multiple climate tipping points (such as breakdowns of major ocean circulation systems). It is not difficult to see that such extreme outcomes present serious threats to tourism and communities that have grown to be dependent on tourism. As Stefan Gössling said before the world's biggest tourism convention, ITB in Berlin in 2025: 'We have already entered the beginning of the age of non-tourism' (Niranjan, 2025).

Yet, despite this reality, progress in the tourism domain is also not promising. Recently, Scott and Gössling have assessed the tourism sector's success in delivering outcomes related to 38 actionable climate pledges identified in existing climate declarations, finding:

> A visible gap exists between pledges and performance, with limited to no demonstrable progress found on 25 climate action pledges. Time is the enemy, and the tourism sector must move beyond ambition to produce results at scale to stabilize and reduce emissions and build climate resilience. (2025: 1)

The role of tourism in climate change is one of the most important before us as a discipline, a practice and an 'industry'. Climate justice deliberations compel us to grapple with the uneven, unjust and invisibilised power dynamics of tourism processes that for too long have remained hidden and unaddressed. Acting to address the threat of climate change demands a solidarity that we must summon to work collectively together. Climate justice analyses spotlight the need for Global North and Global South inequities (and other gross disparities) to be resolved and for us to work towards a shared future together. Tourism has a role to play in this – either contributing to greater injustices or alternatively helping to build the cooperation, understanding and sharing that we need to collaborate for solutions. This book is offered in support of the latter vision. We invite our readers to join us in a shared agenda for transformations in tourism to support climate justice, human thriving and ecological recovery. The key to confronting this challenge to humanity is our collective solidarity and we all must rise to meet it together.

References

Becken, S. and Rastegar, R. (2025) Advancing climate justice in tourism: A critical evaluation of the TPCC Stocktake. *Annals of Tourism Research* 113, 103962. https://doi.org/10.1016/j.annals.2025.103962.

Bianchi, R.V. and Milano, C. (2024) Polycrisis and the metamorphosis of tourism capitalism. *Annals of Tourism Research* 104, 103731. https://doi.org/10.1016/j.annals.2024.103731.

Climate Action Tracker (2025) CAT 2035 Climate Target Update Tracker. See https://climateactiontracker.org/climate-target-update-tracker-2035/ (accessed June 2025).

Fennell, D. and Sheppard, V. (2021) Tourism, animals and the scales of justice. *Journal of Sustainable Tourism* 29 (2–3), 314–335. https://doi.org/10.1080/09669582.2020.1768263.

Gössling, S. and Humpe, A. (2020) The global scale, distribution and growth of aviation: Implications for climate change. *Global Environmental Change* 65, 102194. https://doi.org/10.1016/j.gloenvcha.2020.102194.

Guasca, M., Vanneste, D. and Van Broeck, A.M. (2021) Peacebuilding and post-conflict tourism: Addressing structural violence in Colombia. *Journal of Sustainable Tourism* 30 (2–3), 427–443. https://doi.org/10.1080/09669582.2020.1869242.

Grimwood, B.S.R., Lee, E. and Higgins-Desbiolles, F. (2024) Unsettling geographies of tourism. *Tourism Geographies* 26 (6), 899–916. https://doi.org/10.1080/14616688.2024.2402997.

Fletcher, R., Blanco-Romero, A., Blázquez-Salom, M., Cañada, E., Murray Mas, I. and Sekulova, F. (2021) Pathways to post-capitalist tourism. *Tourism Geographies* 25 (2–3), 707–728. https://doi.org/10.1080/14616688.2021.1965202.

Hickel, J. (2023) The double objective of democratic ecosocialism. *The Monthly Review* (Online). See https://monthlyreview.org/2023/09/01/the-double-objective-of-democratic-ecosocialism/ (accessed June 2025).

Higgins-Desbiolles, F. (2018) Sustainable tourism: Sustaining tourism or something more? *Tourism Management Perspectives* 25, 157–160. https://doi.org/10.1016/j.tmp.2017.11.017.

Higgins-Desbiolles, F. (2022) Subsidiarity in tourism and travel circuits in the face of climate crisis. *Current Issues in Tourism* 26 (19), 3091–3101. https://doi.org/10.1080/13683500.2022.2116306.

Higgins-Desbiolles, F. (2025) On what terms might regenerative tourism proponents justly engage with indigenous knowledges? *Journal of Travel Research* 00472875251341308, https://doi.org/10.1177/00472875251341308.
Higgins-Desbiolles, F. and Bigby, B.C. (eds) (2023) *The Local Turn in Tourism: Empowering Communities*. Channel View Publications.
Higgins-Desbiolles, F. and Everingham, P. (2024) Degrowth in tourism: Advocacy for thriving not diminishment. *Tourism Recreation Research* 49 (1), 215–219. https://doi.org/10.1080/02508281.2022.2079841.
Jamal, T. and Camargo, B.A. (2014) Sustainable tourism, justice and an ethic of care: Toward the just destination. *Journal of Sustainable Tourism* 22 (1), 11–30. https://doi.org/10.1080/09669582.2013.786084.
Larson, R. (2024) *Mastering the Universe: The Obscene Wealth of the Ruling Class, What They Do with Their Money, and Why You Should Hate Them Even More*. Haymarket Books.
Martini, A. (2024) Geographies of mobility justice: post-disaster tourism, recognition justice, and affect in Tohoku, Japan. *Mobilities* 19 (3), 363–378. https://doi.org/10.1080/17450101.2023.2242002.
Morea, J.P. (2021) Environmental justice, well-being and sustainable tourism in protected area management. *Journal of Ecotourism* 20 (3), 250–269. https://doi.org/10.1080/14724049.2021.1876072.
Niranjan, A. (2025) Do heatwaves, wildfires and travel costs signal the end of the holiday abroad? *The Guardian* 23 August. See https://www.theguardian.com/business/2025/aug/23/do-heatwaves-wildfires-and-travel-costs-signal-the-end-of-the-holiday-abroad (accessed September 2025).
Piketty, T. (2014) *Capital in the Twenty-First Century*. Harvard University Press.
Rastegar, R. (2020) Tourism and justice: Rethinking the role of governments. *Annals of Tourism Research* 85, 102884. https://doi.org/10.1016/j.annals.2020.102884.
Rastegar, R. (2022) Towards a just sustainability transition in tourism: A multispecies justice perspective. *Journal of Hospitality and Tourism Management* 52, 113–122. https://doi.org/10.1016/j.jhtm.2022.06.008.
Rastegar, R. (2025) Regenerative justice and tourism: How can tourism go beyond restoration? *Annals of Tourism Research* 111, 103896. https://doi.org/10.1016/j.annals.2025.103896.
Rastegar, R. and Becken, S. (2024) Embedding justice into climate policy and practice relevant to tourism. *Journal of Sustainable Tourism* 33 (10), 2011–2028. https://doi.org/10.1080/09669582.2024.2377720.
Rastegar, R. and Ruhanen, L. (2023) Climate change and tourism transition: From cosmopolitan to local justice. *Annals of Tourism Research* 100, 103565. https://doi.org/10.1016/j.annals.2023.103565.
Rastegar, R., Higgins-Desbiolles, F. and Ruhanen, L. (2023) Tourism, global crises and justice: Rethinking, redefining and reorienting tourism futures. *Journal of Sustainable Tourism* 31 (12), 2613–2627. https://doi.org/10.1080/09669582.2023.2219037.
Rokke, N. (2025) Why war is one of the world's biggest climate threats. *Forbes*, 26 March. See https://www.forbes.com/sites/nilsrokke/2025/03/26/why-war-is-one-of-the-worlds-biggest-climate-threats/ (accessed June 2025).
Scott, D. and Gössling, S. (2025) Beyond ambition: A review of tourism climate change declaration outcomes and prospects from Baku. *Journal of Sustainable Tourism* 1–22. https://doi.org/10.1080/09669582.2025.2508878
Sultana, F. (2022) The unbearable heaviness of climate coloniality. *Political Geography* 99, 102638. https://doi.org/10.1016/j.polgeo.2022.102638.
Shrestha, R.K., Decosta, J.P.L.E., Whitford, M. and Shrestha, R. (2024) 'A place where I belong'-The ambiguous role of the outsider-within identity among indigenous Gurung women tourism entrepreneurs in Nepal. *Journal of Hospitality and Tourism Management* 58, 286–297. https://doi.org/10.1016/j.jhtm.2024.02.002.

Sheppard, V.A. and Fennell, D.A. (2019) Progress in tourism public sector policy: Toward an ethic for non-human animals. *Tourism Management* 73, 134–142. https://doi.org/10.1016/j.tourman.2018.11.017.

Venn, A. (2023) Rendering international human rights law fit for purpose on climate change. *Human Rights Law Review* 23 (1), ngac034. https://doi.org/10.1093/hrlr/ngac034.

UNFCCC (2022) Conflict and climate. See https://unfccc.int/news/conflict-and-climate (accessed June 2025).

UN Tourism (2024) World tourism day 2024: A global message of tourism for peace. See https://www.unwto.org/news/world-tourism-day-2024-a-global-message-of-tourism-for-peace (accessed May 2025).

Warlenius, R. (2017) Decolonizing the atmosphere: The climate justice movement on climate debt. *The Journal of Environment & Development* 27 (2), 131–155. https://doi.org/10.1177/1070496517744593.

Index

adaptation 2, 6, 25, 37, 44, 77, 122, 177, 216, 234, 242, 250, 272
affective solidarity 19, 186–8, 190–1, 193, 198–201, 274, 277, 279
airports 83, 85, 88, 90, 92, 121, 129, 132, 192
air traffic 18, 50, 75–6, 78–83, 85, 87, 91, 94–5
alternative tourism 167–8, 172, 176–9, 273
apartheid 18, 167, 180–1
aviation xii–xiii, 9, 11, 18, 25, 36–7, 40, 45, 47, 49–1, 53, 60, 62–3, 67, 70–2, 76, 79–87, 90–2, 94–5, 144, 235, 249, 273, 277
aviation climate policies 18, 60, 63, 67

Baku Declaration on Enhanced Climate Action in Tourism 25, 235
Bendell, Jem 19, 246, 248, 265
biodiversity xv, 9, 26, 82, 102, 104, 115, 145, 147, 154, 157, 170, 172–3, 178, 210, 219, 249, 252–3, 257
business as usual xxiv, 19, 32, 52, 63, 78, 241, 250, 254–6, 265, 275

capitalism xxiv, 11–12, 15–16, 75, 77, 84, 220, 229, 231–3, 241, 250, 253, 256–7, 259–60, 262, 280
capitalist growth logic 198
capitalist tourism 19, 87, 237, 239
carbon credit 61, 144, 159, 161, 249, 273
carbon emissions xiii, xxiv, 4, 7, 11–12, 25–31, 33–4, 36–7, 40–2, 45–7, 53, 55–6, 61, 67, 69–70, 116, 146, 189, 193, 197, 235, 277, 282
carbon intensity 30–1, 37
carbon-intensive elitism 188
carbon markets 18
carbon offset 14, 61, 76, 78, 142, 144–8, 156, 160–4, 249, 275, 277
carbon project 68, 142, 144–50, 154–60, 163–4
Caribbean xiv, xvi–xvii, 9, 48, 53, 76, 78, 88, 90, 120–4, 126–30, 132, 134–6, 138, 275

climate coloniality, -ism 8, 11, 14, 42, 55, 57, 280
climate crisis 2, 5–7, 13, 15, 17, 19, 40, 50, 55, 75–8, 83, 86, 92–4, 189–90, 198, 227, 229, 234, 241–2, 253, 255–6, 262–3
climate debt 14, 18, 250
climate disruptions 18, 122, 131, 274
climate emergency xvi, xxv, 55, 77–8, 82, 85, 87, 94, 248, 263
climate finance 24, 36–7, 242, 278
climate justice xiv–xv, xxiii–xiv, 1–2, 4–8, 10–11, 14–15, 17–20, 24, 26, 35, 40–5, 47, 51, 53–7, 63, 72, 75–6, 83, 86–7, 101–3, 108, 111–16, 121–2, 124, 126, 137, 148, 170, 172, 176–8, 180, 186–92, 197–201, 205–7, 220–1, 227–30, 233–5, 238, 240–1, 246, 250, 253, 264, 271–4, 277, 280–3
Climate Justice in Tourism report 7–8
climate mobilities 120–2, 125
climate neo-colonialism 249
climate refugees 5
climate washing 14, 242, 249, 264
collapse 19, 57, 64, 75, 77, 94, 123, 136, 162, 234, 241, 246–8, 251–5, 257, 259, 262–5
collective action. -solidarity 95, 186, 189–91, 271–2, 280
common but differentiated responsibilities (CBDR) 14
colonial, -ism 2, 6–8, 11, 13–16, 18–19, 40, 54–5, 57, 76, 87, 93, 120–1, 123, 125–8, 137, 167, 169, 175, 179–80, 186–9, 191–3, 205–7, 209, 212, 214, 216, 218, 220, 228–9, 231, 233, 240, 250, 273–5, 277, 279
community based tourism 41, 102, 114, 281
community-based water management 18, 101, 113
community conservancies 143, 146, 151–2
Conference of the Parties (COP) 1, 6, 25, 26, 37, 177, 228, 234, 235, 242, 249, 263
conservancy 142–6, 148–57, 160, 162, 164
cosmopolitan justice 9–10, 271, 277

Covid-19 25, 60, 63–5, 70, 72, 136–8, 142–3, 211, 252
Corsia 61, 68, 78–9
critical climate justice praxis 1, 4, 15, 238, 264, 274, 281
critical feminism, critical feminist climate justice 15, 234, 238
critical participatory action research (CPAR) 205, 212
cruise 8, 36, 40, 42, 45, 47, 50–1, 54, 92, 120–2, 124, 128–35, 192, 196

Davis, Angela 227, 229, 231, 241
decarbonise, -ation 11, 18, 47, 61–2, 76–8, 86, 92, 94, 126, 251, 257, 279
decolonial 13, 19, 54, 209, 216, 231, 242, 276, 280
decolonisation 1, 11, 13, 192, 206, 209, 212, 215, 220, 276
deep adaptation 19, 246, 248, 251–2, 254–6, 258, 260–1, 264–5, 280
degrowth 1, 11–12, 18, 70, 75–6, 85–7, 94, 262, 271, 273, 278
disaster capitalism 15, 250
distributional effects 60, 63, 69
distribution inequality 26
distributive justice 8, 42–3, 53, 56, 145, 147, 189, 274
drought xiii, 18, 37, 49, 53, 77, 101–2, 105–6, 120–1, 126, 137, 154, 157–8, 187, 195,

ecological justice 9, 11, 188, 250, 257, 264, 277
economic vulnerability 52, 71, 84, 160
ecotourism xv, 102, 113–14, 168–9, 172–3, 175, 211
emissions 4, 7, 14, 16, 24–6, 28, 30–1, 34, 34–7, 45, 47, 61, 67–9, 72, 77–82, 85, 87, 93–5, 126, 144, 159, 163, 176, 273, 280
emissions intensity 30–1
environmental justice 5–7, 15, 78, 137, 145, 167–8, 170, 172–3, 175–8, 181, 188–9, 274
epistemic (in)justice 212–14
Equality in Tourism (EIT) 236
equity 15, 19, 49, 71, 76, 116, 125, 145, 153, 177, 188, 218, 228, 235–6, 264, 278, 281–2
European Trading System (ETS) 67
extreme weather events 25–6, 31, 37, 46, 101, 104–5, 108

Fiji 49, 54–5, 93, 206
financial leakage 48, 53–4, 56

flight shame 12, 62, 67
flood 26, 101, 104–6, 111, 138, 148, 248
fossil fuels 4–5, 11, 61, 77, 80–1, 86, 92, 241, 248–9
friction 120, 122, 127, 131, 133, 135–7, 201
fuel efficiency 79

gender, diverse genders 13, 15, 54, 133, 227–31, 233–4, 236–8, 272, 274, 276
gender justice 15, 50, 227–9, 234–5, 238, 241
genealogical 218–19
glacial lake outburst floods 107, 111
glacier 193, 198
Glasgow Declaration on Climate Action in Tourism xvi, 25, 36, 176, 235
glass-ceiling feminism 227, 229, 234, 236–7, 241
Global North 5, 8–9, 11, 14, 16, 26, 77, 80, 87, 93, 95, 163, 188–9, 231, 233, 242, 255, 259, 263, 272, 275, 277, 283
Global South xiii, 5–, 8, 11–17, 25–6, 41, 46, 50, 53, 55, 77, 80, 82, 85–6, 95, 101–2, 116, 163, 233, 240, 242, 253, 255, 257, 273, 278–9, 283
Global Tourism Vulnerability Index 35
global warming 2, 4, 31–2, 61, 108, 124, 148
Great Barrier Reef 7, 9
Green Climate Fund 37, 112, 263
Greenland 18–19, 77, 186–7, 192–6, 199–201, 274

Hickel, Jason 210

Indigenous 13, 15, 18, 42–3, 50, 54, 57, 75–6, 87, 95, 108–9, 111, 115, 161, 167, 169–70, 177, 179, 181, 187, 189, 192, 196, 205–21, 231, 234–6, 240, 247, 249, 273, 275–6, 278–80
inequality 10, 16, 18, 26, 30, 35–6, 40, 48, 54, 60–4, 70–2, 84, 93, 122, 126, 197, 211, 232, 234, 237, 256, 263, 273
inequity 12, 49, 125, 228, 235
inter-generational equity 162
International Aviation Transport Association (IATA) 78
International Court of Justice (ICJ) 5, 178
intersectionality 16, 232–3
intragenerational equity 16
Israeli occupation 18, 167–8, 170, 180, 276

just transition xiii, xvi, xxiv, 16, 18, 20, 60, 70–2, 75, 76–7, 87, 92, 94–5, 189, 209, 249, 252–3, 257, 263–4, 271, 279

justice xi, xxiv, 1–2, 5, 15, 19, 43, 45, 60, 83, 87, 92, 95, 103, 145, 180, 190, 200, 206–8, 211, 213, 228, 235, 240, 248, 213
justice thinking 19, 228

Kangwaleva 208–9, 215, 217–22
Kenya 18, 143–4, 146, 148–9, 151–3, 273
Kyoto Protocol 14

labour, - exploitation 40–1, 50, 56
land tenure insecurity 160
liberation xiv, 172, 179–80, 229, 232, 234
Liñi, Walter Hadye Prime Minister 206
loss and damage fund 6, 16, 25, 37

Maldives 8–9, 64, 71, 93
marginalisation 14, 16, 42, 52, 54, 57, 108, 193, 230, 274
Maya xvii, 88, 90
melting destinations 19, 186–8, 190–3, 199–201
Mexico xiv, xvii, 18, 29, 75, 87–9, 91, 93, 276
migration 88–9, 121, 126–7, 137, 197, 208, 219, 254, 281
mitigate 2, 14, 76, 102, 106, 110–11, 205, 221, 239, 273, 279
mobility justice xvi, 16, 18, 120–2, 125–7, 135, 137
Mottley, Mia Prime Minister 1
multinational corporations 45, 47–8, 89, 94–5, 258

neocolonial 78, 92, 210, 216, 220, 273
Nepal xi, xiii, xvi–xvii, 18, 101–14, 116, 271, 273, 278
net zero 8, 25, 61, 78, 82, 176, 257
non-government organisations (NGOs) 76, 146

oppression, anti-oppression 2, 14, 19, 163, 170, 178, 180, 188, 191, 200, 228–32, 238, 242, 276

Palestine xiii, xv, 18, 167–74, 176–81, 273, 276
Palestine Institute for Biodiversity and Sustainability (PIBS) 170
Paris Agreement 4, 24, 31, 77–8, 228, 272
pastoral communities 18, 153–4, 162
per capita emissions 29–30, 36
Planetary Boundaries framework 3
pluriversal 13, 211, 257–8

polycrisis 1, 3, 12, 19, 55, 205, 241, 254, 261–2, 264, 272, 277, 280
power xxiv, 2, 8, 12, 16–17, 55–6, 122, 124, 127, 159, 180, 198, 205, 212, 221, 227, 240, 249–50, 253, 260, 281–2
power asymmetries 18, 145, 147, 164, 273
procedural justice 8, 42–3, 54, 57, 145, 189, 275
public-private partnerships 110, 113

racial, -ising 6–7, 13, 16, 121, 229, 241, 276
radical 77, 238, 272
rainwater harvesting 105–7, 110, 112, 114
reclamation 83, 207, 246, 248, 254, 260–1, 265
recognition, - justice 7–8, 14, 40, 43, 45, 104, 145, 189, 276
reconciliation 43, 190, 246, 248, 254, 258–9, 265
regeneration, -ive xxiv, 43, 157, 209–11, 215, 217, 220, 253, 258, 278
Regenerative Vanua xii, xv–xvi, 19, 212, 214–16, 219, 222
Regenerative Vanua stewardship framework 19, 205, 215–16, 276, 278
relinquishment 19, 246, 248, 254–7, 265
reparations 5–6, 16, 18, 87, 93, 242
reparative xxiv, 9, 95, 279
residence-based accounting 28
resilience 25, 42, 44, 101, 104, 110–11, 121, 126, 181, 205, 246, 254–5, 264
resistance xv, xvii, 13, 76, 83–4, 90–2, 121, 160, 173, 181, 273, 276
restoration 19, 43, 52, 114–16, 188, 210, 246, 248, 254, 256–8, 260, 265
restorative justice 8–9, 276
rhythm 120, 122, 127, 131, 133–7
risk 17, 24, 26, 35, 37, 42, 53, 60, 70, 108, 126, 135–6, 152, 154, 156, 160, 194, 264
routes 69, 85, 110–11, 120, 122, 127–8, 132, 134–5, 137

self-determination 19, 172, 194, 206, 273
Seychelles 9, 35, 65
small island developing states (SIDS) 5, 47, 64–5, 71
social justice 5, 11, 15, 42, 45, 75, 92, 94, 112, 137, 188, 197, 210, 215, 220, 233
social movements 186, 188, 198–9, 229, 256
socialising 242
societal collapse 246, 248, 251
socioecological systems 251
socioenvironmental conflicts 76, 78, 84

solidarity 10, 168, 172, 178–9, 187, 190–1, 215, 221, 240, 277, 281, 283
sovereignty 13, 19, 179, 217, 276
spiritual 2, 87, 171, 206, 208, 217–18, 221
stalled mobilities 18, 120, 122, 136–7
Stay Grounded xiii, xvii, 18, 47, 75–6, 83, 87, 90, 94, 256, 273
steward 208, 213, 217–20
steward, -ship 18, 19, 111–16, 178–9, 205, 208–10, 213, 215–20, 239, 261
storian 212–13, 215
sustainable aviation fuels (SAFs) 11, 45, 61, 76, 235, 249
structural justice 15, 18, 54, 57, 227–8, 237–8, 241, 256
Sultana, Farhana 5, 15, 42–3, 54, 57, 126, 188, 189–90, 199, 250, 280
sustainable tourism xi, xiii–xvi, xviii, 9, 31, 46, 49, 101–2, 108–9, 111–14, 116, 167–8, 170–1, 176, 196, 198, 261

taxes 2, 67, 69
Thunberg, Greta 227–8
tourism carbon emissions 26–8, 33–4, 46, 116, 235
tourism demand redistribution 24, 26
tourism dependency 2, 13, 17, 263
tourism-dependent economies (tourism reliant) 67, 135–7, 253
tourism futures 17, 19, 194, 199, 241, 246, 255, 262, 265, 271, 278
tourism governance 18, 42–3, 46, 51–4, 56–7, 152, 278
Tourism Panel on Climate Change (TPCC) 47, 176, 235, 254

tourism governance 18, 42–3, 46, 51–4, 56–7, 278
touristification 75–6, 85, 88–9, 91, 94, 276
traditional water management practices 111
Tyrol 18–19, 186–7, 193, 197–9, 201

unequal mobilities 18, 122, 133
United Nations Framework Convention on Climate Change (UNFCCC) 4, 6, 177, 228
United Nations Sustainable Development Goals (UNSDGs) 197
United Nations World Tourism Organisation (UN Tourism) 36
US Virgin Islands xvii, 18, 120–1, 124, 134, 136–8, 273, 277

Vanua(s) 205–10, 212–22
Vanuatu xii, 5, 205–8, 210, 212–14, 218–19, 222, 271, 273, 276
vulnerability 17, 35, 41, 50, 52, 57, 70, 121, 191

wastewater recycling 110, 112
water, - conservation, -management, - security 18, 101–3, 105–7, 109–16
water security 101, 106, 109–11, 114
wealth extraction 2, 12, 17, 84, 240–1
wetland restoration 114, 116
Whyte, Kyle 42, 178, 212, 228, 247
wildlife 9, 52, 104, 114, 142–4, 146, 151–4, 157–9, 163, 169–70, 279
World Bank xvi, 66, 171

zero emissions tourism 69

For Product Safety Concerns and Information please contact our EU Authorised Representative:

Easy Access System Europe

Mustamäe tee 50

10621 Tallinn

Estonia

gpsr.requests@easproject.com

www.ingramcontent.com/pod-product-compliance
Ingram Content Group UK Ltd.
Pitfield, Milton Keynes, MK11 3LW, UK
UKHW021839210426
5322IPUK00021B/363